OCEANOGRAPHY AND OCEAN ENGINEERING

LAGOONS

HABITAT AND SPECIES, HUMAN IMPACTS AND ECOLOGICAL EFFECTS

OCEANOGRAPHY AND OCEAN ENGINEERING

Additional books in this series can be found on Nova's website
under the Series tab.

Additional e-books in this series can be found on Nova's website
under the e-book tab.

OCEANOGRAPHY AND OCEAN ENGINEERING

LAGOONS

HABITAT AND SPECIES, HUMAN IMPACTS AND ECOLOGICAL EFFECTS

MWINYIKIONE MWINYIHIJA
EDITOR

nova publishers
New York

For permission to use material from this book please contact us:
Telephone 631-231-7269; Fax 631-231-8175
Web Site: http://www.novapublishers.com

Additional color graphics may be available in the e-book version of this book.

Library of Congress Cataloging-in-Publication Data

Lagoons : habitat and species, human impacts and ecological effects / editor, Mwinyikione Mwinyihija, PhD, DSc, FSB, FCIWEM., HSC, CSci., CEnv., CBiol., C.WEM.
pages cm
Includes index.
ISBN: 978-1-62808-092-6 (hardcover)
1. Lagoons. 2. Lagoon ecology. I. Mwinyihija, Mwinyikione.
GB2203.2.L34 2013
577.7'8--dc23

2013018514

Published by Nova Science Publishers, Inc. † New York

'To all those who have dedicated their lives to pursue, evolve and disseminate knowledge to the global audience'

Mwinyihija, 2013

CONTENTS

PREFACE

The study title 'Lagoon: Habitat and Species, Human Impacts and Ecological Effects' was an attempt to provide insights attributed to both natural and anthropogenic influences to the lagoon ecosystems. Human influence related to lagoons is diverse but commonly drawing benefits in areas of transportation, fish aquaculture, sand mining, recreation, waste management etc. Indeed prior to understanding the structure and approach in this book it is imperative to take cognizance of the diverse and very articulate authorship of the chapters herein. This book in its respective chapters draws representation from South America, Africa (including the Mediterranean region) and Europe. Moreover, the chapters individually address and share particular areas of interest that include physical, chemical and biological processes involved in lagoon systems.

To select a few areas for example, in an effort to comprehend the fundamentals of habitat and species of lagoon systems, the eminent conspicuous issue is lack of information related to the composition of micro-biota of the Mediterranean coastal lagoons. Fortunately a few authors have dedicated their chapters to relate and compound on this aspect. In addition, issues related to anthropogenic activities and its ecological impacts have received due attention and foretell the biogeochemistry of applying lagoons in managing waste emanating from such activities. Thusly, the book through its chapters encompasses a scholarly approach (by research, exposure and experience) directed towards comprehending on these critical aspects raised of lagoon ecosystem, in particular depicting;

a. Relationship between benthic vegetation and sediment biogeochemistry
b. Development of micro-algal bloom
c. Aerobic and anaerobic influences
d. Primary producers and benthic fauna relationships etc.

The importance of lagoons, whether open, restricted or in any other form are closely discussed by the authors in this book. Essentially the basis was also to include structural, locational and complex aspects that provided an insight in the biodiversity and characterization of the dynamism of such systems. For example, on a narrow view to illustrate the structural approach in this book, benthic fauna and flora rich in plants like charophytes (including other important aquatic invertebrates and related fauna) are given due attention in some of the chapters explaining the explicit and implicit factors that impact on their threats and decline.

To improve on scope and enrich the discussion, aspects such as eutrophication and pollution is also included in the book. In reaching for the goals stated here-above the span of coverage also enveloped areas touching on shallow waters separated by large bodies of water associated with coastal features to anthropogenic waste water treatments lagoons. The latter's interest lies on the low cost associated to such lagoons and biogeochemistry influencing anaerobic, natural aerobic or aerated systems. Therefore with a more inclusive, depth and varied approach in the study of lagoons, there is no iota of doubt that the book provides a highly enlightening scope of work related to the lagoon ecosystem its habitat, species and ecological impacts relating human influences on all the dimensions therein.

Dr. M. Mwinyihija,
PhD, DSc, FSB, FCIWEM, HSC, CSci.,CEnv.,CBiol.,C.WEM.
Editor

In: Lagoons
Editor: Mwinyikione Mwinyihija
ISBN: 978-1-62808-092-6

Chapter 1

INTERACTIONS BETWEEN SPECIES IN COASTAL LAGOONS: PREDATION ON RECENT BRACKISH OSTRACOD POPULATIONS OF THE LAGOON OF VENICE (NE ITALY)

***F. Ruiz*[*1], *M. Abad*[1], *M. L. González-Regalado*[1], *A. Toscano*[1], *L. Menegazzo*[2], *E. Molinaroli*[2], *G. Rampazzo*[2], *M. Pistolato*[2] *and E. X. García*[3]**

[1]Departamento de Geodinámica y Paleontología, Universidad de Huelva, Huelva, Spain
[2]Departamento de Geología, Universidad de Huelva, Huelva, Spain
[3]Departamento de Botánica y Zoología, Universidad de Guadalajara, México

ABSTRACT

This chapter analyzes the first evidences of predation observed on recent brackish ostracod populations of the Lagoon of Venice. The study of ninety-nine samples and almost 1300 specimens allowed extraction of 12 bored valves. The bioerosion structures (*Oichnus simplex* Bromley) are concentrated in the central area of the valves. Borehole diameter shows a negative but not significant correlation with the dimensions of the predated valves, whereas the number of bored shells is significantly correlated with the ostracod density. Predation affects the most abundant species of this environment (mainly adults and the last juvenile instars).

[*] Departamento de Geodinámica y Paleontología, Universidad de Huelva, Avda. Tres de Marzo, s/n, 21071-Huelva, Spain.

1. Introduction

The Lagoon of Venice (LV) is one of the largest lagoons in the Mediterranean (550 km^2 in area, ~50 km in length, 7-15 km in width). This lagoon is connected to this sea through three inlets (Lido, Malamocco and Chioggia) with variable widths (500-1000 m) and depths (8-15 m). These inlets are situated on the eastern side of the lagoon and permit the water exchanges during the semidiurnal ebb-flood periods (~111 x 10^9 m^3/year; Gacic et al. 2005).

This complex environment has been divided into four sub-basins (Figure 1: Treporti, Lido, Malamocco and Chioggia; Molinaroli et al. 2009), separated by watersheds with minimum residual flows (< 10 cm/s; Cavaleri 1980, Solidoro et al. 2004). The inner areas are constituted by intertidal marshes, intertidal mudflats, submerged mudflats and numerous natural or navigation channels, with a mean depth lower than 1 m (Sarretta et al. 2010). Salinity ranges usually between 28°/₀₀ and 36°/₀₀, although these values diminish near some fluvial mouths (< 7°/₀₀; Marangon 1996).

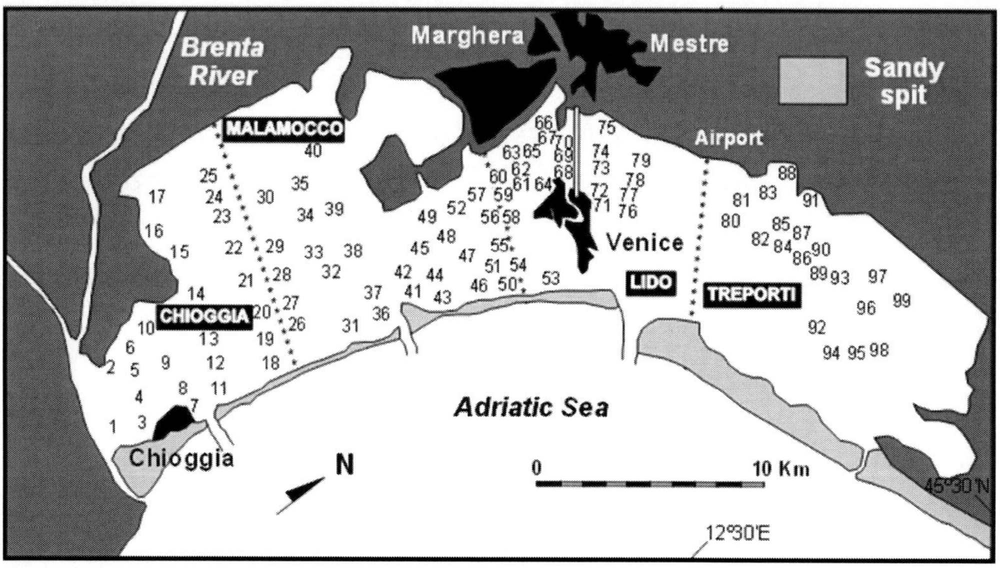

Figure 1. Lagoon of Venice, with the four sub-basins and location of samples.

Bottom sediments are mostly silty (mean mud content ~80%; Molinaroli et al. 2009). These fine sediments are polluted near the industrial zone of Porto Marghera, a 'contaminated site of national interest' (SIN) with high heavy metal concentrations (Zn: 200-3800 mg/kg; Cu: 30-310 mg/kg; Hg: 0.7-15 mg/kg; Pb: 50-280 mg/kg; Zonta et al. 2007). These values decrease remarkably in the remaining areas, although most trace elements (e.g. Cr, Cu, Zn, and Hg) have concentration above background values (Donazzolo et al. 1984, Coccioni et al. 2009).

A number of different homogeneous areas (biotopes) have been determined in this lagoon from the distribution of the microfauna, especially benthic foraminifera (e.g. Albani et al. 2007, 2010, Serandrei-Barbero et al. 2008). Data on species composition of ostracods is comparatively inadequate, with a general overview carried out by Ruiz et al. (2000). *Cyprideis torosa* (Jones, 1850) is the dominant ostracod species, although marine species

(*Pontocythere turbida* (Müller, 1894), *Semicytherura* spp.) are frequent near the three inlets and some species of *Xestoleberis* are well represented near the low hydrodynamic watersheds.

These microcrustaceans represent a food source for numerous groups, such as gastropods (Elewa 2007), carnivorous bivalves (Leal 2008) or fishes (Vaske Junior et al. 2008). Evidences of predation on ostracod shells have been studied almost exclusively in marine environments, both in recent and fossil populations (e.g., Reyment et al. 1966, 1987, Ruiz and González-Regalado 1990; Ruiz et al. 2009).

This paper investigates the evidences of predation on ostracod populations of LV. It represents one of the first systematic analysis of these evidences in coastal lagoons. This study includes: a) the main features of the boreholes; b) a comparison with the external dimensions of the bored valves; c) an estimation of the percentages of bored valves; d) the main characteristics of the different environments with these evidences; and e) a revision of the ontogenetic stages affected by this predatory pressure.

2. Materials and Methods

Ninety-nine samples (Figure 1) were collected from bottom sediments of LV, each weighing 25 g dry weight. All samples were washed using hot tap water and sieved using 0.063 mm mesh width. The material was dried at 70°C and hand picked under a binocular microscope.

If possible, 50 individuals were picked from each sample. The count was then recalculated to yield the total number of ostracods in the whole sample and the ostracod density. This procedure yielded almost 1300 ostracod valves and carapaces. All material was examined for evidence of predation. The collections are housed at the Department of Geodynamics and Paleontology of the University of Huelva, Spain.

Both length and height of each drilled ostracod were measured as well as the diameter of the drilled holes. Pope-diagrams were prepared in order to summarize information on preferred locations of boreholes. The term 'Pope-diagram' for predation maps was introduced by Carriker (1955). These diagrams have been used in previous studies on ostracod predation (e.g. Reyment et al. 1987).

In addition, the external diameter (D) of these evidences was measured and compared with length and height of valves that include them. The Pearson's linear correlation coefficient was used for this purpose.

Finally, material consists of adults and larval shells and consequently the ontogenetic stage of each bored valve or carapace was determined.

3. Taxonomical Accounts

Ichnogenus *Oichnus* Bromley, 1981
(Type ichnospecies : *Oichnus simplex* Bromley, 1981)
Oichnus simplex Bromley, 1981
(Figure 2, a, 1-11)

This ichnospecies is a bioerosion structure observed in the external side of several ostracod valves. All these boreholes present a cylindrical structure, with external and internal openings of about the same size (15-85 μm). In external view, these cylinders show a circular or elongated shape (Figure 2, a), with the presence of scratches near the periphery.

Oichnus paraboloides Bromley, 1981 differs from *O. simplex* in the truncated, cone-shaped structure of the borehole. *O. simplex* implies the total penetration of the valve, whereas *Fossichnus solus* Nielsen, Nielsen and Bromley, 2003 represents an intermediate stage with an incomplete perforation.

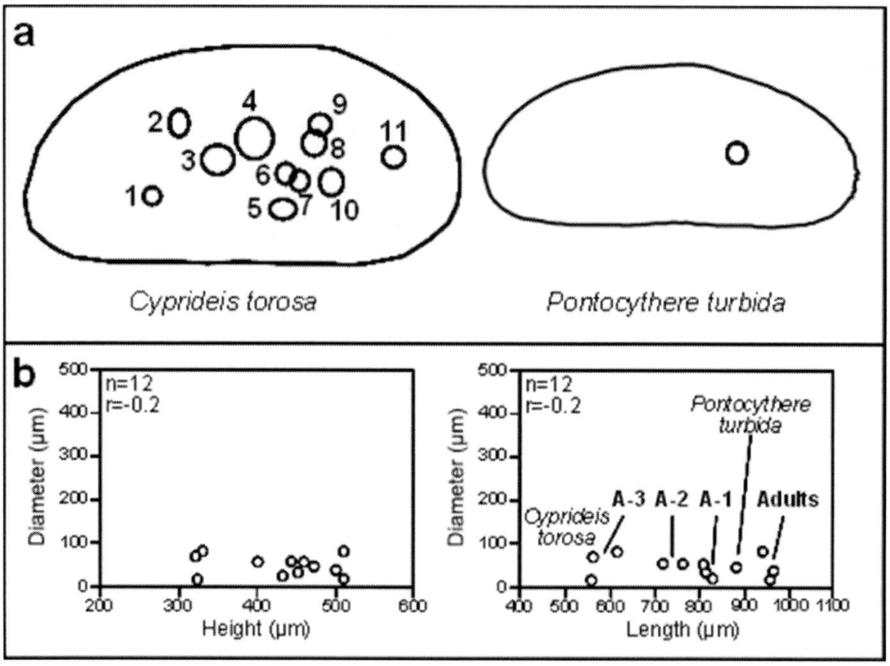

Figure 2. a: Pope-diagrams showing some predation evidences observed in the Lagoon of Venice. Scale not respected. b: Heigth vs diameter and length vs diameter diagrams, with inclusion of the Pearson's correlation coefficient.

4. Results

4.1. Drill Holes

Pope-diagrams (Figure 2, a) indicate a preferential location of the drill holes on the central area of the valves (>83 %), whereas these predation structures are uncommon near the anterior and ventral margins (one specimen in each case). Each bored valve present a single mark.

The analysis of the specimen size *vs* drill hole diagrams (Figure 2, b) shows a negative but no significant correlation between the carapace dimensions and D. The correlation coefficient is very similar (r=-0.2) between D and either length or height.

4.2. Percentages and Distribution of Predation

These evidences of predation are very scarce in LV (12 valves; total mean: 0.86%), with the absence of bored valves in 90 of 99 samples studied. In the nine remaining samples, the predation intensity is very variable (Table 1: 0.4-11.8%), although the number of drilled valves per sample is very low (≤ 3).

Table 1. Ostracoda. Distribution, density and diversity

Samples/Species	*Cyprideis torosa*	*Pontocythere turbida*	Individuals picked	Individuals per 10 g	% Bored shells
VEN-1			3	1	
VEN-2		1	83	33	1.2
VEN-3					
VEN-4					
VEN-5			7	3	
VEN-6			18	7	
VEN-7					
VEN-8			15	6	
VEN-9					
VEN-10					
VEN-11					
VEN-12					
VEN-13					
VEN-14			4	2	
VEN-15					
VEN-16					
VEN-17					
VEN-18					
VEN-19					
VEN-20					
VEN-21					
VEN-22			4	2	
VEN-23					
VEN-24					
VEN-25			25	10	
VEN-26			5	2	
VEN-27					
VEN-28					
VEN-29					
VEN-30					
VEN-31			9	4	
VEN-32			1	< 1	
VEN-33			3	1	

This number presents a positive and significant correlation (r= 0.367; $p < 0.01$) with the ostracod density. Samples with bored individuals were collected mainly near the airport and the innermost areas located between Venice and Porto Marghera.

Table 1. (Continued)

Samples/Species	*Cyprideis torosa*	*Pontocythere turbida*	Individuals picked	Individuals per 10 g	% Bored shells
VEN-34					
VEN-35					
VEN-36					
VEN-37					
VEN-38			6	2	
VEN-39			2	1	
VEN-40			3	1	
VEN-41			48	19	
VEN-42					
VEN-43					
VEN-44					
VEN-45			1	< 1	
VEN-46			8	3	
VEN-47			11	4	
VEN-48					
VEN-49	1		40	16	2.5
VEN-50			1	< 1	
VEN-51			8	3	
VEN-52			9	4	
VEN-53	2		17	7	11.8
VEN-54			10	4	
VEN-55					
VEN-56			6	2	
VEN-57			7	3	
VEN-58			9	4	
VEN-59			1	< 1	
VEN-60	1		13	5	7.7
VEN-61			9	4	
VEN-62					
VEN-63			2	1	
VEN-64	1		20	8	5
VEN-65					
VEN-66					

Samples/Species	*Cyprideis torosa*	*Pontocythere turbida*	Individuals picked	Individuals per 10 g	% Bored shells
VEN-67					
VEN-68					
VEN-69			2	1	
VEN-70					
VEN-71					
VEN-72					
VEN-73					
VEN-74					
VEN-75					
VEN-76					
VEN-77			19	8	
VEN-78			1	< 1	
VEN-79					
VEN-80			10	4	
VEN-81	1		275	110	0.4
VEN-82			10	4	
VEN-83	1		54	22	1.9
VEN-84			8	3	
VEN-85			4	2	
VEN-86			9	4	
VEN-87			112	45	
VEN-88			141	56	
VEN-89			3	2	
VEN-90			2	1	
VEN-91	3		67	27	4.5
VEN-92			6	2	
VEN-93			5	2	
VEN-94	1		62	25	1.6
VEN-95			38	15	
VEN-96			41	16	
VEN-97			5	2	
VEN-98			3	1	
VEN-99			1	1	

4.3. Predated Species

Bored valves belong almost exclusively to *Cyprideis torosa* (11 valves), the most abundant species of LV and subjected to a predatory pressure on (A-3) juvenile instar, at least (Figure 2, b). In addition, a bored (A-3) instar of the marine species *Pontocythere turbida* has been also picked. The measure of the prey dimensions permits to calculate a 'size refuge' (*sensu* Colbath 1985) for the ostracod populations of LV (length < 540 μm; height < 300 μm), beyond which freedom from attack would be assured.

These two bored species present a smooth or slightly punctuated external surface. In addition, one bored specimens of *Cyprideis torosa* collected near the Dese mouth shows a nodded pattern.

5. Discussion

Oichnus simplex is usually attributed to the predatory strategy of muricid or euliminid gastropods (Carriker 1955, Reyment 1963, Donovan and Pickerill 2004) and small turbellarians (Maddocks 1988). The resulting boreholes are generally smaller in diameter than those produced by naticids or other groups of gastropods (Bhatia et al. 1987), which can explain the small size of the boreholes observed. Some possible muricid predators collected in LV are *Hexaplex trunculus* (Linnaeus, 1758), *Nassarius nitidus* (Jeffreys, 1867) or *Nassarius reticulatus* (Linnaeus, 1768) (Berto et al. 2006, 2007).

The central concentration of these boreholes has been described by Reyment (1963, 1966) in both fossil and recent ostracod populations. Nevertheless, some elongated species show a different pattern, with the boreholes restricted to the anterior or posterior areas of the carapace (e.g. *Bythocypris*; Reyment 1966), whereas other studies indicate a random arrangement of drilling sites (Aranki 1987).

The negative correlation coefficient measured between hole dimensions and prey size is unusual, because this parameter is always positive and sometimes statistically significant in previous investigations (Reyment et al. 1987, Ruiz et al. 2009). This disparity may be due to: a) the low number of data; or b) the attack of different ontogenetic stages of ostracods by juvenile specimens of gastropods with similar sizes.

Predation intensity of LV is very low, with a very high number of samples without bored shells. These low values have been also detected on ostracod populations of Holocene sequences (Ruiz et al. 2010), but they contrast with the frequent record of boreholes in fossil and recent shallow marine environments (Ruiz et al. 2009, 2011).

The significant correlation between the ostracod density and the number of bored shells was not observed in previous papers on ostracod predation (Reyment 1966). Nevertheless, this data should be used with caution, owing to the scarcity of drilled valves in LV.

Predators attack the last juvenile instars (A-1 to A-3) and adults. This preference has been also observed in the western Niger delta (Reyment 1966). Intermediate stages (A-4 and A-5) are also present in LV, although they are not predated (Ruiz et al. 2000).

Conclusion

Predation is extremely rare on ostracod populations of LV. Bored valves are found mainly in the innermost areas of the lagoon, where adults and the last juvenile instars are predated probably by muricid gastropods. A positive, significant correlation was found between the ostracod density and the number of drilled valves.

Boreholes are concentrated in the central area of bored valves, with a negative correlation between the hole diameter and the prey size. The unusual results derived from these statistical analyses (hole diameter *vs* prey dimensions; ostracod density *vs* predation intensity) are conditioned, at least partially, by the low number of drill shells.

ACKNOWLEDGMENTS

This work was funded by a Spanish DGYCIT Project (CGL2010-15810), an Excellent Project of the Andalusia Board funded by FEDER (SEJ-4770) and a Research Group of the Andalusia Board (RNM-238).

REFERENCES

Albani, A. D., Serandrei-Barbero, R., Donnici, S. (2007). Foraminifera as ecological indicators in Lagoon of Venice, Italy. *Ecol. Indicat.*, 7, 239-253.

Albani, A. D., Donnici, S., Serandrei-Barbero, R. (2010). Seabed sediments and foraminifera over the Lido Inlet: Comparison between 1983 and 2006 distribution patterns. *Cont. Shelf Res.*, 30, 847-858.

Aranki, J. F. (1987). Marine Lower Pliocene Ostracoda of southern Spain with notes on the Recent fauna. *Pub. Palaeont. Inst. Univ. Uppsala*, 13, 1-134.

Berto, D., Giani, M., Covelli, S., Boscolo, R., Cornello, M., Macchia, S., Massironi, M. (2006). Mercury in sediments and *Nassarius reticulatus* (Gastropoda, Prosobranchia) in the southern Venice Lagoon. *Sci. Tot. Environ.*, 368, 298-305.

Berto, D., Giani, M., Boscolo, R., Covelli, S., Giovarnardi, O., Massironi, M., Grassia, L. (2007). Organotins (TBT and DBT) in water, sediments and gastropods of the southern Venice lagoon (Italy). *Mar. Poll. Bull.*, 55, 425-435.

Bhatia, B., Sirjana, S., Sunila, B. (1989). Predation by gastropod drills on Early Lutetian ostracods from Kutch, Western India. *Geobios*, 22, 387-394.

Carriker, M. R. (1955). Critical review of biology and control of oyster drills *Urosalpinx* and *Eupleura*. *U.S. Dept. Interior Fish and Wildlife Serv., Spec. Sci. Rep.-Fisheries*, 148, 1-150.

Cavaleri, L. (1980). Sediment transport in shallow lagoons. *Nuovo Cimento 3C*, 5, 527-540.

Coccioni, R., Frontalini, F., Marsili, A., Mana, D. (2009). Benthic foraminifera and trace element distribution: A case study from the heavily polluted lagoon of Venice (Italy). *Mar. Poll. Bull.*, 59, 257-267.

Colbath, S. L. (1985). Gastropod predation and depositional environments of two molluscan communities from the Miocene Astoria Formation at Beverly beach state park, Oregon. *J. Paleont.*, 59, 849-869.

Donazzolo, R., Orio, A. A., Pavoni, B., Perin, G. (1984). Heavy metals in sediments of the Venice lagoon. *Oceanol. Acta*, 7, 25-32.

Donovan, S. K., Pickerill, R. K.. (2004). Traces of cassid snails upon the echinoids from the Middle Miocene of Poland: Comments on Ceranka and Zlotik (2003). *Acta Palaeont. Pol.*, 49, 483-484.

Elewa, A. M. T. (2007). Predation due to changes in environment: Ostracod provinciality at the Paleocene-Eocene thermal maximum in North and West Africa and the Middle East.

In: Elewa, A. M. T. (ed.). *Predation in organisms-A Distinct Phenomenon*. Springer Verlag, Berlin and Heidelberg, pp. 7-26.

Gacic, M., Kovacevic, V., Mancero Mosquera, I., Mazzoldi, A., Cosoli, S. (2005). Water fluxes between the Venice Lagoon and the Adriatic Sea. In: Fletcher, C. A., Spencer, D. T. (eds.), *Flooding and Environmental Problems of Venice and Venice Lagoon: State of Knowledge*. Cambridge University Press, Cambridge. pp. 431-444.

Leal, J. H. (2008). A remarkable new genus of carnivorous, sessile bivalves (Mollusca: Anomalodesmata: Poromyidae) with descriptions of two new species. *Zootaxa*, 1764, 1-18.

Maddocks, R. F. (1988). One hundred million years of predation on ostracods. In: Hanai, T. T., Ikeya, N., Ishizaki, K. (eds.), *Evolutionary biology of Ostracoda*. Elsevier, Amsterdam. pp. 637-657.

Marangon, L. (1996). Studio dei diversi ambienti lagunari tramite l'integrazione di osservazioni sui sedimenti e campionamenti delle acque con modelli idrodinamici diffusivi. Tesi di laurea, Università Ca'Foscari degli studi di Venezia, Venezia, Italy.

Molinaroli, E., Guerzoni, S., Sarretta, A., Masiol, M., Pistolato, M.. (2009). Thirty-year changes (1970-2000) in bathymetry and sediment texture recorded in the lagoon of Venice sub-basins, Italy. *Mar. Geol.*, 258, 115-125.

Reyment, R. A. (1963). Studies on Nigerian Upper Cretaceous and Lower Tertiary Ostracoda, part 2: Danian, Paleocene and Eocene ostracods. *Stockholm Cont. Geol*, 10, 1-28.

Reyment, R. A. (1966). Preliminary observations on gastropod predation in the western Niger delta. *Palaeogeog., Palaeoclimatol., Palaeoecol.*, 2, 81-102.

Reyment, R. A., Honigstein, A. (1987). Predation by boring gastropods on Late Cretaceous and Early Palaeocene ostracods. *Cret. Res.*, 8, 189-209.

Ruiz, F., González-Regalado, M. L. (1990). Bioerosión en ostrácodos. *Geogaceta*, 6, 87-90.

Ruiz, F., González-Regalado, M. L., Baceta, J. I.,Menegazzo-Vitturi, L., Pistolato, M., Rampazzo, G., Molinaroli, E.. (2000). Los ostrácodos actuales de la laguna de Venecia (NE de Italia). *Geobios*, 33, 447-454.

Ruiz, F., Abad, M., González-Regalado, M. L., Civis, J., González-Delgado, J. A., García, E. X., Toscano, A.. (2009). Predation on Neogene ostracods of southwestern Spain. *Riv. Ital. Paleont. Stratig.*, 116, 149-157.

Ruiz, F., Abad, M., González-Regalado, M. L., Toscano, A., Rodríguez Vidal, J., Cáceres, L. M., Pozo, M., Carretero, M. I., García, E. X. (2010). Evidencias de depredación en ostrácodos holocenos del Parque Nacional de Doñana (SO de España). *Rev. Esp. Micropaleont.*, 42, 267-273.

Ruiz, F., Abad, M., González-Regalado, M. L., Tosquella, J., García, E. X., Toscano, A., Muñoz, A., Pendón, J. G. (2011). Predation on Recent marine ostracod populations of southwestern Spain. *Ameghiniana*, 48, 113-121.

Sarretta, A., Pillon, S., Molinaroli, E., Guerzoni, S., Fontolan, G.. (2010). Sediment budget in the Lagoon of Venice, Italy. *Cont. Shelf Res.*, 30, 34-949.

Serandrei-Barbero, R., Albani, A. D., Donnici, S. (2008). *Atlante dei Foraminiferi della Laguna di Venezia*. Istituto Veneto di Scienze, Lettere ed Arti, Venezia.

Solidoro, C., Canu, D., Cucco, A., Umgiesser, G. (2004). A partition of the Venice Lagoon based on physical properties and analysis of general circulation. *J. Mar. Syst.*, 51, 147-160.

Vaske Júnior, T. R., Lessa, R. P., Barbosa, T. M., Tolotti, M. T., Ribeiro C. B. (2008). Stomach contents of the Caribbean pomfret *Brama caribbea* (Mead, 1972) from stomach contents of great pelagic predators from southwestern equatorial Atlantic. *Bol. Inst. Pesca*, 34, 241-249.

Zonta, R., Botter, M., Bassin, D., Pini, R., Scattolin, M., Zaggia, L. (2007). Sediment chemical contamination of a shallow water area close to the industrial zone of Porto Marghera (Venice Lagoon, Italy). *Mar. Poll. Bull.*, 55, 529-542.

In: Lagoons
Editor: Mwinyikione Mwinyihija

ISBN: 978-1-62808-092-6

Chapter 2

DYNAMICS OF HARMFUL DINOFLAGELLATES AND CILIATES IN THE GHAR EL MELH LAGOON (TUNISIA)

Amel Dhib,[1,2*] ***Souad Turki,***[1] ***Fourat Akrout***[1] ***and Lotfi Aleya***[2]
[1]Institut National des Sciences et Technologies de la Mer (INSTM), Laboratoire Milieu Marin, Centre la Goulette, Tunisie
[2]Université de Franche-Comté, Laboratoire de Chrono-Environnement, France

ABSTRACT

Mediterranean lagoons are among the most vulnerable ecosystems. The Ghar El Melh lagoon is a Mediterranean water body, located in north-eastern Tunisia. It's Ramsar site suffers from various natural and anthropogenic interferences, such as pollution and silting. To investigate the water quality of the Ghar El Melh lagoon, an annual survey of abiotic (physical and chemical parameters) and biotic factors (phytoplankton and ciliate) was carried out. The lagoon ranges from euryhaline to hyperhaline (26.6-51.2 psu) and can be considered hypereutrophic with outstanding concentrations of total phosphorus (0.74-6.93 μmol l^{-1}) and dissolved inorganic nitrogen (0.83-22.38 μmol l^{-1}). The phytoplankton community was dominated by dinoflagellates (54.87 %) and diatoms (38.85 %). Algae proliferations were recorded in the spring and summer with a maximal abundance of about 4.51 10^5 cells l^{-1}. Among 32 dinoflagellate species identified in this study, 19 were reported in the literature as harmfuls. For ciliate assemblage, a list of 28 species was compiled and was largely dominated by the Spirotrichea group (80.66 %) in which Choreotrichida represented 44.16 % followed respectively by Tintinnida (34.15 %) and Strombidiida (16.39 %). High concentrations of the Spirotrichea species were recorded in spring season.

Keywords: Ghar El Melh lagoon, water quality, abiotic and biotic parameters

---

[*] Corresponding author: Amel Dhib. E-mail: amell.dhib@gmail.com.

INTRODUCTION

The coastlines of the Mediterranean Sea comprise about six hundred lagoons. As a semi enclosed system, coastal lagoons are strongly dependent on the watershed inputs (Valiela et al. 1997; De Wit et al. 2005) and are particularly vulnerable ecosystems (Cloern, 2001; Verlaque, 2001; Lloret et al. 2008).

The Ghar El Melh lagoon is a Mediterranean water body, located in north-eastern Tunisia, and it was included within a recently (November 2007) designated Ramsar site (Ramsar Bureau, 2007a). The site fulfills the Ramsar criteria for internationally important wetlands that are representative of lagoons within the Southern Mediterranean, its habitat for migrating fish, especially in winter, and as a wintering and breeding site for many birds including grebes, herons, sterns, flamingos and gulls. It also qualifies as internationally important as its population of Collared Pratincole (*Glareola pratincola*) exceeds 1 % of the total Palaeoarctic population. Unfortunately, during the last few years the Ghar El Melh lagoon has been affected by the progressive restriction of seawater exchange (Menif, 2000). This factor, coupled with various natural and anthropogenic interferences, such as pollution and silting, contributed to the alteration of the water and sediment quality, resulting in a decrease in biodiversity (SCET-ERI, 2000).

To meet this situation, an annual survey of abiotic (physical and chemical parameters) and biotic factors (phytoplankton and ciliates) was carried out. To our knowledge, this work is the first investigation of the planktonic ciliated communities in this site. In Tunisia, studies addressing the distribution of ciliate assemblages was recently reviewed for the south-east coast (Elloumi et al. 2006; Hannachi et al. 2009; Kchaou et al. 2009; Drira et al. 2009; Rekik et al. 2011) and they were very scarce in northern part (Hannachi et al. 2011). The Ghar El Melh Lagoon remains relatively unexplored for HAB bloom formation for phytoplankton in comparison to other Tunisian coastal areas. The only occasional data from the lagoon were those reporting fish mortality in 1995–1996, mainly caused by *Alexandrium minutum* (Romdhane et al. 1998) and in 2007 due to dinoflagellate *Kryptoperidinium foliaceum* (Turki et al. 2007).

In summary, the results of seasonal changes of phytoplankton and ciliates coupled with physical and chemical parameters at five stations in the Ghar El Melh lagoon are analyzed and our aim is to investigate the water quality of this area.

MATERIAL AND METHODS

1. Study Site

The Ghar El Melh Lagoon (37° 06′N, 10° 11′E) is coastally located in the northern part of the Gulf of Tunis, between Ras Tarf (Cape Farina) and the estuary of the Medjerda River (Figure 1). The lagoon is shallow with an average depth of 0.8 m and influenced by the regional water circulation (Ben Ismail et al. 2012). It is biologically very rich, harboring fish, shellfish, plants, and several sea birds, with nearly 20,000 ducks (Isenmann et al. 2005). This lagoon is a complex of three smaller lagoons that cover an area of about 35 km^2: the main lagoon or El Bhira (26.7 km^2), Sabkhet El Ouafi (5.2 km^2) in the southeast and Sabkhet Sidi

Ali El Mekki (3.7 km^2) in the northeast. The principal lagoon is connected to the Mediterranean Sea through a permanent connection 70 m wide, identified as El Boughaz and associated with coastal sand bars. Other low-lying sections of these dunes provide intermittent connections to the sea (Ayache et al. 2009). The region presents a subhumid to humid climate (mean temperature, 17 °C; annual rainfall, 580 mm; precipitation spread from October to April, with a maximum in January), with dominant northwesterly winds exceeding 15 ms^{-1} (Thompson et al. 2009). The lagoon is supplied with relatively small volumes of drainage water from the Utica floodplain (which lies in the lower valley of the Mejerda River to the west), but is also known to have several springs (northeast of the main lagoon) and to benefit from the seasonal runoff from the Djebel Nadhour and Demina hills to the north. It has a mean salinity of 37 psu (range, 27–48).

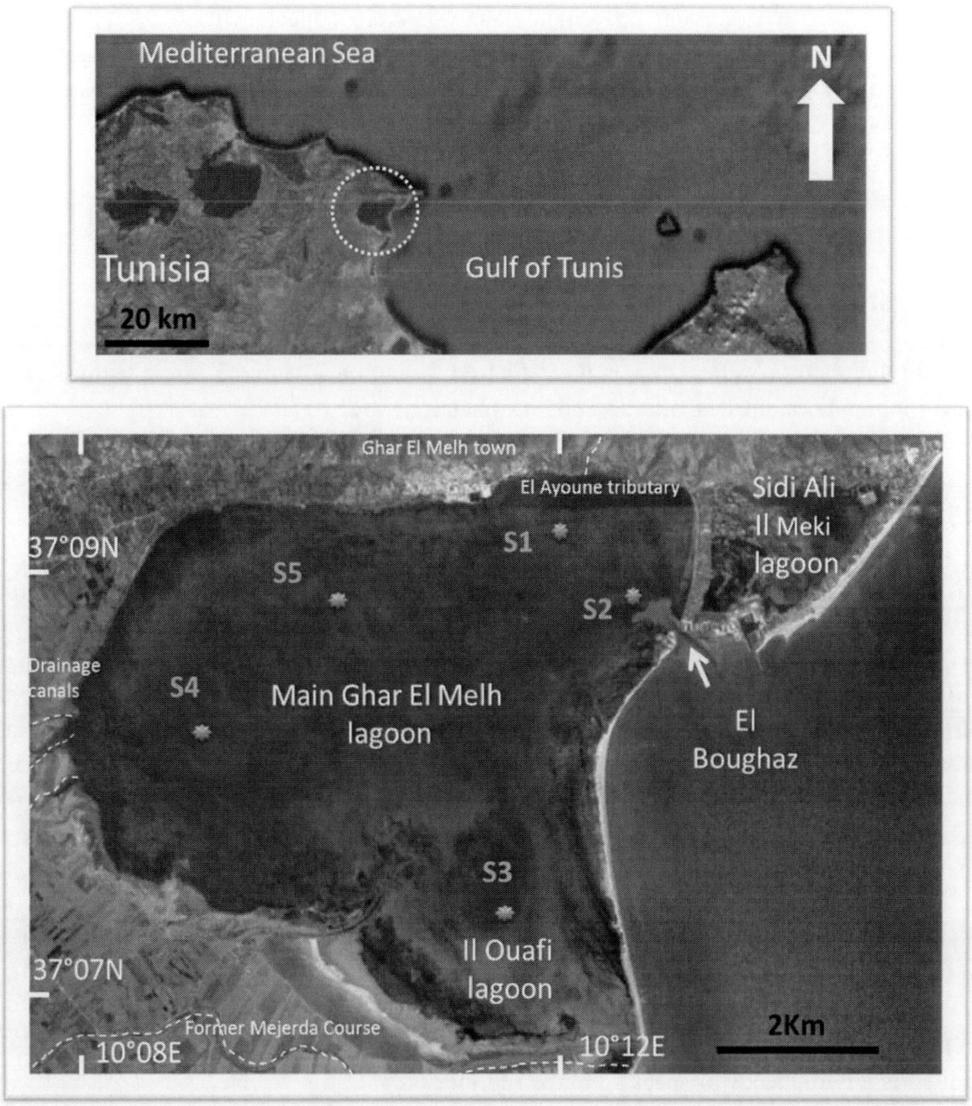

Figure 1. Geographic situation of the Ghar El Melh Lagoon and location of the sampling stations from S1 to S5.

2. Sampling

Samples (N=110) were collected twice a month, from January 2011 to January 2012, at five stations (S1 to S5) (Figure 1). Station 1 is located in the north east of GML. Station 2 faces El Boughaz, the one permanent 70 m wide connection with the Gulf of Tunis. Station 3 is located in the south east of the lagoon in the middle of Sabkhet El Ouafi. Station 4, in the south west of GML, affected by industrial discharges. Station 5, in the north of the lagoon, affected by agricultural and urban spills.

3. *In situ* Measures

Samples for phytoplankton and ciliates analysis were collected from the whole water column using a polyvinylchloride (PVC) tube. Environmental variables were measured in the field using a WTW multiparameter for water temperature (°C) and salinity (psu). Depth and transparency was measured using a Secchi disk.

For analysis of chemical variables, water samples were collected in 1000-ml polypropylene bottles, 30–50 cm under the surface water.

4. Laboratory Analysis

In the laboratory, water samples were filtered through a 47 mm/45 µm Whatman GF/C filter for ammonium, nitrites, nitrates and phosphates.

5. Ciliate and Phytoplankton Enumeration

Sub-samples (25 ml) of ciliate and phytoplankton were counted under an inverted microscope after fixation with lugol solution (final concentration 1% v/v) and settling for 48 hours using the Utermôhl method (1958).

Identification of algal taxa was achieved according to various keys of Dodge (1982, 1985); Sournia (1986), Nezan and Piclet (1996) and Hasle and Syversten (1997). Ciliate identification was made using the keys of Corliss (1961); Petz (1999) and Struder-Kypke and Montagnes (2002).

Tintinnids were identified using lorica morphology and species description according to Balech (1959) and Kofoid and Campbell (1929, 1939). Cell numbers were expressed as cells l^{-1}.

The level of community structure was assessed with Shannon and Weaver's (1949) H′ diversity index.

$$H' = - \sum ni / N * \log2 \; ni / N$$

ni/N: is the frequency of species i in the sample

N: number of species in the community

HYDROLOGIC PARAMETERS

1. Temperature

The mean water temperature increased in spring and reached its maximum in the summer. The highest temperature (27.6 °C) was recorded at station 5 in January 2011, while the lowest (11.1 °C) at station 4 in January 2012. Thermal stratification doesn't exist because of the shallowness of the lagoon (< 2m) (Figure 2).

2. Salinity

The water salinity varied from 51.2 psu in July in station 4 to 26.6 psu in December in station 3. Salinity peaked in the summer. This may be ascribed to the high temperature, water evaporation and the weak inflow of incoming freshwater to the lagoon (Figure 3).

3. Nutriments

Total phosphorus (T-P) concentrations were higher (0.74-6.93 μmol l^{-1}). Orthophosphate concentrations represented only 8.1% of the total phosphorus and reached prominent values (0.05-0.74 μmol l^{-1}). Total nitrogen (T–N) concentrations showed regular trends (mean± s.d.= 19.27 ± 2.97). Nitrogen consisted mainly of the dissolved organic form (68.5 %).

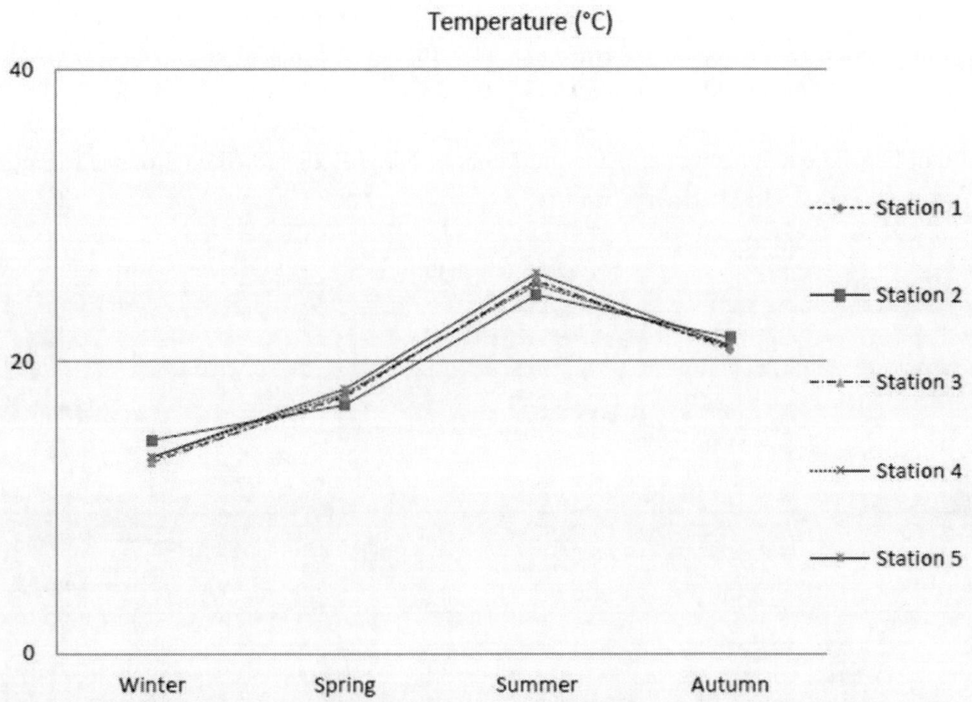

Figure 2. Seasonal variations of temperature water in five stations of Ghar El Melh lagoon.

The dissolved inorganic form (DIN: NO_2+NO_3+NH_4) represented only 31.5 % of the total nitrogen. Ammonium concentrations accounted for 53.79 % of total DIN followed by nitrate 41.92 % and nitrite 4.28 %. DIN values were higher (0.83-22.38 µmol l^{-1}). The nitrate/ammonium ratio was low and peaked in winter with a maximum of 360.62 in February at station 5. The total azote and total phosphorus showed similar trends (Figure 4).

BIOLOGIC PARAMETERS

1. Phytoplankton Community

The different combinations of phytoplankton groups consisted of Dinophyceae (32 species), Bacillariophyceae (22 species), Cyanobacteriae (5 species), Chlorophyceae (one species) and Euglenophyceae (one species) (Table 1).

Dinoflagellates (54.88 %) and diatoms (38.86 %) were the dominant groups, as known in other Northern Tunisian ecosystems (Daly Yahia et al. 1997; Ramdani et al. 2004; Dhib et al. 2012) and Mediterranean regions (Estrada, 1985; Bravo et al. 2010).

Dinoflagellate group was represented mainly by 5 genera: *Prorocentrum* (7 species), *Scrippsiella* (3 species) *Gymnodinium* (2 species), *Dinophysis* (3 species) and *Protoperidinium* (6 species) (Figure 5).

Two peaks of dinoflagellate abundance were recorded in spring (3.68 10^5 cells l^{-1} in station 5) and early summer (3.95 10^5 cells l^{-1} in station 4) (Figure 6) only awarded to *P. micans* proliferations. Another two high concentrations were recorded; the first one was in the summer and only ascribed to *S. trochoidea* (1.78 10^5 cells l^{-1} in station 3); the latter was during the winter season which was mainly awarded to the proliferation of *P. micans* (8.4 10^4 cells l^{-1} in station 1), *Prorocentrum triestinum* (1.7 10^4 cells l^{-1} in station 1), *D. sacculus* (1.4 10^4 cells l^{-1} in station 1) and *S. trochoidea* (1.3 10^4 cells l^{-1} in station 1).

The majority of dinoflagellates species recorded in our study (19 species; 94.55 % of total dinoflagellate abundance) are taxa indicators of a highly disturbed environment (Figure 7).

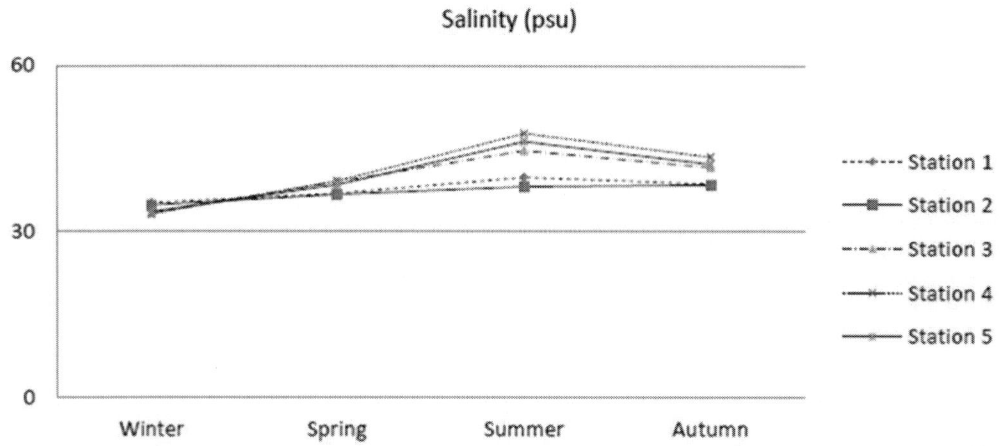

Figure 3. Seasonal variations of salinity in five stations of Ghar El Melh lagoon.

Table 1. List of the phytoplankton species found in the Ghar El Melh lagoon

Phytoplankton community			
Groups	Order	Genus	Species
Dinophyceae	Gymnodiniales	*Gymnodinium*	*Akashiwo sanguinium*
			Gymnodinium catenatum
	Prorocentrales	*Prorocentrum*	*Prorocentrum micans*
			Prorocentrum spp
	Dinophysales	*Dinophysis*	*Dinophysis sacculus*
			Dinophysis spp
		Oxyphysis	*Oxyphysis oxytoxoides*
	Peridiniales	*Alexandrium*	*Alexandrium catenella*
		Ceratium	*Neoceratium* spp
		Gonyaulax	*Lingulodinium polyedra*
			Gonyaulax spinifera
		Protoperidinium	*Protoperidinium diabolum*
			Protoperidinium spp
		Coolia	*Coolia monotis*
		Diplopsalis	*Diplopsalis lenticula*
		Scrippsiella	*Scrippsiella trochoidea*
			Scrippsiella spp
		Pyrophacus	*Pyrophacus horologium*
	Pyrocystales	*Pyrocystis*	*Pyrocystis noctiluca*
Bacillariophyceae	Naviculales	*Navicula*	*Navicula* spp
		Gyrosigma	*Gyrosigma* sp.
		Diploneis	*Diploneis* sp.
		Plagiotropsis	*Plagiotropsis* sp.
	Bacillariales	*Nitszchia*	*Nitszchia* spp
		Pseudo-nitszchia	*Pseudo-nitszchia* sp.
		Bacillaria	*Bacillaria* sp.
		Denticulopsis	*Denticulopsis* sp.
		Cylindrotheca	*Cylindrotheca closterium*
	Fragilariales	*Thalassionema*	*Thalassionema nitzshioides*
		Striatella	*Striatella unipunctata*
		Climacosphenia	*Climacosphenia* sp.
	Chaetocerotales	*Chaetoceros*	*Chaetoceros* spp
	Licmophorales	*Licmophora*	*Licmophora gracilis*
	Rhizosoleniales	*Guinardia*	*Guinardia* sp.
		Rhizosolenia	*Rhizosolenia* sp.
	Achnanthales	*Cocconeis*	*Cocconeis* sp.
	Leptocylindrales	*Leptocylindrus*	*Leptocylindrus danicus*
Cyanobacteriae	Oscillatoriales	*Oscillatoria*	*Oscillatoria salina*
			Oscillatoria sp.
		Spirulina	*Spirulina* sp.
		Lyngbya	*Lyngbya* sp.
	Nostocales	*Anabaenea*	*Anabaenea* sp.
Chlorophyceae	Chlorococcales	*Pediastrum*	*Pediastrum simplex*
Euglenophyceae	Euglenales	*Euglena*	*Euglena* sp.

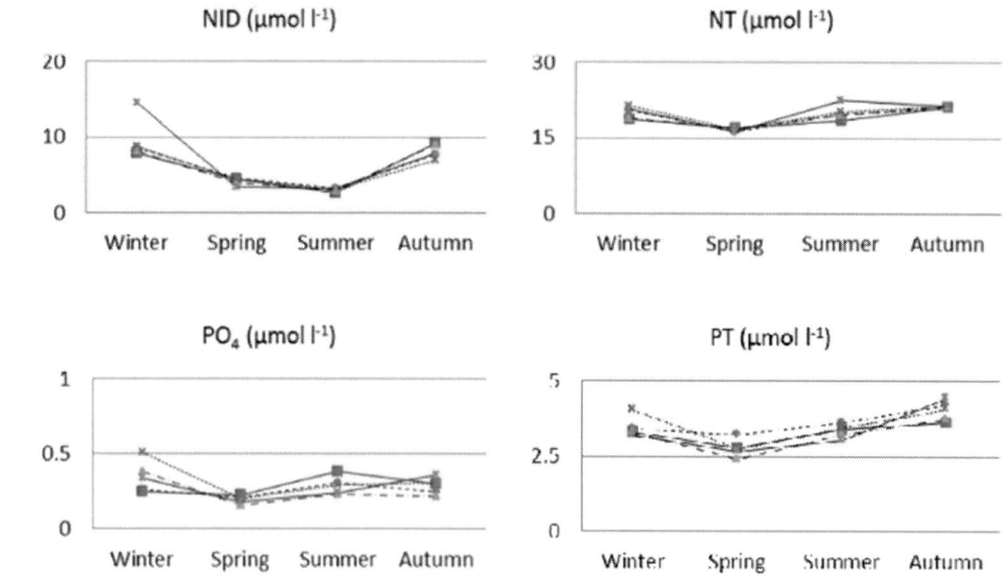

Figure 4. Seasonal variations of nutriments in five stations of Ghar El Melh lagoon.

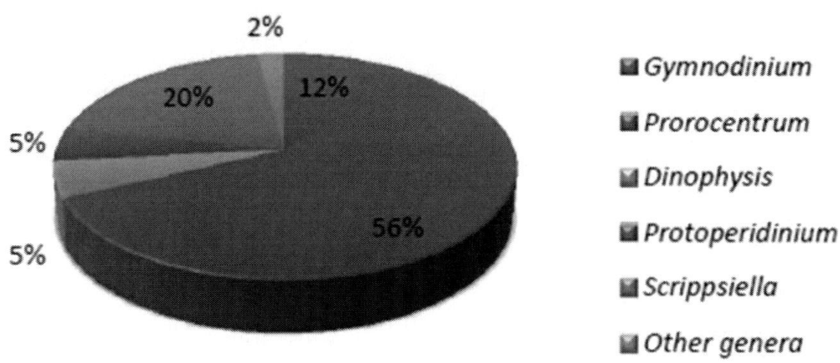

Figure 5. Relative contribution of different genera to dinoflagellate community.

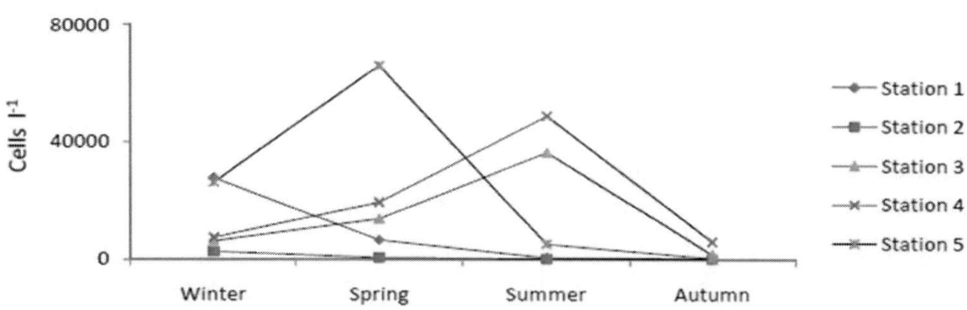

Figure 6. Seasonal changes of dinoflagellate community in five stations of Ghar El Melh lagoon.

Many papers have reported the harmful effects of the dominant taxon *P. micans* (52.74%) which blooms and forms a red tide in many parts of the world (Taylor and Seliger 1979; Anderson et al. 1985; Graneli et al. 1990; Fukuyo et al. 1990; Faust et al. 1999).

Early reports on *P. micans*, indicating that this strain produces a paralytic shellfish poison (Pinto and Silva 1956), remain unconfirmed since recent incidents involving shellfish mortality have been attributed to oxygen depletion (Lassus and Berthome 1988). Toxic effects were reported for *Prorocentrum lima* (Marr et al. 1992), *Prorocentrum concavum* (Hu et al. 1993), *Prorocentrum minimum* (Olenina et al. 2010), *Prorocentrum mexicanum* (An et al. 2010), *D. sacculus* (Boni et al. 1993), *Dinophysis accuminata* (Kat, 1985), *Dinophysis rotundata* (Lee et al. 1989), *Coolia monotis* (Holmes et al. 1995), *Gymnodinium catenatum* (Mee et al. 1986), *Alexandrium catenella* (Hallegraeff, 1991; Taylor et al., 1995), *Lingulodinium polyedra* (Bennouna et al. 2002) and *G. spinifera* (Rhodes et al. 2006).

Also, formation of a bloom and/or red tide have been reported for *Prorocentrum gracile* (Alvial and Garcia 1986), *P. triestinum* (Glibert et al. 2012), *S. trochoidea* (Marshall 2009; Tang et al. 2010), *Akashiwo sanguineum* (Robinson and Brown 1983; Bockstahler and Coats 1993), *Neoceratium furca* (Machida et al. 1999) and *Neoceratium fusus* (Baek et al. 2009).

In contrast to previous reports (Romdhane et al. 1998; Turki et al. 2007), it is noteworthy that no apparent bloom effects (fish mortalities or human contaminations) were detected during our study in the Ghar El Melh lagoon.

Concerning the diatom group, it was dominated by 6 genera: *Navicula* (3 species), *Nitzschia* (2 species), *Gyrosigma* (one species), *Cylindrotheca* (one species), *Thalassionema* (one species) and *Climacosphenia* (one species) (Figure 8)

Two peaks of diatoms were detected in the summer in order to 4.4 10^5 cells. l^{-1} and 3.90 10^5 cells. l^{-1} both in station 5 (Figure 9). The diatoms abundance was characterised by blooms of the following species: *Navicula duerrenbergiana, Navicula phyllepta, Cylindrotheca closterium* and *Nitzschia linearis.* The diversity index of phytoplankton assemblages was close between the different stations (2.23- 2.92 bits $cells^{-1}$).

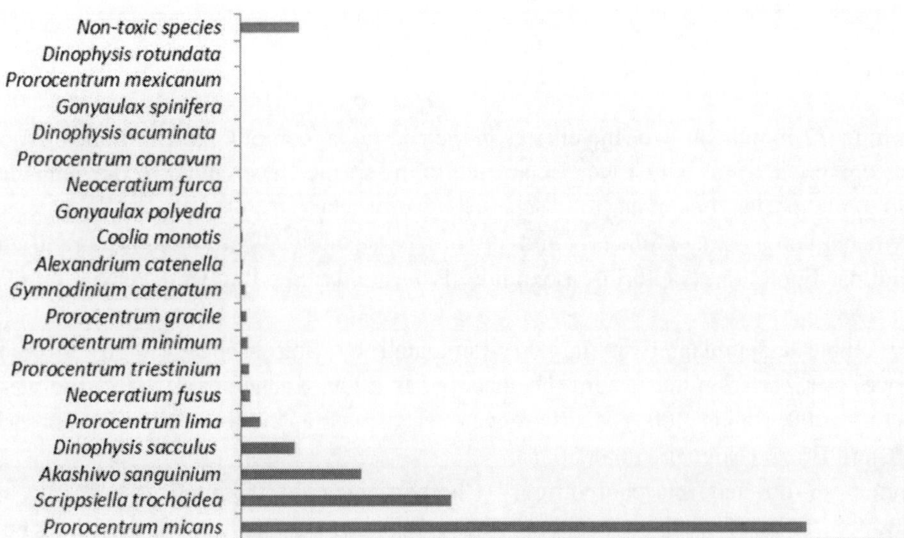

Figure 7. Relative contribution of the harmful species to dinoflagellate community.

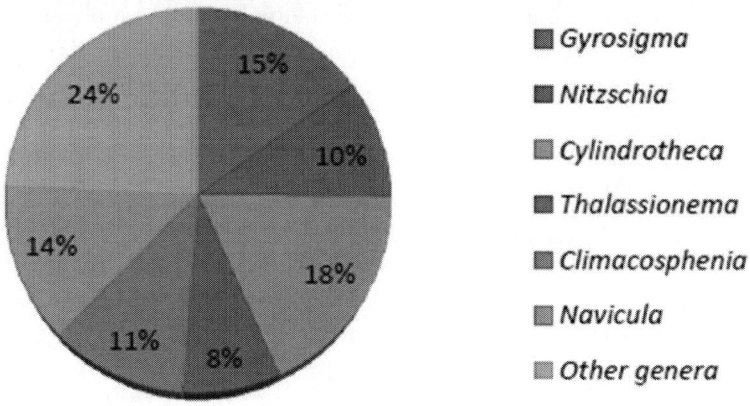

Figure 8. Relative contribution of different genera to diatom community.

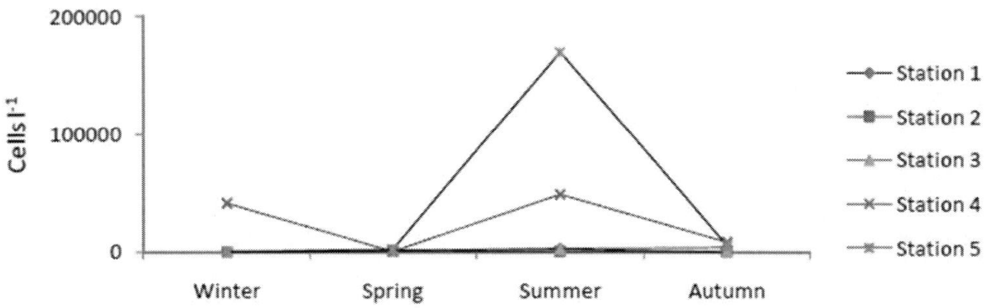

Figure 9. Seasonal changes of diatom community in five stations of Ghar El Melh lagoon.

2. Ciliate Community

During a 12-month study of the ciliates in the coastal lagoon of Ghar El Melh, a list of 28 taxa was compiled, 25 of which were identified to the species level and 3 to the genus level. The different species belonged to 5 classes (Spirotrichea, Litostomatea, Karyorelectea, Oligohymenophorea and Ciliatea) and 9 orders, namely Tintinnida, Choreotrichida, Strombidiida, Euplotida, Haptorida, Loxodida, Philasterida, Sessilida and Peniculida (Table 2).

The ciliate assemblage was largely dominated by Spirotrichea (80.66 % of total abundance) as reported in the eastern (Hannachi et al. 2009; Kchaou et al. 2009) and western (Pérez et al. 2000; Balkis, 2004) Mediterranean Sea, as well as in other environments such as north-eastern Brazil (Nogueira et al. 2005).

Among spirotrichs, the choreotrichs (Choreotrichida + Tintinnida) were the most abundant (63.17 % of total abundance). This group has usually been reported to be the dominant planktonic ciliates in most environments (Carlough and Meyer 1989, Dupuy et al. 2007; Hannachi et al. 2009; Rekik et al. 2011).

Only 12 ciliate species made a considerable contribution to the total ciliate abundance (88.92 %) (Figure 10).

Table 2. List of the ciliate species found in the Ghar El Melh lagoon

Ciliate community			
Class	Subclass	Order	Species
Spirotrichea	Oligotrichia	Strombidiida	*Strombidium conicum*
			Strombidium acutum
			Strombidium chlorophilum
	Choreotrichia	Choreotrichida	*Strobilidium neptuni*
			Strobilidium spiralis
			Leegaardiella sol
			Lohmanniella oviformis
		Tintinnida	*Tintinnopsis* spp
			Eutintinnus tubulosus
			Eutintinnus macilentus
			Codonella aspera
			Favella serrata
			Favella azorica
			Petalotriccha ampula
			Amphorides amphora
			Codonellopsis morchella
			Amphorellopsis acuta
			Metacylis mediterranea
	Hypotrichia	Euplodida	*Euplotes charon*
Litostomatea			*Mesodinium rubrum*
Karyorelictea			*Aspidisca* sp.
Ciliatea			*Paramecium* sp.
Oligohymenophorea			*Uronema marinum*
			Vorticella sp.

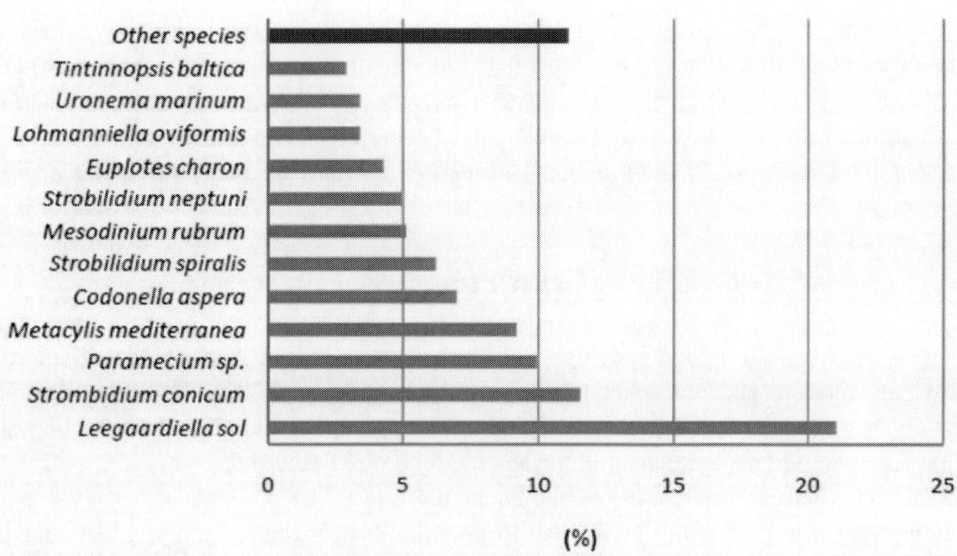

Figure 10. Relative contribution of different ciliate species.

We note that naked ciliates dominated the ciliate community of the Ghar El Melh lagoon (72.47 %) in concordance with the results found in the Gulf of Hamamet (Hannachi et al. 2011) and Mediterranean regions where the loricate ciliate contribute from 5 to 10 % to the ciliate community (Margalef 1963, Dolan and Marassé, 1995; Dolan 2000). Nevertheless, tintinnids dominated in terms of the number of species (15 species) and accounted for 43.6 %, 34.15 % and 27.54 % of total choreotrichs, spirotrichs and total ciliate abundances respectively. The dominance of tintinnids was reported in the Gulf of Tunis (Hannachi et al. unpublished data), the Gulf of Gabès (Hannachi et al., 2009; Kchaou et al., 2009) and other areas (Revelante and Gilmartin 1987) and was mainly related to pollution from the surroundings. In addition, loricate ciliates possess higher adaptability to eutrophic environments than other ciliates (Krsinic et al. 1987; Cordeiro et al. 1997). Our survey demonstrating the importance of tintinnids confirms these findings because Ghar El Melh lagoon is the subject of eutrophication and pollution from anthropogenic activities.

Total ciliate abundances ranged from 0 to 5.35 10^3 cells l^{-1} (mean ± s.d.= 3.8 10^2 ± 6.8 10^2 cells l^{-1}) showing remarkable increase in spring season (mean ± s.d.= 4.76 10^2 cells l^{-1} ± 2.13 10^2 cells l^{-1}) mainly ascribed to the high densities of Spirotrichea species including *Strombidium conicum* (maximal abundance = 2.6 10^3 cells l^{-1} in S5), *Leegaardiella sol* (maximal abundance = 1.64 10^3 in S1) *Metacylis mediterranea* (maximal abundance = 10^3 cells l^{-1} in S4), *Codonella aspera* (maximal abundance = 720 cells l^{-1} in S5), *Strobilidium spiralis* (maximal abundance = 520 cells l^{-1} in S5) and *Tintinnopsis baltica* (maximal abundance = 200 cells l^{-1} in S3).

Another peak of ciliate abundance occurred in September only awarded to the typical ciliate species *Paramecium* sp. (maximal abundance = 5. 36 10^3 cells l^{-1} in S5). This species is especially common in scums and can be considered then as a bioindicator of water quality.

Overall, Choreotrichida, Tintinnida and Strombidiida have the larger abundances in spring season while the remaining groups have the larger abundance and biomass in autumn season mainly awarded to *Paramecium* sp. which has the most important abundance recorded in this study (Figure 11). It has been reported that, in spring season, ciliates are the major group of microzooplankton grazers as they respond rapidly to increasing phytoplankton availability and dinoflagellates follow the ciliate bloom with some delay (Löder et al. 2011).

The diversity index of ciliate assemblages was very close between the different stations (2.02-2.36 bits $cells^{-1}$). In this study, we recognized some species as mixotrophic (*S. conicum* and *Lohmanniella oviformis*), autotrophic (*Mesodinium rubrum*) and benthic (*Aspidisca* sp.).

Conclusion

The Ghar El Melh lagoon is a highly disturbed area suffered from an exacerbation of salinity, temperature and nutriments gradients which is the main case of algae blooming especially harmful dinoflagellates. In addition, ciliates, as a valuable bio-indicator of water quality, are an important component of the aquatic food web in the site.

Such situation can easily weaken the economic interest of the lagoon. So, it becomes absolutely urgent to protect this ecosystem from all kinds of pressures and to adopt concrete measures such as the creation of a sewage wastewater and the dredging areas close to the pass “El Boughaz” the only connection of the lagoon to the sea.

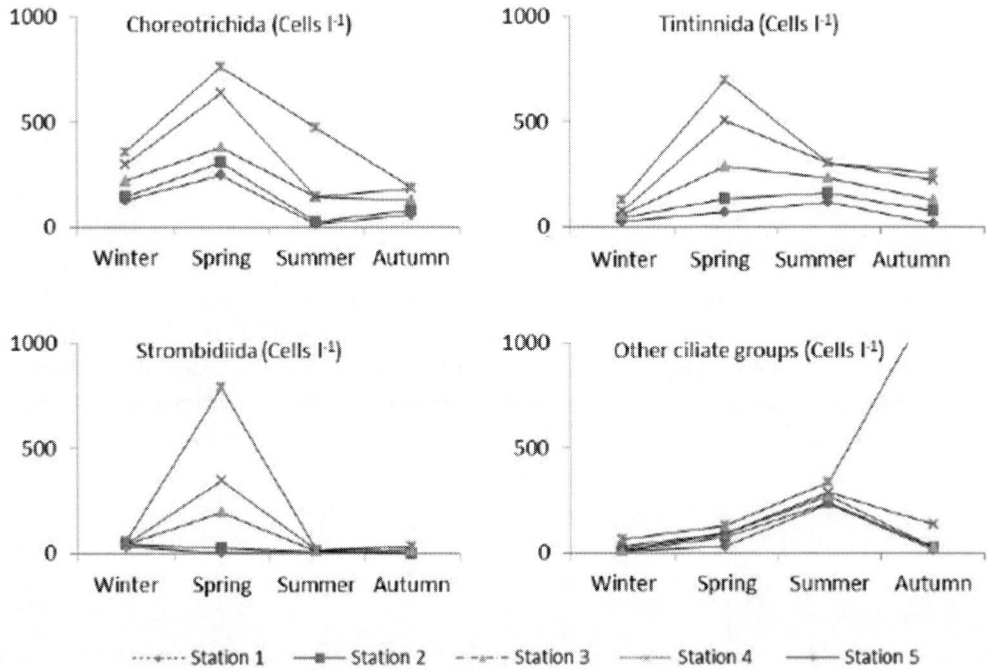

Figure 11. Seasonal changes of ciliate groups in five stations of Ghar El Melh lagoon.

REFERENCES

Alvial, M. A. and Garcia, B. J. (1986). Dynamics of a red tide caused by *Prorocentrum gracile* Schütt at Iquique. *Chile Review of Marine Biology* 22, 97–123.

An, T., Winshell, J., Scorzetti, G., Fell, J. W., and Rein, K. R. (2010). Identification of okadaic production in the marine dinoflagellate *Prorocentrum rhathymum* from Florida Bay. *Toxicon* 55, 653-657.

Anderson, D. M., White, A. W. and Baden, D. G. (1985). *Toxic dinoflagellates*. New York: Elsevier, 561 p.

Ayache, F., Thompson, J. R., Flower, R. J., Boujarra, A., Rouatbi, F., and Makina, H. (2009). Environmental characteristics, landscape history and pressures on three coastal lagoons in the Southern Mediterranean Region: Merja Zerga (Morocco), Ghar El Melh (Tunisia) and Lake Manzala (Egypt). *Hydrobiologia* 622, 15–43.

Baek, S. H., Shimode, S., Kyoungsoon, S., Han, M., and Kikuchi, T. (2009). Growth of dinoflagellates, *Ceratium furca* and *Ceratium fusus* inSagami Bay, Japan: the role of vertical migration and cell division. *Harmful Algae*, 8, 843–856.

Balkis, N. (2004). Tintinnids (Protozoa: Ciliophora) of the Büyükçekmece Bay in the Sea of Marmara. *Sci. Mar.* 68, 33–44.

Ben Ismail, S., Sammari, C., Gasparini, G. P., Bèranger, K., Brahim, M., and Aleya, L., (2012). Water masses exchanged through the Channel of Sicily: Evidence for the presence of new water masses on the Tunisian side of the channel. Deep-Sea Research I, 63, 65–81. Cloern, J. E., 2001. Our evolving conceptual model of the coastal eutrophication problem. *Marine Ecology Progress Series* 210, 235-265.

Bennouna, A., Berland, B., El Attar, J., and Assobhei, O. (2002). *Lingulodinium polyedrum* (Stein) Dodge red tide in shellfish areas along Doukkala coast (Moroccan Atlantic). *Oceanologica* Acta 25, 159–170.

Bockstahler, K. R. and Coats, D. W. (1993). Grazing of the mixotrophic dinoflagellate *Gymnodinium sanguineum* on ciliate populations of Chesapeake Bay. *Marine Biology* 116, 477–487.

Boni, L., Milandri, A., Poletti, R., and Pompei, M. (1993). DSP cases along the coasts of Emilia-Romagna (Northwestern Adriatic Sea). In: T. J. Smayda and Y. Shimizu (Eds.), *Toxic phytoplankton blooms in the Sea* (pp. 475–481). Amsterdam: Elsevier.

Bravo, I., Figueroa, R. I., Garcés, E., Fraga, S., and Massanet, A. (2010). The intricacies of dinoflagellate pellicle cysts: the example of *Alexandrium minutum* cysts from a bloomrecurrent area (Bay of Baiona, NW Spain). *Deep Sea Research* II 57, 166–174.

Carlough, L. A., Meyer, J. L. (1989). Protozoans in two southeastern blackwater rivers and their importance to trophic transfer. *Limnol. Oceanogr*. 34, 163–177.

Cordeiro, T. A., Brandini, F. P. and Martens, P. (1997). Spatial distribution of the Tintinnina (Ciliophora, Protista) in the North Sea, spring of 1986. *J. Plankton Res*., 19, 1371–1383. Corliss, J. O. (1961) the ciliated Protozoa: characterization, classification and guide to the literature.

Daly Yahia, M. N., Daly Yahia-Kefi, O. and Romdhane, M. S. (1997). Le plancton des lagunes de Ghar El Melh et du lac Nord de Tunis. (p. 90) *Projet National Mobilisateur code 50*. Faculté des Sciences de Tunis.

De Wit, R., Leibreich, J., Vernier, F., Delmas, F., Beuffe, H., Maison, Ph., Chossat, J.-C., Laplace-Treyture, C., Laplana, R., Clavé, V., Torre, M., Auby, I., Trut, G., Maurer, D., Capdeville, P. (2005). Relationship between land-use in the agro-forestry system of les Landes, nitrogen loading to and risk of macro-algal blooming in the Bassin d'Arcachon coastal lagoon (SW France). *Estuarine, Coastal and Shelf Science* 62, 453-465.

Dhib, A., Frossard, V., Turki, S., Aleya, L. (2012). Dynamics of harmful dinoflagellates driven by temperature and salinity in a northeastern Mediterranean lagoon. *Environmental Monitoring and Assessment*. DOI 10.1007/s10661-012-2797-4.

Dodge, J. D. (1982). *Marine Dinoflagellates of the British Isles*. (p. 303) Her Majesty's Stationery Office. London.

Dodge, J. D. (1985). *Atlas of Dinoflagellates*. Farrand Press, London, 119 p.

Dolan, J. (2000). Tintinnid ciliate diversity in the Mediterranean Sea: longitudinal patterns related to water column structure in late spring–early summer. *Aquat. Microb. Ecol*. 22, 69–78.

Dolan, J. R. and Marassé, C. (1995). Planktonic ciliate distribution relative to a deep chlorophyll maximum: Catalan Sea, NW Mediterranean, June 1993. *Deep-Sea Res*. 42, 1965–1987.

Drira, Z., Hamza, A., Belhassen, M., Ayadi, H., Bouaïn, A., Aleya, L. (2008a). Dynamics of dinoflagellates and environmental factors during the summer in the Gulf of Gabes (Tunisia, Eastern Mediterranean Sea). *Sci. Mar*. 72, 59–71.

Dupuy, C., Ryckaert, M., Le Gall, S., Hartmann, H. J. (2007). Seasonal Variations in Planktonic Community Structure and Production in an Atlantic Coastal Pond: The Importance of Nanoflagellates. *Microb. Ecol*. 53, 537–548.

Elloumi, J., Carrias, J. F., Ayadi, H., Sime-Ngando, T., Boukhris, M., and Bouain, A. (2006). Composition and distribution of planktonic ciliates from ponds of different salinity in the solar saltwork of Sfax, Tunisia. *Estuarine, Coastal and Shelf Science* 67, 21–29.

Estrada, M. (1985). Deep phytoplankton and chlorophyll maxima in theWestern Mediterranean. In: M. Moraitou-Aposto- Poulou and V. Kiortis (Eds.), *Marine Mediterranean ecosystems* (pp. 247–277). New York: Plenum Press.

Glibert, P. M., JoAnn, M. B. and Todd, M. K. (2012). Recent insights about relationships between nutrient availability, forms, and stoichiometry, and the distribution, ecophysiology, and food web effects of pelagic and benthic *Prorocentrum* species. *Harmful Algae* 14, 231–259.

Faust, M. A., Larsen, J. and Moestrup O. (1999). Potentially toxic phytoplankton. 3. Genus *Prorocentrum* (Dinophyceae), (p. 23) *ICES Identification Leaflets for Plankton. Leaflet 184.*

Fukuyo, Y., Takano, H., Chihara, M., and Matsuoka, K. (1990). Red tide organisms in Japan. *An illustrated taxonomic guide* (p. 407). Tokyo: Uchida Rokakuho.

Graneli, E., Sundstrom, B., Edler, L., and Anderson, D. M. (1990). *Toxic marine phytoplankton*. New York: Elsevier, 554 p.

Hallegraeff, G. M. (1991). Aquaculturists Guide to Harmful Australian Microalgae. Fishing Industry Training Board of Tasmania/CSIRO Division of Fisheries, *Hobart*, 111 p.

Hasle, G. R., Syversten, E. E. (1997). Marine diatoms. In: Tomas, C. R. (ed) *Identifying marine phytoplankton.* Academic, St. Petersburg 5–385.

Hannachi, I., Drira, Z., Bel Hassen, M., Hamza, A., Ayadi, H., Bouain, A., and Aleya, L. (2009). Abundance and biomass of the ciliate community during a spring cruise in the Gulf of Gabès (East Mediterranean Sea, Tunisia). *Acta Protozoologica* 47, 293–305.

Hannachi, I., Drira, Z., Bel hassen., M, Hamza., A., Ayadi, H., and Aleya, L. (2011). Species composition and spatial distribution of abundances and biomass of phytoplankton and ciliates during summer stratification in the Gulf of Hammamet (Tunisia). *Journal of the Marine Biological Association of the United Kingdom.* Doi:10.1017/S00253154100020 92.

Holmes, M. J., Lewis, R. J., Jones, A., and Wong Hoy, A. W., (1995). Cooliatoxin, the first toxin from *Coolia monotis* (Dinophyceae). *Nat. Tox.* 3, 355-362.

Hu, T., deFreitas, A. S. W., Doyle, J., Jackson, D., Marr, J., Nixon, E., Pleasance, S., Quilliam, M. A., Walter, J. A., and Wright, J. L. C. (1993). New DSP toxin derivatives isolated from toxic mussels and the dinoflagellates, *Prorocentrum lima* and *Prorocentrum concavum*. In: T. J.

Isenmann, P., Gaultier, T., El Hili, A., Azafzaf, H., Dlensi, H., and Smart, M. (2005). *Birds of Tunisia*. (p. 600) Société d'études ornithologiques de France.

Kat, M. (1985). Dinophysis acuminata blooms, the distinct cause of Dutch mussel poisoning. In: D. M. Anderson, A. W. White and D. G. Baden (Eds.), *Toxic dinoflagellates* (pp. 73–78). New York: Elsevier.

Kchaou, N., Elloumi, J., Drira, Z., Hamza, A., Ayadi, H., Bouain, A., and Aleya, L. (2009). Distribution of ciliates in relation to environmental factors along the coastline of the Gulf of Gabès, Tunisia. Estuarine, *Coastal and Shelf Science* 83, 414–424.

Krsinic, F. (1987). Tintinnines (Ciliophora, Oligotrichida, Tintinnina) in eastern Adriatic Bays. Estuar. *Coast. Shelf Sci.* 24, 527–538.

Lassus, P. and Berthome, J. P. (1988). *Status of 1987 algal blooms in IFREMER*. ICES/ annex, III C.M 33, 5–13.

Lee, J.-S., Igarashi T., Fraga, S., Dahl, E., Hovgaard, P., and Yasumoto, T. (1989). Determination of diarrhetic toxins in various dinoflagellate species. *J. Appl. Phycol.* 1, 147-152.

Lloret, J., Marin, A., Marin-Guirao, L. (2008). Is lagoon eutrophication likely to be aggravated by global climate change? *Estuarine, Coastal and Shelf Science* 78, 403-412.

Loder, M. G. J., Meunier, C., Wiltshire, K. H., Boersma, M., Aberle, N. (2011). The role of ciliates, heterotrophic dinoflagellates and copepods in structuring spring plankton communities at Helgoland Roads, North Sea. *Mar. Biol.* 158, 1551–1580.

Machida, M., Fujitomi, M., Hasegawa, K., Kudoh, T., Kai, M., Kobayashi, T., and Kamiide, T. (1999). Red tide of *Ceratium furca* along the Pacific coast of central Japan in 1997. *Nippon Suisan Gakkaishi* 65, 755–756.

Margalef, R. (1963). Rôle des ciliés dans le cycle de la vie pélagique en Méditerranée. *Rapp Comm. Int. Mer Medit.* 17, 511–512.

Marr, J. C., Jackson, A. E. and McLachlan, J. L. (1992). Occurrence of *Prorocentrum lima*, a DSP toxin-producing species from the Atlantic coast of Canada. *Journal of Applied Phycology* 4, 17–24.

Marshall, H. G. (2009). Phytoplankton of the York river. *Journal of Coastal Research* 57, 59–65.

Mee, L. D., Espinosa, M. and Diaz, G. (1986). Paralytic shellfish poisoning with a *Gymnodinium catenatum* red tide on the Pacific coast of Mexico. *Mar. Environ. Res.* 19, 77-92.

Menif, D. (2000). Les Gobiidés des côtes tunisiennes: Morphologie et biologie de *Zosterisessor ophiocephalus* (Pallas, 1811) et Gobius niger Linnaeus, 1758. Thèse de Doctorat, Université de Tunis.

Nezan, E. and Piclet, G. (1996). *Surveillance du phytoplancton marin: Les principales classes hormis les diatomophycées.* (p. 57) Manuel illustré adapté à la formation des analystes.

Nogueira, E. M. S., Sassi, R., Cordeiro, T. A. (2005). The Rhabdonellidae (Tintinnina: Oligotrichida) from Atol Das Rocas and Fernando De Noronha Archipelago, Southwestern Atlantic, Brazil. *Arq. Ciên. Mar.* 38, 93–104.

Olenina, I., Wasmund, N., Hajdu, S., Jurgensone, I., Gromisz, S., Kownacka, J., Toming, K., Vaicuiute, D., and Olenin, S. (2010). Assessing impacts of invasive phytoplankton: the Baltic Sea case. *Marine Pollution Bulletin* 60, 1691–1700.

Pérez, M. T., Dolan, J. R., Vidussi, F., Fukai, E. (2000). Diel vertical distribution of planktonic ciliates within the surface layer of the NW Mediterranean (May 1995). *Deep-Sea Res.* 47, 479–503

Petz, W. (1999). Ciliophora. In: Boltovsky, D. (Ed.), *South Atlantic Zooplankton*, vol. 2. Backhuys Publishers, Leiden, The Netherlands 265–319.

Pinto, J. S. and Silva, E. S. (1956). The toxicity of *Cardium edule L.* and its possible relation to the dinoflagellate Prorocentrum micans Ehr. *Notas Instituto di Bioliologia Marina*, 12, 1–20.

Ramdani, M., Elkhiati, N., Flower, R. J., Thompson, J. R., Kraiem, M. M., Ayache, F., and Ahmed, M. H. (2004). Environmental influences on the qualitative and quantitative composition of phytoplankton and zooplankton in North African coastal lagoons. *Hydrobiologia* 622, 113–131.

Ramsar Bureau, 2007a. The List of Wetlands of International Importance. Ramsar Bureau, Gland.

Rekik, A., Drira, Z., Guermazi, W., Elloumi, J., Maalej, S., Aleya, L., and Ayadi, H. (2011). Impacts of an uncontrolled phosphogypsum dumpsite on summer distribution of phytoplankton, copepods and ciliates in relation to abiotic variables along the near-shore of the southwestern Mediterranean coast. *Marine Pollution Bulletin* 64, 336– 346.

Revelante, N., Gilmartin, M. (1987). Seasonal Cycle of the Ciliated Protozoan and Micrometazoan Biomass in a Gulf of Maine Estuary. *Estuar. Coast. Shelf Sci.* 25, 581–598.

Rhodes, L., McNabba, P., de Salasb, M., Briggsc, L., Beuzenberga, V., and Gladstonea, M. (2006). Yessotoxin production by *Gonyaulax spinifera*. *Harmful Algae* 5, 148–155.

Robinson, M. G. and Brown, L. N. (1983). A recurrent red tide in a British Columbia coastal lagoon. *Canadian Journal of Fisheries and Aquatic Sciences* 40, 2135– 2143.

Romdhane, M. S., Eilertsen, H. C., Daly Yahia, O., and Daly Yahia, M. N. (1998). Toxic Dinoflagellate blooms in Tunisia lagoons: causes, and consequences for aquaculture. *Harmful Algae* 80–83.

SCET-ERI, (2000). *Diagnostic de la situation actuelle, Etude de l'amélioration de la qualité des eaux dans la lagune de Ghar El Melh, phase I-partie III,* (texte principale), Ministére de l'Agriculture, Direction Générale de la Pêche et d'Aquaculture, 293 p.

Shannon, C. E. and Weaver, G. (1949). *The mathematical theory of communication*. Urbana, Chicago, IL: University of Illinois Press, 118 p.

Sournia, A. (1986). Atlas du phytoplancton marin. Cyanophycées, Dictyochophycées, *Raphidophycées*. (p. 219) Edition du CNRS.

Struder-Kypke, M.C., Montagnes, D. J. S. (2002). Development of web-based guides to planktonic protists. *Aquatic Microbial Ecology* 27, 203–207.

Tang, X. H., Yu, R. C., Zhang, Q. C., Wang, Y. F., Yan, T., and Zhou, M. J. (2010). Molecular phylogenetic analysis of dinoflagellate *Scrippsiella trochoidea* isolated from the East Asian waters. *Chinese Journal of Oceanolology and Limnology* 28, 323–328.

Taylor, D. L. and Seliger, H. H. (1979). *Toxic dinoflagellate blooms*. New York: Elsevier, 505 p.

Taylor, F. J. R., Fukuyo, Y. and Larsen, J. (1995). Taxonomy of harmful dinoflagellates. In: G. M. Hallegraeff, D. M. Anderson and A. D. Cembella (eds), *Manual on Harmful Marine Microalgae, IOC Manuals and Guides No. 33*. UNESCO, France 283-317.

Thompson, J. R., Flower, R. J., Ramdani, M., Ayache, F., Ahmed, M. H., Rasmussen, E. K., and Petersen, O. S. (2009). Hydrological characteristics of three North African coastal lagoons: insights from the MELMARINA project. *Hydrobiologia* 622, 45–84.

Turki, S., Balti, N. and Ben Salah, C. (2007). First detection of *Kryptoperidinium foliaceum* (Stein1883) in Tunisian waters. *Harmful Algae News*, 9, 10.

Utermohl, H. (1958). Zur Vervollkommung der quantitative phytoplankton. Methodik. Mitt. *Intern. Ver. Limnol.* 9, 1-38.

Valiela, I., Collins, G., Kremer, J., Lajtha, K., Geist, M., Seely, B., Brawley, J., Sham, C. H., (1997). Nitrogen loadings from coastal watersheds to receiving estuaries: new method and application. *Ecological Applications* 7, 358-380.

Verlaque, M. (2001). Checklist of the macroalgae of Thau lagoon (Herault, France), a hot spot of marine species introduction in Europe. *Oceanologica Acta* 24, 29-49.

In: Lagoons
Editor: Mwinyikione Mwinyihija
ISBN: 978-1-62808-092-6

Chapter 3

NEW CALEDONIA LAGOON: A THREATENED PARADISE UNDER ANTHROPOGENIC PRESSURE?

C. Grenz,[1,*] R. Le Borgne,[2,†] J.-P. Torréton[3,‡] and R. Fichez[1,4,§]

[1]Aix Marseille Université, Université du Sud Toulon-Var, CNRS/INSU, IRD, MIO, UM 110, Marseille, Cedex, France
[2]Centre IRD de Nouméa, Nouméa Cédex, New Caledonia,
[3]ECOSYM, Van Phuc Diplomatic Compound Kim Ma, Ba Dinh, Hanoi, Vietnam
[4]IRD, Universidad Autonoma Metropolitana, Departamento de Hidrobiología, Iztapalapa, México D.F., Mexico

ABSTRACT

The French archipelago of New Caledonia hosts one of the three most extensive reef systems in the world and was partly added to the UNESCO World Heritage List in 2008. In the very specific context of rapidly growing land-based mining extraction and ore processing activities, industrial managers, political decision makers, lagoon users and citizens have been expressing an increasing demand for the implementation of efficient environmental monitoring programs dedicated to the effects related to mining activities. As a result, New Caledonia lagoons were thoroughly studied particularly during the last two decades through national research projects. This chapter aims at reviewing the state-of-the-art of knowledge about the general features of New Caledonia geography and climatology, lagoon environmental drivers including hydrodynamics, pelagic and benthic processes, and the potential stress of terrestrial and anthropogenic inputs on marine communities.

* Corresponding author: C. Grenz. Aix Marseille Université, Université du Sud Toulon-Var, CNRS/INSU, IRD, MIO, UM 110, 13288 Marseille, Cedex 09, France. E-mail: christian.grenz@univ-amu.fr; phone: +33 (0)4 91041641; fax: +33 (0)4 91041643.

† R. Le Borgne: Centre IRD de Nouméa, B.P. A5, 98848 Nouméa Cédex, New Caledonia.

‡ J.-P. Torréton: IRD - UMR 238 ECOSYM, Van Phuc Diplomatic Compound Appt. 202, Bldg. 2G - 298, Kim Ma, Ba Dinh, Hanoi, Vietnam.

§ R. Fichez: Aix Marseille Université, Université du Sud Toulon-Var, CNRS/INSU, IRD, MIO, UM 110, 13288 Marseille, Cedex 09, France. IRD, Universidad Autonoma Metropolitana, Departamento de Hidrobiología, Av. San Rafael Atlixco 186, Col. Vicentina C.P. 09340, Iztapalapa, México D.F., Mexico.

Insights addressing economic, ecological and political issues related to the complex status of self-governed New Caledonia are discussed with respect to living resources and environmental management.

INTRODUCTION

New Caledonia is a tropical archipelago located between latitudes 19 and 23° South and longitudes 163 and 168° East, just North of the Tropic of Capricorn, in the South Pacific Ocean (Figure 1). This system hosts one of the largest reef structures worldwide which was inscribed by UNESCO on its World Heritage site in 2008. New Caledonia marine area (EEZ and Territorial Waters) extends over 1,450,000 km^2. The main island, Grande Terre (*ca.* 400 x 50 km), is a high island surrounded by a lagoon of approximately 22,000 km^2 (Andréfouët et al., 2009), delimited by a 1100 km-long barrier reef. Being close to the Philippines marine biodiversity hot spot, New Caledonia lagoon is known for its high biodiversity level. Remoteness of the islands further favoured the appearance of endemic species amounting to 5 % of all marine species. Within groups with low dispersal capacities like sponges, up to 71 % of the identified species are endemic (Levi et al., 1998). New Caledonia hosts multiple habitats like coral reefs, mangroves, seagrass and green algae beds which provide homes to fish, marine invertebrates, marine mammals (such as whales and dugongs), and reptiles such as sea snakes and turtles.

New Caledonia has been intensively studied, from 1774 on, when the British explorer James Cook and his colleagues drew the first cartography of the main island reefs and lagoons (Fromaget and Richer de Forges, 1992; Payri and Richer de Forges, 2007). Later in 1946 the Institut Français d'Océanie (IFO) Research Centre in Nouméa was founded and developed research on geomorphology, geology and marine ecology (*e.g.* Catala, 1948). The IFO Centre was subsequently renamed as the ORSTOM (Office de la Recherche Scientifique et Technique Outre-Mer) and now IRD (Institut de Recherche pour le Développement) Centre.

Intensive exploration on New Caledonia reefs and lagoons took place in 1960-1963 sponsored by the Singer-Polignac Foundation (Taisne, 1965). Research on lagoon hydrology and water circulation, pelagic properties and ecosystem functioning extended during the 70's (Dandonneau et al., 1981; Rougerie, 1986) and the 80's thanks to ship facilities (R/V Dawa, Vauban and Alis) and new scientific programmes specifically dedicated to the lagoons (SMIB (Substances Marines d'Intérêt Biologique) and LAGON.

Several years later, integration of lagoon research activities and development of knowledge on anthropogenic input effects led to the ECOTROPE programme (1996-2001) and the CAMELIA research unit (2001-2009), both of which strongly contributed to the PNEC (Programme National Environnement Côtier) study of the New Caledonia pilot site. Goals of the first PNEC New Caledonia programme (1999-2003) were "(1) to identify and model the processes involved in the transport and transformation of the main inputs from terrestrial and anthropogenic origin and (2) to understand the effects of these inputs on various functional characteristics of the ecosystem". The main research activities included water circulation and particle transport modelling, the description of biochemical processes, the evaluation of temporal and spatial variations of lagoon organisms in terms of biomass and activities, and the exploration for ecosystem bioindicators (Fichez and Grenz, 2005).

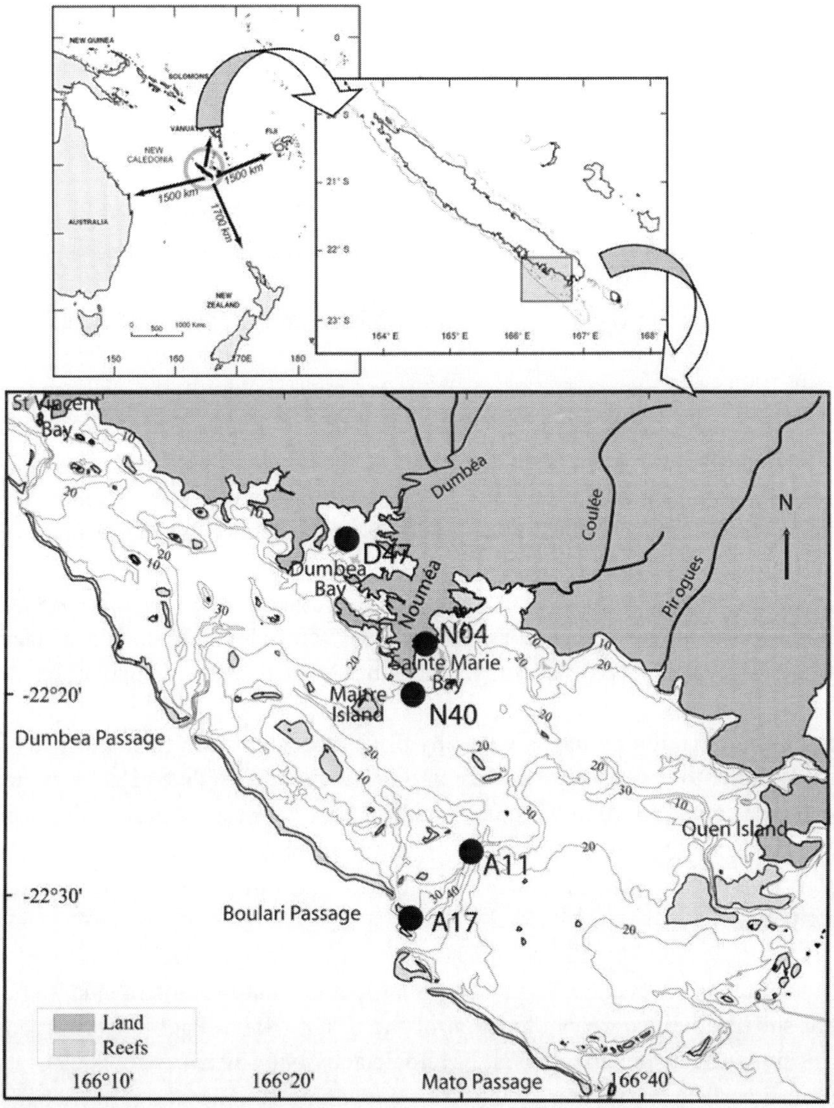

Figure 1. Map of the South West lagoon of New Caledonia and location, along a reef to near-shore transect of selected stations representative of the different trophic characteristics encountered.

The second PNEC programme (2004-2007) step up the above findings with studies on larval fish development, the origin and extent of New Caledonia upwelling, the nature of metal contaminants and their effects on benthic communities (including corals), the impacts of shrimp farm waste water discharges into the lagoons and the assessment of human use of the lagoon. Following the PNEC programme, more recent studies considered more specific issues, such as the mangrove ecosystem functioning. This paper reviews and synthesizes scientific literature including findings acquired during the 8 year-long PNEC project, on the lagoon structure, functioning and uses.

It evaluates the main anthropogenic pressures that potentially threaten New Caledonia lagoon. It refers mainly to the New Caledonia SW lagoon and uses, when available, results from other areas.

General Features of New Caledonia Geography and Climatology

New Caledonia archipelago includes the main island, the Loyalty and Bélep archipelagos, the Isle of Pines and scattered small coral islands, such as Surprise islands (Figure 1). The main island, Bélep and part of the Isle of Pines are made of basalt, peridotites, metamorphic and sedimentary rocks, while the other islands are made of limestone from coral origin. All of them are bordered with fringing reefs and most of them have a barrier reef with a lagoon of variable importance. Standing close to the Capricorn tropic, New Caledonia climate is considered as semi-tropical with two marked seasons generated by the displacement of the South Pacific Convergence Zone (SPCZ). In summer (November-March) ocean temperature classically range from 26 to 27 °C. Tropical depressions and cyclones originate more Northerly related to higher temperature regimes but affect New Caledonia with an average frequency of 1.5 cyclones per year. In winter (June-August), ocean temperatures decrease to 22-23 °C.

Inter-season transition periods are usually drier especially from September to November. Trade winds from the south-east are common and more frequent (250 days per year in Noumea) and sustained (5-10 $m.s^{-1}$) in the southern part whereas westerly gales accompanied by rainfall, occasionally occur in winter. The latitudinal extent of New Caledonia, leads to a northward positive temperature gradient and a higher cyclonic frequency in the Northern part of the archipelago.

Finally, due to its longitudinal location in the Western Pacific (163-168°E), New Caledonia is influenced by ENSO (El Niño-Southern Oscillation) interannual variations and subsequent shift in the geographic location of the SPCZ (South Pacific Convergence Zone), that significantly affects precipitation and cyclonic activity (Vincent et al., 2009).

The ocean surface circulation around New Caledonia is under the influence of the south equatorial current, but superficial currents generated by the dominant trade winds partly oppose to this general circulation scheme.

During El Niño events the ITCZ (Intertropical Convergence Zone) shifts along the equator and the associated water mass known as the Western Pacific Warm Pool is displaced eastward resulting in a decrease in average temperatures by 1 to 1.5 °C (Delcroix and Lenormand, 1997), and long drought periods in the south west Pacific.

Lefèvre et al. (2010) identify four types of weather conditions in summer, using an objective classification of the wind regimes. They show that the mountain range of New Caledonia is hydrodynamically steep and as a result of trade-wind obstruction by the mountainous island, the flow is shaped by coastally trapped mesoscale responses with a spatial scale of about 150 km. Combined to orographic blocking and flow splitting, the strong diurnal cycles along the shoreline add to the complexity of the wind patterns that constitutes a key factor to understand the dynamics of the lagoon circulation and its impact on ecosystems.

Like all neritic zones, lagoons are subjected to the double influence of terrestrial and oceanic inputs. The respective contribution of both effects relies on the lagoon size and the water flow intensity between the open ocean and the lagoon, either through passes or above the barrier reef (Le Borgne et al, 2010). Such contributions are indeed quite variable in the New Caledonia archipelago, so no general pattern may be suggested.

Effect of oligotrophic oceanic waters entering the lagoons reduces the continental influence and gets lower and lower shoreward. While oligotrophy is the prevailing trophic status of oceanic water masses, temporary enrichment processes have been observed recently in the SW region of the main island, with the "Caledonian" upwelling (Hénin and Cresswell, 2005) and the uprising of the hydrographical structure between the two horns located north of the isle of Pines (Neveux et al., 2010). In both cases, a sustained trade wind, 16-30 knt velocity, is required and leads to a 2-4 °C temperature decrease of superficial waters and a 1.5-2.3 fold increase of chlorophyll *a* (Chl a) and phytoplankton abundance (Neveux et al., 2010). Marchesiello et al. (2010) show that upwelling processes, which bring deep waters originating from the west, are weaker than expected. According to the authors, the divergence between the Ekman drift and the continental shelf is counterbalanced by a geostrophic convergence and transport of water masses drifting from the north along the shelf.

Lagoon Environmental Drivers

Environmental forcing parameters are essential drivers to ecological organisation and functioning of pelagic and benthic systems and a proper definition of these constraining factors is necessary to establish ecological cause-effect relationships. Previous work on the SW lagoon hydrodynamics demonstrated that the semi-diurnal tide, with amplitudes of 1.5 m at spring tide and 0.3 m at neap tide, propagated from the south to the north; its influence on lagoon circulation being largely dependent on the geomorphology (Douillet, 1998). As a generality, the lagoon is subject to strong oceanic inputs mostly due to the influence of trade winds and recent estimates using modelling suggested an average flushing time of 11 days for the south-west part of the lagoon, corresponding to a total area of circa 2000 km^2 (Bujan et al., 2000; Jouon et al., 2006).

Water transfer over the reef crest could significantly contribute to hydrodynamic budgets and impact ecological functioning and it is only recently that a preliminary study was conducted on tidal effects (Bonneton et al., 2007). Investigating particulate matter aggregation, and hydrodynamically driven suspended particulate transport and sediment resuspension processes is also essential to improve our understanding of major environmental issues such as light limitation induced by water turbidity or the respective fate of contaminants bound to fine particles (Ouillon et al., 2004, 2005; Fernandez et al., 2006; Mari et al. 2007, Migon et al., 2007, Jouon et al., 2008, Mari et al., 2009).

High tropical islands are subjected to heavy seasonal rainfall so mechanical erosion of soils is a general feature. Rainfall in New Caledonia varies a lot according to elevation and wind exposure, between 500 and 3000 mm per year. Highest values are recorded along the central mountain area and lowest on the leeward western coast of Grande Terre. Erosion processes and related sediment inputs are amplified by human pressure related to bush fire, agriculture and land cleaning, by fire and mining activities.

Laganier (1994) showed that erosion in Ouenghi basin lies between 8 and 200 t yr^{-1} per km^2 of catchment area, up to 1000 t yr^{-1} in 1974 when a new open-cast mine came into service along with the construction of its main road. Combined to cyclone floods, these huge sediment inputs cause severe degradation to the lagoon ecosystem and the fringing reefs as in the Great Barrier Reef (GBR) and should be considered as a priority in marine conservation planning (Brodie et al., 2001). In order to map the soil sensitivity to erosion at a catchment scale in New Caledonia, Dumas et al. (2010) described 2 methodologies by means of remote sensing and GIS (Geographical Information System) tools. Using 2 models based on a cognitive multi criteria approach and on a Universal Soil Loss Equation respectively, the authors could spatialize and quantify the potential areas of erosion at the catchment basin scales. Combined to an ICZM (Integrated Coastal Zone Management) framework these tools could be useful to identify risk areas by building specific scenarios related to changes in land use.

Due to the significance of ocean water intrusion in the lagoon, trophic status is basically known as oligotrophic with low concentrations of nutrients and phytoplankton biomass (Table 1). The build up of phytoplankton biomass with water aging however is a well known natural process at the root of the definition of eutrophication and it is therefore difficult to discriminate between natural and anthropogenic eutrophication. Ouillon et al. (2010) presented an extensive synthesis on recent numerical modelling approaches dealing with hydrodynamics, fine suspended sediment transport, wind-wave, and small-scale atmospheric circulation that are compared to *in situ* measurements. Beyond the essential determination of circulation patterns in the lagoon, hydrodynamic modelling conducted in the New Caledonia lagoon further permitted to calculate different types of commonly called "water residence times" (Jouon et al., 2006) which proved to be very powerful indicators of the interactions between physics and plankton ecology (Torréton et al., 2007), and metals distribution (Migon et al 2007). They provide new data on the hydrodynamics generated by swell-break over the reef crest and validated a wind-wave model in a fetch-limited environment. The analysis of spatial and temporal variability of suspended matter concentration over short and long time-scales showed aggregates to be more abundant near or around reefs and the possible biological origin on this aggregation process were highlighted. A combination of *in situ* (monochromatic measurements) and remote (surface spectral measurements and satellite observations) optical approaches was used to assess the distribution of suspended particulate material and to provide indirect calibration and validation of a suspended sediment transport model including resuspension processes. Community structure alterations due to anthropogenic modification of the environment are rarely confronted to environmental status assessment (Fichez et al., 2010). These authors, therefore, described the distribution of parameters related to the trophic status of the coral reef lagoon of New Caledonia under a large range of environmental conditions. They identified the main factors influencing lagoon water status and provide a typological classification of lagoon waters. Combined ACP-cluster analysis appears as an interesting tool to discriminate between natural trophic enrichment along the offshore-inshore gradient and anthropogenically induced eutrophication.

The authors provided evidence that water typology could be used to establish relationships between environmental drivers and benthic or pelagic community structure (Dumas et al., 2007; Sasal et al., 2007).

Table 1. Means, standard deviations (SD) and ranges of some chemical and biological variables in representative stations of the South West lagoon of New Caledonia. Sampling depth is 3 m, Station numbers are indicated on Figure 1 and the sampling period extended from July 1997 to June 2003

		Salinity	NH_4 µM	NO_3+NO_2 µM	PO_4 µM	Si(OH)4 µM	Tot DN µM	Tot DP µM	POP µM	POC µM	PN µM	CHLa µg/l	Phaeo µg/l
Reef passage (A17)	Mean	35.53	0.21	0.04	0.05	1.33	6.00	0.27	0.17	5.80	0.75	0.20	0.17
	SD	0.15	0.24	0.05	0.03	0.44	1.19	0.09	0.52	1.35	0.26	0.07	0.08
	Maximum	35.73	0.87	0.15	0.13	2.79	8.86	0.52	2.39	8.56	1.37	0.35	0.38
	Minimum	35.02	0.00	0.00	0.00	0.84	4.17	0.14	0.02	3.41	0.36	0.08	0.06
Outer lagoon (A11)	Mean	35.55	0.23	0.02	0.04	1.44	5.95	0.25	0.20	6.07	0.75	0.23	0.18
	SD	0.16	0.22	0.02	0.03	0.41	0.93	0.07	0.68	1.57	0.23	0.15	0.11
	Maximum	35.77	0.83	0.05	0.11	2.96	7.67	0.46	3.08	9.73	1.11	0.79	0.42
	Minimum	35.02	0.01	0.00	0.00	0.92	4.38	0.14	0.01	3.29	0.30	0.06	0.01
Inner lagoon (N40)	Mean	35.51	0.21	0.03	0.03	2.15	6.32	0.25	0.13	7.49	1.08	0.31	0.29
	SD	0.31	0.15	0.04	0.03	0.66	1.16	0.07	0.26	1.76	0.69	0.13	0.25
	Maximum	35.97	0.53	0.14	0.08	3.55	9.15	0.42	1.18	11.45	3.81	0.75	1.05
	Minimum	34.42	0.02	0.00	0.00	1.28	4.45	0.12	0.03	5.09	0.49	0.10	0.07
Embayment (D47)	Mean	35.36	0.23	0.08	0.05	5.98	7.22	0.27	0.18	9.86	1.32	0.52	0.42
	SD	0.40	0.14	0.14	0.04	2.89	1.22	0.08	0.39	2.45	0.43	0.28	0.20
	Maximum	36.29	0.63	0.63	0.12	12.87	9.59	0.52	1.83	14.21	2.26	1.21	0.81
	Minimum	34.52	0.04	0.00	0.00	1.66	4.46	0.13	0.05	5.39	0.51	0.10	0.13
Eutrophised (N04)	Mean	35.42	0.76	0.32	0.27	3.34	9.39	0.62	0.32	19.65	2.33	1.56	0.89
	SD	0.48	0.60	0.70	0.21	1.86	1.47	0.22	0.59	12.70	1.07	1.32	0.59
	Maximum	36.15	2.31	2.75	0.77	6.51	12.88	1.14	2.78	66.71	5.25	5.39	3.00
	Minimum	34.02	0.02	0.00	0.05	0.81	6.71	0.28	0.05	6.79	0.69	0.24	0.17

NH4: ammonium; NO_3 + NO_2: sum of nitrate and nitrite, PO_4: soluble reactive phosphorus; $Si(OH)_4$: silicate; POP: particulate organic phosphorus; POC: particulate organic carbon; PN: particulate nitrogen; CHLa: chlorophyll a; Phaeo: phaeopigments.

The authors further presented and discussed the variability of lagoon water biogeochemistry at various time scales, showing that high frequency variability and seasonal variability are of a similar order of magnitude especially in embayment where the displacement of water bodies may significantly and rapidly modify onsite environmental conditions. Despite the importance of short term variability evidence of a reinforcement of oligotrophic conditions during ENSO occurrences are provided, a factor that could be of significance to ENSO related biological events such as coral bleaching. Considering the potential use of biogeochemical parameters in environmental surveys they considered that high temporal variability in dissolved inorganic nutrients concentrations strongly hindered their use as indicators of environmental status whereas particulate organic material and more specifically Chl a, are much more robust indicators.

As briefly stated before, occurrence of upwelling events may occasionally bring nutrients to the surface oceanic waters and boost lagoon primary production. Ganachaud et al. (2010) observed contrasting water properties and planktonic composition during two cruises, performed during distinct physical conditions. During the December 2004 cruise, strong trade winds conditions triggered the occurrence of a significant upwelling event. In December 2005, relatively low trade winds conditions allowed for the formation of a south-eastward current, christened by the authors as the "Alis Current of New Caledonia", which resulted in vertically stratified oceanic waters. Increase in primary and secondary productivities under upwelling conditions however was far lower than what can be observed in typical upwelling systems worldwide. Fuchs et al. (2012) developed a 3D coupled biogeochemical-hydrodynamic model used on-line to study La Nina events on the South West lagoon. The same model was applied to infer the impact of upwelling which events derived from Sea Surface Temperature (SST) indexes were found to occur between 16 and 28% of the time (Fuchs et al., 2013). Using a forward Lagrangian transport analysis, upwelled waters from the upper nutricline were found to be able to reach the South West lagoon and phytoplankton responded promptly to these nutrient inputs with a 50 % increase in Chl a concentrations in three days. Even occasional and of limited potential in terms of new nutrient inputs, upwellings may temporarily contribute to increase their trophic status in the specific context of the prevailing strongly oligotrophic nature of the New Caledonia oceanic surface waters.

LAGOON PELAGIC PROCESSES

Ocean waters in the New Caledonia basin area are nutrient depleted and Chl a concentrations are below 0.1 $\mu g.l^{-1}$ (Dandonneau and Gohin, 1984). Lagoons are generally subject to strong oceanic inputs and prevailing lagoon environmental conditions are therefore near oligotrophy. Water temperature in the lagoon tends to be 1° to 2°C lower in winter and 1° to 2°C higher in summer than in the ocean (Rougerie, 1986). Salinity is generally close to ocean salinity but rainfall can occasionally decrease the local salinity of the surface water in protected embayments. The South West lagoon of New Caledonia is an area bounded to the north by the southern portion of the Saint-Vincent Bay and to the south by a line running from Ouen Island to Mato Pass. This area covers around 2000 km^2, 5% of which is occupied by reef structures and the remainder by soft bottom structures (Testau and Conand 1983). Unlike atoll lagoons this coastal lagoon is influenced by terrestrial inputs.

The predominant characteristics of the lagoon waters can be resumed by their oligotrophic to mesotrophic status with low particle concentrations. Even if planktonic biomasses and activities are low, temporal variability has been recognized as significant at time scales ranging from hours to years. This has been observed either for nutrient, phytoplankton or zooplankton standing stocks (Dolan et al., 2006; Rodier and Le Borgne, 2008; Le Borgne et al., 2010) or in the case of planktonic functioning (Briand et al., 2004; Renaud et al. 2005; Jacquet et al., 2006; Mari et al., 2007; Conan et al., 2008; Rochelle-Newall et al., 2008a; Pringault et al., 2009; Le Borgne et al., 2010; Torréton et al., 2010). Spatial variability is mainly driven by localized nutrient inputs fueling pelagic production or by large scale hydrodynamic processes dispersing the water masses inside the lagoon (Douillet, 1998), leading to significant coast-sea gradients of nutrient, particles, and plankton biomass, production and diversity (Jacquet et al., 2006; Mari et al., 2007;Torréton et al., 2007; Weinbauer et al., 2010).

One of the most important characteristics of the pelagic compartment is probably the "trophic gradient", due to the decreasing continental influence and the increasing oceanic one from the shore to the open ocean. Consequences of the gradient may be easily seen on the biomass and production values. The annual mean for Chl a is 4 times larger (0.25 vs. 0.97 μg l^{-1}) in Ste Marie Bay than in the middle of the lagoon and the same ratio is observed for primary production (8.1 vs. 45.3 mg C m^{-3} d^{-1}) (Jacquet, 2005). A similar gradient may be observed on zooplankton dry weights (Binet and Le Borgne, 1996) and on planktonic size structure, with a greater contribution of large size classes as nutrient availability gets higher shoreward. A potential consequence of these shifts in size classes is an expected non-linear response of upper trophic levels to biomass and productivity enrichments since transfer efficiencies along food webs typically increase with prey sizes (Legendre and Rassoulzadegan, 1995). Another consequence is the export capacity of the planktonic ecosystem since vertical fluxes of particulate organic matter are also dependent upon the size structure of primary producers (Legendre and Rassoulzadegan, 1996). On the short term scale, the weather regimes will affect features of the pelagic environment. Tropical cyclones represent a special case of short term events and Neveux et al. (2009) report on the Erica cyclone effects in March 2003. During the cold season, when the wind regime often has lower amplitude with more westerlies, two weather regimes may be defined (Anonymous, 1981). Precipitation, which results in high terrestrial inputs, and wind, affecting mixing of the water column, are the main two factors of variability on the short term scale.

As shown by both Le Borgne et al., (2010) and Torréton et al., (2010), seasonal variations of the phytoplanktonic biomass, as described by Chl a concentrations, display a maximum in June and two minima in February and September-October. Mesozooplankton (*i.e.* the 200-2000 μm size class) follows an almost inverse pattern, with a minimum in August and two maxima in March and November. Highest zooplankton biomasses with specific compositions are found in estuarine bays, allowing their predators, such as larval fish, to concentrate there as shown by Carassou et al. (2010). Concerning long term variability, Le Borgne et al. (2010) analysing a 10-year long time series, showed that zooplankton biomass displays a 3-4 year long periodicity, which does not appear to be significantly related to the SOI (Southern Oscillation Index).

One of the reasons is that this equatorial index may poorly reflect variations that occur in the geographical position of the SPCZ, which sets amounts of precipitation and, therefore, amounts of continental inputs to lagoons.

Amplitudes of the short term and seasonal signals are similar for planktonic variables such as Chl a, primary, and bacterial production (Torréton et al., 2010). In the Noumea area, Chl a concentrations increased by a factor of 5 to 6 linked mainly to diatom production. They showed that the contribution of the >2µm Chl a size class to total is 60% in Sainte Marie Bay in Noumea, while it is only 28% in the middle of the lagoon, near Maître Islet. Similar contributions were calculated for primary production: 61% and 19%, respectively. Inversely, the bacterial production represents a lower percentage of primary production in Ste Marie Bay (21%) than near Maître Islet (34%). Except for a few sites, eutrophicated by urban wastes in the region of Noumea, planktonic production is characteristic of mesotrophic systems nearshore and never gets high. The reason is likely nitrogen limitation, which can, indeed, be alleviated by diazotrophy, the physiological process that allows cyanobacteria, mainly, to use molecular nitrogen. Based on direct measurements, Rodier and Le Borgne (2010) provide an average of 5% for the contribution of *Trichodesmium* spp. (filamentous cyanobacteria) diazotrophy to total phytoplankton nitrogen needs. Of course, such a contribution is highly variable because *Trichodesmium* are able to bloom at times. Recent diazotrophy estimates based on indirect methods would indicate, however, that small cyanobacteria of the nano- and picoplankton size classes would be responsible for as much as 10 times that of *Trichodesmium* (Biegala and Raimbault, 2008). If such results were to be confirmed, this would mean that planktonic diazotrophy could play a major role in the lagoon ecosystem functioning and alleviate nitrogen limitation. In that case, iron or soluble reactive phosphorus (SRP) would become limiting. Iron is probably not, taking its measured concentrations into account. SRP, however, would be the limiting element as it is already for *Trichodesmium* in the region. One of the arguments for such phosphorus limitation is based on the very low available phosphorus concentrations found in lateritic soils and rivers (Tenorio et al., 2005), which dominate in the main island. It remains organic phosphorus compounds could be used by primary producers instead of SRP, as highlighted by the authors. At this stage, the debate between nitrogen and phosphorus limitation in the SW New Caledonia lagoon remains open and the answer probably depends on the area involved, considering the nature of its continental inputs. Aside from the temporal variability described above, *Trichodesmium spp* can bloom during the period extending from September to April. The two main species, *T. erythraeum* in the SW lagoon and *T. thiebautii* in the SE one, may bloom 3-7 days after a nutrient concentration increase provided environmental conditions are satisfied: sea temperature >26°C or lower temperature with a calm sea. The only difference between the two species behaviour is a much higher buoyancy for *T. erythraeum* which generates important surface accumulations of this species. Both species have been found toxic, with ciguatoxin-like and paralysing toxins (Kerbrat et al., 2010), and could be responsible for some cases of ciguatera.

In order to infer drivers of the spatial and temporal variabilities in the lagoon, a coupled 3D hydrodynamic - biogeochemical model was used in a diagnostic mode (Faure et al., 2010a, 2010b). Similar to observations, the model predicted strong and stable cross-shore gradients both in Chl a biomass and in bacterial production. Key conclusions were that the impact of terrestrial inputs, in terms of elevated Chl a and nutrients is restricted to inshore regions and bays near river mouths.

High residence times of water masses combined to nutrient enrichment inshore explain this phenomenon. In parallel the zone of increased bacterial production seemed to be considerably larger than the zone of increased Chl a as shown by a larger alongshore strip of increased ammonium contents. Satellite imagery is a promising tool to collect additional data either to infer variability or for model validation. Dupouy et al. (2010) tested this method for Chl a concentrations in the lagoon and surrounding ocean open waters. As a first attempt they showed that best results were obtained for deep waters mostly because of turbidity effects in the shallowest zones.

Lagoon Benthic Processes

A number of studies in benthic ecology have been conducted on the New Caledonian lagoon from the early sixties (Salvat 1964, 1965; Thomassin and Coudray 1981; Faure et al. 1981; Chardy and Clavier, 1988). From these literature data it can be established that the benthic biomass of the New Caledonian lagoon is among the highest values recorded in the tropical zone and is near the middle of the range of values for the northern and temperate zones. Sediment beds are by far the most common habitat in the lagoon of Grande Terre. Chardy et al. (1988) determined the structure and functioning of the soft bottom communities of the New Caledonian lagoon combining various criteria such as sediment composition, benthic community composition, or metabolic activity. Using additional data from a quantitative survey of the macrobenthos and several sedimentological studies (Dugas and Debenay 1982, Debenay 1985), Chardy et al. (1988) defined three major subsystems stretching from the land to the rear of the barrier reef. Muddy substrates (MS) are located close to the shore and largely influenced by terrigeneous inputs; grey sands (GS) cover the middle of the lagoon; white sands (WS) are confined to the vicinity of the rear of the barrier reef. The largest subsystem GS covers 50 % of the area. MS is the second largest area (35 %) and occupies bays, depressions and submarine valleys. Finally WS accounts for the remainder 15 % and refers to the inner reef zones. It has been demonstrated that lagoon sediments are essentially of native origin and that, even for the white sands, coral originated carbonates only contribute to a small proportion of deposited sediments (Chevillon 1996).

Process oriented studies were performed in order to assess the overall metabolism in New Caledonia lagoon sediments with special attention paid to microbenthic mineralization, benthic primary production and, oxygen and nutrient fluxes at the sediment water interface (Boucher and Clavier, 1990; Boucher et al., 1994; Clavier and Garrigue, 1999; Clavier et al., 1994; Grenz et al., 2003; Clavier et al., 2005, Pringault et al. 2008). Metabolic flux measurements indicated that average benthic respiration rates were higher than gross primary production, indicating that the whole lagoon was net heterotrophic. In order to infer the seasonal variability of benthic metabolic fluxes, Grenz et al. (2010) made two land to reef transects representing most of sediment types encountered in the lagoon, during the dry and wet seasons.

Measurements of sediment to water nutrient and oxygen fluxes were achieved along the transects. Inter-sample variability was as high as seasonal differences so that no seasonally driven temperature effect could be observed on benthic nutrient fluxes in the temperature range.

Compared to pelagic primary production rates, the authors could highlight the tight coupling existing between the benthic and pelagic compartments and calculated that oxygen sediment consumption and related carbon and nutrient fluxes could balance nearly all net pelagic primary production in the lagoon. These biogeochemical estimates point to the functional importance of sediment biogeochemistry in the lagoon of New Caledonia.

Coral reefs have been developing in New Caledonia waters since the Miocene to form major barrier reef structures surrounding the main island but made also the Loyalty Islands up and more remote islands, atolls or submerged reefs. Coral reefs may be locally exposed to terrestrial inputs, effects of inland mining activities, discharge of urban and industrial waste waters or anchor damage.

In spite of these local damages, coral reefs of New Caledonia are still considered as sufficiently pristine. Coral reefs are among the most productive ecosystems of the planet (Gattuso et al., 1998), higher than the adjacent lagoon sediments per surface unit. When considering the size of each system, coral reef and sediment productions are equivalent at the whole system scale of SW lagoon (Clavier and Garrigue, 1999). The possibility in the face of climate change or other stressors that coral reefs will prove to be resilient primarily because of their inherent complexity has been hypothesized in a review recently authored by Rogers (2012).

New Caledonia lagoons are still preserved from threats like coral bleaching compared to other reefs as the Great Barrier Reef in Australia and Fidji Islands (http://reefgis.reefbase.org). Nevertheless highly degraded fringing reefs can be observed within bays in New Caledonia close to the main inland as shown by Adjeroud et al. (2010).

These authors studied the spatial distribution of juvenile and adult corals from land to barrier reefs. The data showed that composition and abundances of coral assemblages were related to substrate types but not to water quality or metal concentrations in sediments. Moreover they suggested that the coral populations were structured by recruitment processes, keeping in mind that the coral larvae can be supplied by less disturbed nearby reefs. Nevertheless, densities of juvenile were relatively low thus indicating that replenishment capacities and potential for recovery following major disturbances in this lagoon are probably limited.

Mangrove swamps are characterized by 24 mangrove species in New Caledonia covering an estimated area of 250 km2 (Duke 2006; Virly 2006). Mangroves traditionally used for wood, medicinal plants, and as fishing grounds have been affected by urbanisation. Mangroves suffered from both deliberate destruction and indirect detrimental effects of other nearby land use.

However, there is a growing acknowledgement of the economic and environmental value of mangrove swamps resulting in the reinforcement of protective regulations. A study of heavy metals and organic matter distributions in mangrove sediments of various sites representative of live and dead forests and intertidal unvegetated areas showed that dense Rhizophora stands favour the retention of toxic heavy metals, making them less available to biota (Marchand et al., 2011).

Seagrasses are represented by 12 different species, building dense beds located in shallow waters (1 to 6 m) on the leeward side of the patch reefs and on the landward face of the barrier reef (Payri, 2007). Green algae of the *Caulerpa* and *Halimeda* genera also form dense beds, calcareous skeletons from the various *Halimeda* species significantly contributing to lagoon sediment carbonate deposition.

POTENTIAL EFFECT OF TERRESTRIAL AND ANTHROPOGENIC INPUTS

Since 2 decades, New Caledonia has experienced a rapidly growing land-based mining extraction and ore processing activity as the island is home to about a quarter of the world's known nickel resources. As a consequence, industries, political decision makers, lagoon users and the people have been expressing an increasing demand for the implementation of efficient environmental monitoring programs dedicated to register and control the effects of mining activities. Such programs should strongly rely on the use of biomonitoring species, as already implemented in temperate areas in the frame of US and EU Mussel Watch programs, biomonitoring presenting the very valuable combined advantages of providing a direct and time-integrated assessment of the metal fraction that is actually available to organisms (Phillips, 1991). However, and despite historical knowledge on natural and anthropogenic metal contamination in New Caledonia (Ambatsian et al., 1997) and reported high levels of metals in coastal marine sediments (Fernandez et al., 2006), it was only very recently that significant data on metal transfer in marine organisms was made available (Dalto et al., 2006; Hédouin et al., 2008, 2009; Metian and Warnau, 2008; Chouvelon et al., 2009). Aside from direct accumulation of metals in commercially important species, metal inputs to the New Caledonia South West lagoon also presents the potential to perturb the planktonic foodweb structure and the carbon transfer. For instance, Zn additions at concentrations lower than the clean water limits set by environmental protection agencies for tropical waters were shown to reduce the relative proportion of diatoms, increase the relative proportions of other phytoplankton, notably dinoflagellates, and affect phytoplankton-bacteria relationship (Rochelle-Newall et al. 2008b). Pringault et al. (2010) using microcosm experiments addressed Ni^{2+} adsorption in marine sediments and concluded that, in the context of the New Caledonia lagoon, the structural effect of Ni^{2+} on microbial population might be limited since prevailing anoxic conditions in coastal sediments favor its accumulation into an innocuous form. A much better assessment of bioaccumulation processes in potential sentinel organisms however was still needed. Hédouin et al. (2010 a) used radiotracer techniques to characterize bioaccumulation of Ag, Cd, Co, Cr and Zn originating from seawater, food and sediment in the oysters *Isognomon isognomon* and *Malleus regula*. Except for Ag and, to a lesser extent, Cd, both bivalves accumulated dissolved metals in a very similar way. The uptake kinetics indicated that Ag, Cd, and Zn were concentrated faster and more efficiently than Cr and Co in both oysters.

The strong retention of Ag, Cd, and Zn observed during the depuration experiment also suggested efficient detoxification processes based on metal storage in soft tissues. Ag accumulation was significantly higher in *I. isognomon* than in *M. regula* evidencing differences in sequestration/detoxification strategies. Overall, the authors provide evidence that sediment acted as a major source of Co and Cd, whereas food was the major source of Zn. Regarding Ag, seawater was the main source for *I. isognomon* (86%), whereas sediment was the predominant route for *M. regula* (92%). The influence of dissolved concentrations of As, Cd, Co, Cr, Mn, and Zn on uptake and depuration kinetics, tissue and subcellular distribution in clams and oysters was investigated by Hédouin et al. (2010 b). Metals were retained in the tissues of the clam *Gafrarium tumidum* and the oyster *Isognomon isognomon* in the following order of Cd > Cr ≈ Zn > Mn > Co > for *G. tumidum* and Zn > Mn, Co > As ≈ Cd ≈ Cr for *I. isognomon*.

The body distribution of metals at the end of the depuration period was generally homogeneous in *G. tumidum* except for Co concentrations that were significantly higher in the gills and lower in the foot. In *I. isognomon* most of the metals were associated with the visceral mass, except for Co which was mainly concentrated in the mantle. The general similarity of uptake and depuration rates in both bivalve species over a large range of concentrations confirms that *G. tumidum* and *I. isognomon* are able to rapidly reveal the extent of the ambient dissolved contamination and to preserve the information over relatively long periods of time (several weeks or months). However this simple relationship between metal concentration in the organisms and in seawater observed under the laboratory conditions can be more complex as it depends on exposure conditions driven by variable food-related bioaccumulation.

Using radiotracers again, Hédouin et al. (2010 c) quantified the assimilation efficiency of Mn, Co and Zn as a function of food quality and quantity (Co concentration, phytoplankton species, cell density). Exposure to phytoplankton with increasing Co concentration did not significantly affect the assimilation efficiency and the relative strength for both bivalves indicating that metal concentration in their tissue would be directly proportional to ambient metal concentration in the organic particulate material. They demonstrated that incorporation of these metals in the oyster *I. isognomon* and the clam *G. tumidum* was strongly depending on phytoplankton species and varied strongly from one metal to the other. Metals were generally better assimilated when bivalves were fed *Isochrysis galbana* with *Emiliana huxleyi* or *Heterocapsa triquetra*, a discrepancy hence appearing to be unrelated to phytoplankton size. Species-related selectivity and/or dietary preferences together with specific differences in metal speciation such as storage of the metal under bioavailable forms in the cytoplasm of the phytoplankton cells are presented as the most likely explanation for the specific differences observed in assimilation efficiency. Finally the authors showed that assimilation was significantly influenced by food availability and demonstrated that the two bivalve species were able to adjust their feeding strategies according to the food conditions prevailing in the environment. This original set of data led to the conclusion that a better knowledge of assimilation efficiency is essential to refining bioaccumulation predictive model as well as data interpretation from field surveys and biomonitoring programmes. Another common inhabitant of the lagoon, the Pacific blue shrimp *Litopenaeus stylirostris* has been studied through a combination of *ex situ* and *in situ* experiments designed to assess metal bioaccumulation and tissue distribution by Metian al. (2010). Laboratory experiments were conducted to determine bioaccumulation kinetics and body distribution of five selected radiotracers (^{110m}Ag, ^{109}Cd, ^{57}Co, ^{51}Cr, and ^{65}Zn) in shrimps exposed to dissolved or food sources of metals. *L. stylirostris* efficiently accumulated dissolved Ag, Cd, Co, Cr and Zn. Apart from Ag and Cd, the other metals reached steady-state equilibrium after 24 hour-long incubation. Biological half life in shrimp tissues increased from ≤ 35 d for Ag, Cr and Zn, to 152 d for Co and 796 d for Cd. Food source contamination/decontamination kinetics was studied by exposing shrimps to radiolabelled mussel tissues for 3 h and radiotracing depuration for 39 d. Ag, Cd, and Co were the most efficiently assimilated metals but their half life did not exceed 32 d. The field study provided a surprisingly missing baseline survey of metal and metalloid concentrations in *L. stylirostris* organs and a first assessment of the risk for human consumers. The authors claim that the concentrations in Cd, Cr, Mn, and Ni that they measured in the intestine and moreover in the hepatopancreas of the New Caledonian *L. stylirostris* were among the highest reported for shrimps or more generally for crustaceans.

On the contrary, metal and metalloid concentrations in the abdominal muscle of farmed *L. stylirostris* were in the range of concentrations reported for the same, or closely related, shrimp species hence suggesting that metals present in the New Caledonia environment are of limited bioavailability and/or that *L. stylirostris* efficiently regulates metal bioconcentration. Taking into account metal concentrations in the edible part of *L. stylirostris,* preliminary risk assessment for the human consumer considering maximum thresholds as recommended by international organizations (Food and Agriculture Organization FAO and World Health Organization WHO) indicated that consumption of shrimp did not present any significant risk. For example, in the case of arsenic, a 50-kg adult would have to eat more than 180 shrimps per week to reach the WHO maximum recommended intake, hence being much more likely to suffer from seafood indigestion than from metal poisoning. The development of tropical shrimp (*Litopenaeus spp*) aquaculture in New Caledonia has been very early linked with environmental awareness and scientific information is essential to enforce and improve measures ensuring sustainable development (Labrosse et al., 2000). One of the top priorities refers to human security and in the specific context of New Caledonia, solid information on metal bioaccumulation processes, organ tropism and accumulation versus depuration kinetics in aquaculture shrimps was strongly needed.

Living Resources and Environmental Management

As already stated, New Caledonian lagoon has been little affected by natural hazards like coral bleaching (Wantiez, 2008) so the major stress drivers of the environment are related to the lack of regulations framing uses of marine resources and to mining activities. For instance, New Caledonia is recognized as an "Oversea Country" of France which leads to an extensive autonomy regarding national laws but with less restrictive regulatory tools for spatial planning at the level of the Island. Due to erosion processes, the open air mining activity generates huge fluxes of terrestrial waste material that end up in the lagoon (Danloux, 1987). Nickel mining is the main economic production which contributes substantially to the Gross Domestic Product and should increase threefold in the coming 15 years (Lagadec et al., 2002). Apart from amplified sediment inputs the foreseen developments of new plants will have two direct consequences in terms of environment pressure (1) urban expansion and associated increase in used water inputs to the lagoon, and (2) intensification of subsistence fishing for local market with potential stock destruction and impact on reef diversity. Extensive research has been carried out on commercial important species particularly on molluscs (*e.g.* Bour, 1987, Baron, 1992; Clavier et al., 1989), echinodermata (*e.g.* Conand, 1981), crustaceans (*e.g.* Delathière, 1990), pelagic, demersal or reef fishes (*e.g.* Conand and Boely, 1982, Kulbicki and Wantiez, 1990, Letourneur et al., 1998). A recent study by Jollit et al (2010) described the spatial structure and characteristics of the informal fisheries based on four geosystems. They estimated that the informal fishery based on a fleet of around 3200 boats represented 2.6 times the commercial fishery activity in New Caledonia, most of which located in the Southern lagoon. Consequently this activity exerts a higher fishing pressure on coral reef resources than professional fishermen. Despite the complexity of the distribution of fishing activity that was observed, informal fishery covered 4 geosystems based on practices and geographical patterns, *i.e.* Melanesian, rural, peri-urban and urban geosystems.

Facing the brutal economic and demographical changes related to future mining projects like Koniambo in the Northern lagoon, Chabanet et al. (2010) analysed the spatio-temporal variability in reef fish assemblages of a fishing area prior to the mining project. Temporal variations in reef fish structure were consistent among biotopes. This similarity was still high when considering diversity and trophic regimes. Results indicated that functional structure prevails over taxonomic structure. They demonstrated that the industrial project of Koniambo will probably induce major perturbations in terms of hyper-sedimentation originated by channel dredging and sediment inputs in this relatively preserved environment. A second impact refers to the intensification of fishing pressure and change in fishing practices linked to population increase in this zone (doubling foreseen by 2015). Continuous monitoring of environmental data based on a Before-After-Control-Impact Paired Series (BACIPS) design for example (Stewart-Oaten, 1996), is needed in order to assess the impact of mining activities on the New Caledonia lagoon but also to develop sustainable management plans for marine resources. Regarding this aspect, Andréfouët and Wantiez (2010) discussed a strategy to characterize the diversity of coral reef habitats and fish communities based on the selection of sampling sites using high resolution remote sensing images. Bases on different criteria, clusters were defined for which the habitats were characterized. The analysis of clustering of the habitat data identified four main structures both for coral habitats and fish community structures. The protocol they used offers a diverse and representative view of the patterns present in New Caledonia lagoon and can be applied anywhere in the frame of further World Heritage site nominations or conservation programs.

Additionally, considering the worldwide examples of shrimp farming detrimental effects (Páez-Osuna, 2001, Pham Thi et al., 2010), one of the key issues to be addressed dealt with the environmental impact of shrimp farms generally located inland just behind mangrove forests in protected bays. A common environmental stress related to shrimp farming is the potential eutrophication risk. Aquaculture production mainly through shrimp farms is expected to double in the next decade (Andrier, 2004) but the impact remains spatially focused in terms of hyper-sedimentation or eutrophication near the discharge areas (Burford et al., 2003). A study on the water composition and, phytoplankton biomass and community structure on the effluents from semi-intensive and intensive shrimp farms was published by Thomas et al. (2010). They showed that suspended particulate nitrogen and phosphorous were significantly higher in the intensive farm effluents than in the semi intensive farm and most exported nutrients were in the form of soluble or particulate organic matter. The authors demonstrated also that anthropogenic eutrophication was most severe in the proximate surrounding of the intensive farm, biomass increase being accompanied by significant change in phytoplankton composition. Even though localized and limited to 2 months beyond the effluent release periods, it is claimed that some negative impact might affect sensitive ecosystems such as seagrass beds in the innermost part of the bays where intensive shrimp farming occurs.

Inscription of lagoon and reef areas of New Caledonia in the UNESCO World Heritage List in 2008, together with a local growing concern about marine environments have brought the need for taking into consideration sustainability of corals reefs in the process of emerging collective actions (David et al., 2010). This tropical lagoon is increasingly affected by serious anthropogenic and environmental changes with related ecological and social consequences and some significant actions have been taken to promote the sustainable development of the island.

On the specific topic of mining impacts, the National Centre for Research and Technology (CNRT) on "Nickel and its environment" was officially created in 2009 as an original scientific structure partly funded by mining companies but altogether ensuring the development of independent and peer reviewed research work on mining activities. Beside some strategic plans developed towards ICZM (Integrated Coastal Zone Management), the lagoon actors need to set up a large-scale observational system for the long term survey of key indicators, as recently initiated by OEIL (Observatoire de l'Environnement en Province Sud) for the SW lagoon in 2009. Only continuous monitoring of environmental data will help to separate effects of climatic and anthropogenic changes from natural fluctuations. Regular monitoring of the water quality and the condition of the inshore coral reefs delivers information on how the lagoon is changing over time and whether some regions are more affected by land runoff than others.

Moreover, combined with other initiatives in neighboring systems, a modern network of environmental data providers, exploiting the most recent observation techniques and technologies, could lead to the development of edge-cutting information products to end-users. A first initiative was taken in 2010 at a national scale by launching the GOPS (Grand Observatoire du Pacifique Sud) which coordinates the resources and scientific research of thirteen French research institutes and universities in the region. Even if the economic, ecological and political issues related to New Caledonia are complex, the various actors involved seek a broad consensus in favor of adopting an ICZM plan for the New Caledonia coastal areas.

Conclusion

New Caledonia's soils, derived largely from ultramafic rocks, are unusually rich in iron, magnesium, and phytotoxic heavy metals such as nickel. Native terrestrial flora species have adapted to such severe conditions for a long time as well as marine species of the surrounding lagoon, very likely. Thus New Caledonia has a hot and humid climate which induces heavy rainfall generating huge runoff of terrestrial particles disseminated from land and redistributed all over the lagoon.

These events are intensified during tropical storms or cyclone floods which happen frequently. Up to now, the New Caledonia lagoon has been little affected by natural hazards and the main pressures come from human activities. Localized threats in terms of eutrophication derived from urban discharge of untreated waste and shrimp farming become a real issue since a doubling of the aquaculture production area and extended urban projects associated to the threefold increase in metal production as foreseen for the next decade.

Impacts on the marine environment can be split into two categories: 1) open air mining induces destruction of native vegetation and stripping of topsoil leaving deposits of terrestrial material. These can be swept by runoff directly to the lagoon and induce hypersedimentation, and 2) population increase around the future plants and their resulting enhancement of the use of marine resources, may cause localized risks of overexploitation and even change fishing practices.

More than ever a strategy based on monitoring is needed to maintain the lagoon of New Caledonia at the lowest degree of contamination.

Acknowledgments

The authors acknowledge the participants who contributed through the French Program on Coastal Environments (PNEC) and the various institutes who financially supported the decadal investigation on New Caledonia. We are particularly grateful to Institut de Recherche pour le Développement (IRD), Centre National de la Recherche Scientifique (CNRS) through their Research Institutes: Institut National des Sciences de l'Univers (INSU) and Institut Ecologie et Environnement (INEE), and the different Universities including Aix-Marseille University and Lille University of Science and Technology. On a more local but as important scale, substantial supports was provided by ZoNeCo (Zone économique de Nouvelle-Calédonie), a multi-disciplinary program launched jointly by the Government of France, the Territory of New Caledonia and the three Provinces in association with the research institutes.

References

Adjeroud, M., Fernandez, J-M., Carroll, A. G., Harrison, P. L., Penin, L., 2010. Spatial patterns and recruitment processes of coral assemblages among contrasting environmental conditions in the southwestern lagoon of New Caledonia. *Marine Pollution Bulletin* 61, 375-386.

Ambatsian, P., Fernex, F., Bernat, M., Parron, C., Lecolle, J., 1997. High metal inputs to closed seas: the New Caledonia lagoon. *Journal of Geochemical Exploration* 59, 59–74.

Andréfouët, S., Wantiez, L., 2010. Characterizing the diversity of coral reef habitats and fish communities found in a UNESCO World Heritage Site: the strategy developed for Lagoons of New Caledonia. *Marine Pollution Bulletin* 61, 612-620.

Andréfouët, S., Cabioch, G., Flamand, B., Pelletier, B., 2009. A reappraisal of the diversity of geomorphological and genetic processes of New Caledonian coral reefs: a synthesis from optical remote sensing, coring and acoustic multibeam observations. *Coral Reefs* 28, 691-707.

Andrier, P., 2004. Une vision globale de l'aquaculture de crevettes et de son marché. In: *Styli 2003. Trente ans de crevetticulture en Nouvelle-Calédonie*. Ed. Ifremer, Paris, Actes colloq. n° 38, 280 p.

Anonymous, 1981. Types de temps et cyclones. In: *Atlas de la Nouvelle-Calédonie et dépendances,* ORSTOM Paris.

Baron, J., 1992. Reproductive cycles of the bivalves molluscs *Atactodea striata* (Gmelin), *Gafrarium tumidum* (Röding) and *Anadara scapha* (L.) in New Caledonia. *Australian Journal of Marine and Freshwater Research* 43, 393-402.

Biegala, I., Raimbault, P., 2008. High abundance of diazotrophic picocyanobacteria (<3μm) in a Southwest Pacific coral lagoon. *Aquatic Microbial Ecology* 51, 45-53.

Binet, D., Le Borgne, R., 1996. La station côtière de Nouméa. Dix ans d'observations sur l'hydrologie et le pelagos du lagon sud-ouest de Nouvelle-Calédonie. *Archives Sciences de la Mer Biologie Marine n°2*, Editions IRD, Nouméa.

Bonneton, P., Lefebvre, J. P., Bretel, P., Ouillon, S., Douillet, P., 2007. Tidal modulation of wave-setup and wave-induced currents on the Aboré coral reef, New Caledonia. *Journal of Coastal Research*, SI 50, 762-766.

Boucher, G., Clavier, J., 1990. Contribution of benthic biomass to overall metabolism in New Caledonia lagoon sediments. *Marine Ecology Progress Series* 44, 271– 280.

Boucher, G., Clavier, J., Garrigue, C., 1994. Oxygen and carbon dioxide fluxes at the water–sediment interface of a tropical lagoon, *Marine Ecology Progress Series* 107, 185–193.

Bour, W., 1987. *The trochus resource in New Caledonia*. Naga, The ICLARM Q. 10, 3-4.

Briand, E., Pringault, O., Jacquet, S., Torréton, J.-P., 2004. The use of oxygen microprobes to measure bacterial respiration for determining bacterioplankton carbon growth efficiency along trophic gradients in a coral reef lagoon. *Limnology And Oceanography: Methods* 2, 406-416.

Brodie, J., Christie, C., Devlin, M., Haynes, D., Morris, S., Ramsay, M., Waterhouse J., Yorkston, H., 2001. Catchment management and the Great Barrier Reef. *Water Science and Technology* 43 (9): 203–211.

Bujan, S., Grenz, C., Fichez, R., Douillet, P., 2000. Evolution saisonnière du cycle biogéochimique dans le lagon sud-ouest de Nouvelle-Caledonie. Application d'un modèle compartimental. Compte Rendu Académie des Sciences Paris, Life Sciences, *Ecology*, 323, 225-233.

Burford, M. A., Costanzo, S. D., Dennison, W. C., Jackson, C. J., Jones, A. B., McKinnon, A. D., Preston, N. P., Trott, L. A., 2003. A synthesis of dominant ecological processes in shrimp ponds and adjacent coastal environments in NE Australia. *Marine Pollution Bulletin* 46, 1456-1469.

Carassou, L., Le Borgne, R., Rolland, E., Ponton, D., 2010. Spatial and temporal distribution of zooplankton related to the environmental conditions in the coral reef lagoon of New Caledonia, Southwest Pacific. *Marine Pollution Bulletin* 61, 367-374.

Catala, R., 1948. Contribution à l'étude des poissons de la Nouvelle-Calédonie. Observations sur *Coris angulata*, Lacépède. *Doc. Multigraph. Institut Français d'Océanie*, Nouméa, 3p.

Chabanet, P., Guillemot, N., Kulbicki, M., Vigliola, L., Sarramegna, S., 2010. Baseline study of the spatio-temporal patterns of reef fish assemblages prior to a major mining project in New Caledonia (South Pacific). *Marine Pollution Bulletin* 61, 598-611.

Chardy, P., Chevillon, C., Clavier, J., 1988. Major benthic communities of the south-west lagoon of New Caledonia. *Coral Reefs* 7, 69-75.

Chardy, P., Clavier, J., 1988. Biomass and trophic structure of the macrobenthos in the south-west lagoon of New Caledonia. *Marine Biology* 99, 195-202.

Chevillon, C., 1996. Skeletal composition of lagoonal modern sediments in New Caledonia: Coral, a minor constituent. *Coral Reefs* 15 (3), 199-207.

Chouvelon, T., Warnau, M., Churlaud, C., Bustamante, P., 2009. Hg concentrations and related risk assessment in coral reef crustaceans, molluscs and fish from New Caledonia. *Environmental Pollution* 157, 331-340.

Clavier, J., Baron, J., Lefort, Y., 1989. Etude des bivalves dans le lagon de Nouvelle-Calédonie. Etat d'avancement des travaux au 30 novembre 1989. Nouméa: ORSTOM. *Conventions Sciences de la Mer: Biologie Marine* 3, 31p.

Clavier, J., Boucher, G., Chauvaud, L., Fichez, R., Chifflet, S., 2005. Benthic response to ammonium pulses in a tropical lagoon. Implications for coastal environment functioning. *Journal of Experimental Marine Biology and Ecology* 316, 231-241.

Clavier, J., Boucher, G., Garrigue, C., 1994. Benthic respiratory and photosynthetic quotients in a tropical lagoon, *Compte Rendu Académie des Sciences*, Paris 217, 937–942.

Clavier, J., Garrigue, C., 1999. Annual sediment primary production and respiration in a large coral reef lagoon (SW New Caledonia). *Marine Ecology Progress Series* 191, 79–89.

Conan, P., Joux, F., Torréton, J.-P., Pujo-Pay, M., Rochelle-Newall, E., Mari, X., 2008. Impact of solar ultraviolet radiation on bacterio- and phytoplankton activity in a large coral reef lagoon (SW New Caledonia). *Aquatic Microbial Ecology* 52, 83-98

Conand, C., 1981. Sexual cycles of three commercially important holothurians species (Echinodermata) from the lagoon of New Caledonia. *Bulletin of Marine Science* 31, 523-543.

Conand, F., Boely, T., 1982. Résultats des campagnes de prospection d'appât vivant en Nouvelle-Calédonie. *Pêche maritime* 1247, 96-101.

Dalto, A. G., Grémare, A., Dinet, A., Fichet, D., 2006. Muddy-bottom meiofauna responses to metal concentrations and organic enrichment in New Caledonia South-West Lagoon. Estuarine, *Coastal and Shelf Science* 67, 629-644.

Dandonneau, Y., Dugas, F., Fourmanoir, P., Magnier, Y., Rougerie, F., Debenay, J.-P., 1981. Le lagon de la Grande Terre. Présentation d'ensemble, sédimentologie et hydrologie du sud-ouest. In: *Atlas de la Nouvelle-Calédonie et dépendances*, ORSTOM Paris.

Dandonneau, Y., Gohin, F., 1984. Meridional and seasonal variations of the sea surface chlorophyll concentration in the southwestern tropical Pacific (14 to 32°S, 160 to 175°E). *Deep Sea Research* 31, 1377-1393.

Danloux, J., 1987. *Contribution à l'étude des transports solides en Nouvelle-Calédonie: les transports solides en suspension sur les bassins de la Ouenghi et de la Pouembout.* ORSTOM Editions, Paris, 121 p.

David, G., Leopold, M., Dumas, P. S., Ferraris, J., Herrenschmidt, J. B., Fontenelle, G., 2010. Integrated coastal zone management perspectives to ensure the sustainability of coral reefs in New Caledonia. *Marine Pollution Bulletin* 61, 323-334.

Debenay, J. P., 1985. Recherches sur la sédimentation actuelle et les thanatocénoses des foraminifères de grande taille dans le lagon sud-ouest et sur la marge insulaire sud de Nouvelle Calédonie. *Thèse d'Etat,* Univ. Aix-Marseille, 200p.

Delathière, S., 1990. Biologie et exploitation du crabe de palétuviers Scylla serrata en Nouvelle-Calédonie. *Thèse Université Bretagne Occidentale, océanographie biologique*, 291p.

Delcroix, T. and Lenormand, O., 1997. ENSO signals in the vicinity of New Caledonia, South Western Pacific. *Oceanologica Acta* 20, 481-491.

Dolan, J. R., Jacquet, S. and Torréton, J.-P., 2006. Comparing taxonomic and morphological biodiversity of tintinnids (planktonic ciliates) of New Caledonia. *Limnology and Oceanography* 51, 950-958.

Douillet, P., 1998. Tidal dynamics of the south-west lagoon of New Caledonia: observations and 2D numerical modelling. *Oceanologica Acta* 21, 69-79.

Dugas, F., Debenay, J. P., 1982. *Carte sédimentologique et carte annexe du lagon de Nouvelle Calédonie A 1/50000-, feuille "Nouméa".* ORSTOM, notice explicative 95, 45.

Duke, N. C., 2006. Mangrove biodiversity in New-Caledonia In «Typologies et Biodiversité des mangroves de Nouvelle-Calédonie» *ZoNéCo program report*, 213 pp.

Dumas, P., Kulbicki, M., Fichez, R., Chifflet, S., Ferraris, J., 2007. Environmental factors influencing urchin spatial distributions on New Caledonian disturbed coral reefs. *Journal of Experimental Marine Biology and Ecology* 344, 88-100.

Dumas, P., Printemps, J., Mangeas, M., Luneau, G., 2010. Developing Erosion Models for Integrated Coastal Zone Management: a Case Study of The New-Caledonia West Coast. *Marine Pollution Bulletin* 61, 519-529.

Dupouy, C., Neveux, J., Ouillon, S., Frouin, R., Murakami, H., Hochard, S., Dirberg, G., 2010. Inherent optical properties and satellite retrieval of chlorophyll concentration in the lagoon and open ocean waters of New Caledonia. *Marine Pollution Bulletin* 61, 503-518.

Faure, G., Thomassin, B. A., Vasseur, P., 1981. Reef coral assemblages on windward slopes in the Nouméa lagoon (New Caledonia). *Proc. 4th Int. Coral Reef Symp.* 2, 293-301.

Faure, V., Pinazo, C., Torréton, J.-P., and Jacquet, S., 2010 (a). Modelling the spatial and temporal variability of the SW lagoon of New Caledonia I: a new biogeochemical model based on microbial loop recycling. *Marine Pollution Bulletin* 61, 465-479.

Faure, V., Pinazo, C., Torréton, J-P., Douillet, P., 2010 (b) Modelling the spatial and temporal variability of the SW lagoon of New Caledonia II: realistic 3D simulations compared with in situ data. *Marine Pollution Bulletin* 61, 480-502.

Fernandez, J. M., Ouillon, S., Chevillon, C., Douillet, P., Legendre, R., Fichez, R., 2006. A combined modelling and geochemical study of the fate of terrigeneous inputs from mixed natural and mining sources in a coral reef lagoon (New Caledonia). *Marine Pollution Bulletin*, 52, 320-331.

Fichez, R., Grenz, C., 2005. Chapter V, New Caledonia worksite. In: *PNEC The French Coastal Environment Research Programme, Overview of 1999-2003 activities.* Ifremer Editions, Brest, 107-133.

Fichez, R., Chifflet, S., Douillet, P., Gérard, P., Gutierrez, F., Jouon, A., Ouillon, S., Grenz C., 2010. Biogeochemical typology and temporal variability of lagoon waters in a coral reef ecosystem subject to terrigeneous and anthropogenic inputs (New Caledonia). *Marine Pollution Bulletin* 61, 309-322.

Fromaget, M., Richer de Forges, B., 1992. *Catalogue indexé du milieu marin de Nouvelle-Calédonie.* 2ème édition. Nouméa : ORSTOM. Cat. : Sci. Mer., 274 p.

Fuchs, R., Dupouy, C., Douillet, P., Caillaud, M., Mangin, A., Pinazo, C., 2012. Modelling the impact of a La Niña event on a South West Pacific Lagoon. *Marine Pollution Bulletin* 64, 1596–1613.

Fuchs, R., Pinazo, C., Douillet, P., Fraysse, M., Grenz, C., Mangin, A., Dupouy, C., 2013. Modelling Ocean-Lagoon Interaction during Upwelling Processes in the South West of New Caledonia. *Estuarine, Coastal and Shelf Science*.

Ganachaud, A., Vega, A., Rodier, M., Dupouy, C., Maes, C., Marchesiello, P., Eldin, G., Ridgway, K., Le Borgne, R., 2010. Observed impact of upwelling on water properties and biological activity off the Southwest coast of New Caledonia. *Marine Pollution Bulletin* 61, 449-464.

Gattuso, J.-P., Frankignoulle, M., Wollast, R., 1998. Carbon and carbonate metabolism in coastal aquatic ecosystems. *Annual Review of Ecology and Systematics* 29, 405-434

Grenz, C., Denis, L., Boucher, G., Chauvaud, L., Clavier, J., Fichez, R., Pringault, O., 2003. Spatial variability in sediment oxygen consumption under winter conditions in a lagoonal system in New Caledonia (South Pacific). *Journal of Experimental Marine Biology and Ecology* 285-286, 33-47.

Grenz, C., Denis, L., Pringault, O., Fichez, R., 2010. Spatial and seasonal variability of sediment oxygen consumption and nutrient fluxes at the sediment water interface in a sub-tropical lagoon (New Caledonia). *Marine Pollution Bulletin* 61, 399-412.

Hédouin, L., Bustamante, P., Fichez, R., Warnau, M., 2008. The tropical brown alga Lobophora variegata as bioindicator of mining contamination in the New Caledonia lagoon: a field transplantation study. *Marine Environmental Research*, 66, 438-444.

Hédouin, L., Bustamante, P., Churlaud, C., Pringault, O., Fichez, R., Warnau, M., 2009. Trends in concentrations of selected metalloid and metals in two bivalves from the SW lagoon of New Caledonia. *Ecotoxicology and Environmental Safety*, 72, 372-381.

Hédouin, L., Metian, M., Teyssié, J. L., Fichez, R., Warnau, M., 2010 (a). Delineation of heavy metal contamination pathways (seawater, food and sediment) in tropical oysters from New Caledonia using radiotracer techniques. *Marine Pollution Bulletin* 61, 542-553.

Hédouin, L., Batista, M. G., Metian, M., Buschiazzo, E., Warnau, M., 2010 (b). Metal and metalloid bioconcentration capacity of two tropical bivalves for monitoring the impact of land-based mining activities in the New Caledonia lagoon. *Marine Pollution Bulletin* 61, 554-567.

Hédouin, L., Metian, M., Lacoue-Labarthe, T., Fichez, R., Bustamante, P., Warnau, M., 2010 (c). Influence of food on the assimilation of selected metals in tropical bivalves from the New Caledonia lagoon: qualitative and quantitative aspects. *Marine Pollution Bulletin* 61, 568-575.

Hénin, C., Cresswell, G., 2005. Upwelling along the western barrier reef of New Caledonia. *Marine and Freshwater Research* 56, 105-1010.

Jacquet, S., 2005. Influence des nutriments sur le réseau trophique microbien planctonique dans le lagon SW de Nouvelle-Calédonie. *Ph.D. thesis Paris 6 University*, 278p.

Jacquet, S., Delesalle, B., Torréton, J. P., Blanchot, J., 2006. Response of phytoplankton communities to increased anthropogenic influences (southwestern lagoon, New Caledonia). *Marine Ecology Progress Series* 320, 65-78.

Jollit, I., Léopold, M., Guillemot, N., David, G., Chabanet, P., Lebigre, J. M., Ferraris, J., 2010. Geographical aspects of informal reef fishery systems in New Caledonia. *Marine Pollution Bulletin* 61, 585-597.

Jouon, A., Douillet, P., Ouillon, S., Fraunié, P., 2006. Calculations of hydrodynamic time parameters in a semi-opened coastal zone using a 3D hydrodynamic model. *Continental Shelf Research* 26, 1395-1415.

Jouon, A., Ouillon, S., Douillet, P., Lefebvre, J.-P., Fernandez, J. M., Mari, X., Froidefond, J.-M., 2008. Spatio-temporal variability in suspended particulate matter concentration and the role of aggregation on size distribution in a coral reef lagoon. *Marine Geology* 256, 36-48.

Kerbrat, A-S., Darius, H. T., Pauillac, S., Chinain, M., Laurent, D., 2010. Detection of ciguatoxin-like and paralysing toxins in *Trichodesmium* spp. from New Caledonia lagoon. *Marine Pollution Bulletin* 61, 360-366.

Kulbicki, M., Wantiez, L., 1990. Variations in the fish catch composition in the Bay of St. Vincent, New Caledonia, as determined by experimental trawling. *Australian Journal marine and freshwater research* 41, 121-144.

Labrosse, P., Fichez, R., Farman, R., Adams, T., 2000. New Caledonia. In: *Seas at the Millenium, an environmental evaluation.* Sheppard, C. (ed.), Elsevier, Amsterdam. Vol. 2, 723-736.

Lagadec, G., Perret, C., Pitoiser, A. 2002. Nickel et développement en Nouvelle Calédonie, In: Perret, C. (Ed.), *Perspectives de développement pour la Nouvelle-Calédonie*. PUG, Grenoble, chap.1, 21-42.

Laganier, R., 1994. Contribution à l'étude des processus d'érosion et des risques naturels dans les îles du Sud-Ouest Pacifique (Nouvelle-Calédonie et îles Salomon), *PhD thesis report, Univ. Sci. Techn.* Lille-Flandres-Artois, Orstom.

Le Borgne, R., Douillet, P., Fichez, R., Torréton, J-P., 2010. Hydrography and plankton temporal variabilities at different time scales in the southwest lagoon of New Caledonia: a review. *Marine Pollution Bulletin* 61, 297-308.

Lefèvre, J., Marchesiello, P., Jourdain, N., Menkes, C. E., Leroy, A., 2010. Weather regimes and orographic circulation around New Caledonia. *Marine Pollution Bulletin* 61, 413-431.

Legendre, L., Rassoulzadegan, F., 1995. Plankton and nutrient dynamics in marine waters. *Ophelia* 41, 153-172.

Legendre, L., Rassoulzadegan, F., 1996. Food-web mediated export of biogenic carbon in oceans: Hydrodynamic control. *Marine Ecology Progress Series* 145, 179-193.

Letourneur, Y., Kulbicki, M. and Labrosse, P., 1998. Length-weight relationships of fishes from coral reefs and lagoons of New Caledonia - an update. *Naga* 21, 39-46.

Levi, C., Laboute, P., Bargibant, G., Menou, J.-L., 1998. *Sponges of the New Caledonian lagoon*. Orstom, Paris, 214 p.

Marchand, C., Allenbach, M., Lattier-Vergès, E., 2011. Relationships between heavy metals distribution and organic matter cycling in mangrove sediments (Conception Bay, New Caledonia). *Geoderma* 160, 444–456.

Marchesiello, P., Lefèvre J., Vega, A., Couvelard, X., Menkes, C., Ganachaud, A., 2010. Coastal upwelling, circulation and heat balance around New Caledonia's barrier reef. *Marine Pollution Bulletin* 61, 432-448.

Mari, X., Rochelle-Newall, E., Torréton, J.-P., Pringault, O., and Jouon, A. 2007. Water residence time: a regulatory factor of the DOM to POM transfer efficiency. *Limnology and Oceanography* 52, 808-819.

Mari, X., Migon, C., Nicolas, E., 2009. Reactivity of transparent exopolymeric particles: A key parameter of trace metal cycling in the lagoon of Noumea, New Caledonia. *Marine Pollution Bulletin* 58, 1874-1879

Metian, M., Laetitia Hédouin, L., Eltayeb, M. H., Lacoue-Labarthe, T., Teyssié, J-L., Mugnier, C., Bustamante, P., Warnau, M., 2010. Metal and metalloid bioaccumulation in the Pacific blue shrimp *Litopenaeus stylirostris* (Stimpson) from New Caledonia: laboratory and field studies. *Marine Pollution Bulletin* 61, 576-584.

Metian, M., Warnau, M., 2008. The tropical brown alga *Lobophora variegata* (Lamouroux) Womersley: a prospective bioindicator for Ag contamination in tropical coastal waters. *Bulletin of Environmental Contamination and Toxicology*, 81, 455-458.

Migon, C., Ouillon, S., Mari, X., Nicolas, E., 2007. Geochemical and hydrodynamic constraints on the distribution of trace metal concentrations in the lagoon of Noumea, New Caledonia. *Estuarine Coastal and Shelf Science* 74, 756-765

Neveux, J., Lefebvre, J-P., Le Gendre, R., Dupouy, C., Gallois, F., Courties, C., Gérard, P., Fernandez, J-M., Ouillon, S. (2010) Phytoplankton dynamics in the southern New Caledonian lagoon during a southeast trade winds event. *Journal Marine Systems* 82, 230-244

Neveux, J., Tenorio, M. M. B., Jacquet, S., Torréton, J-P., Douillet, P., Ouillon, S., Dupouy, C., 2009. Chlorophylls and Phycoerythrins as markers of environmental forcings including cyclone Erica effect (March 2003) on phytoplankton in the Southwest lagoon of New Caledonia and oceanic adjacent area. *International Journal of Oceanography*, doi: 10.1155/2009/232513.

Ouillon, S., Douillet, P., Andrefouet, S., 2004. Coupling satellite data with in situ measurements and numerical modeling to study fine suspended sediment transport: a study for the lagoon of New Caledonia. *Coral Reefs* 23, 109-122.

Ouillon, S., Douillet, P., Fichez, R., Panché, J-Y., 2005. Enhancement of regional variations in salinity and temperature in a coral reef lagoon, New Caledonia. C.R. *Geoscience* 337, 1509-1517.

Ouillon, S., Douillet, P., Lefebvre, J. P., Le Gendre, R., Jouon, A., Bonneton, P., Fernandez, J. M., Chevillon, C., Magand, O., Lefèvre, J., Le Hir, P., Laganier, R., Dumas, F., Marchesiello, P., Bel Madani, A., Andréfouët, S., Panché, J. Y., Fichez, R., 2010. Circulation and suspended sediment transport in a coral reef lagoon: the southwest lagoon of New Caledonia. *Marine Pollution Bulletin* 61, 269-296.

Páez-Osuna, F., 2001. The environmental impact of shrimp aquaculture: a global perspective. *Environmental Pollution*, 112, 229-231.

Payri, C., 2007. Revised checklist of marine algae (Chlorophyta, Rhodophyta and Ochrophyta) and seagrasses (Marine Angiosperma) of New Caledonia. In: C. Payri, B. Richer de Forges *Compendium of marine species from New Caledonia*, pp. 95-112. Documents Scientifiques et Techniques Centre IRD de Nouméa II, 7.

Payri, C., Richer de Forges, B., 2007. *Compendium of marine species from New Caledonia*. Documents Scientifiques et Techniques Centre IRD de Nouméa II 7, 391p.

Pham Thi, A., Kroeze, C., Bush, S. R., Mol, A. P. J., 2010. Water pollution by intensive brackish shrimp farming in south-east Vietnam: Causes and options for control. *Agricultural Water Management*, 97, 872-882.

Phillips, D. J. H., 1991. Selected trace elements and the use of biomonitors in subtropical and tropical marine ecosystems. *Review Environmental Contaminant Toxicology*, 120, 105-128.

Pringault, O., Viret, H. and Duran, R., 2010. Influence of Microorganisms on the Removal of Nickel in Tropical Marine Sediments (New Caledonia). *Marine Pollution Bulletin* 61, 530-541.

Pringault, O., Duran, R., Jacquet, S., Torréton, J.-P., 2008. Temporal variations of microbial activity and diversity in marine tropical sediments. *Microbial Ecology* 55, 247-258.

Pringault, O., Tesson, S. V. and Rochelle-Newall, E., 2009. Respiration in the light and bacterio-phytoplankton coupling. *Microbial Ecology* 57, 321-334.

Renaud, F., Pringault, O.,Rochelle-Newall, E., 2005. Effects of the colonial cyanobacterium *Trichodesmium Spp.* on bacterial activity. *Aquatic Microbial Ecology* 41, 261-270.

Rochelle-Newall, E. J., Torréton, J.-P., Mari, X., and Pringault, O., 2008a. Phytoplankton-bacterioplankton coupling in a sub-tropical South Pacific coral reef lagoon. *Aquatic Microbial Ecology* 50, 221-229.

Rochelle-Newall, E. J., Mari, X., Rouchon, C., Torréton, J.-P., Pringault, O., 2008b. Zinc induces shifts in microbial carbon flux. *Aquatic Microbial Ecology* 52, 57–68.

Rodier, M., Le Borgne, R., 2008. Population dynamics and environmental conditions affecting *Trichodesmium* spp. (filamentous cyanobacteria) blooms in the south-west lagoon of New Caledonia. *Journal of Experimental Marine Biology and Ecology*, doi:10.1016/j.jembe.2008.01.016.

Rodier, M., Le Borgne, R., 2010. Population and trophic dynamics of *Trichodesmium thiebautii* in the SE lagoon of New Caledonia. Comparison with *T. erythraeum* in the South-west. *Marine Pollution Bulletin* 61, 349-359.

Rogers, C. S., 2013. *Coral Reef Resilience through Biodiversity.* ISRN Oceanography in press.

Rougerie, F., 1986. Le lagon sud-ouest de Nouvelle-Calédonie: spécificité hydrologique, dynamique et productivité. *Etudes et thèses*, ORSTOM Paris, 234p.

Salvat, B., 1964. Prospections faunistiques en Nouvelle Calédonie dans le cadre de la mission d'études des récifs coralliens. *Cahiers Pacifique* 6, 77-119.

Salvat, B., 1965. Etude préliminaire de quelques fonds meubles du lagon calédonien (additif). *Cahiers Pacifique* 7, 101-106.

Sasal, P., Mouillot, D., Fichez, R., Chifflet, S., Kulbicki, M., 2007. The use of fish parasites as biological indicators of anthropogenic influences in coral-reef lagoons: a case study of Apogonidae parasites in New-Caledonia. *Marine Pollution Bulletin*, 54, 1697-1706.

Stewart-Oaten, A. 1996. Problems in the analysis of environmental monitoring data. In: R. J. Schmitt and C. W. Osenberg (eds), *Detecting ecological impacts: Concepts and application in coastal habitats,* pp. 109-131. Academic Press, San Diego.

Taisne, B., 1965. Organisation et Hydrographie. In: *Expédition Française sur les récifs coralliens de la Nouvelle-Calédonie*. Editions de la Fondation Singer-Polignac, Paris, 5-131.

Tenório, M. M. B., Le Borgne, R., Rodier, M., Neveux, J., 2005. The impact of terrigeneous inputs on the Bay of Ouinné (New Caledonia) phytoplankton communities: a spectrofluorometric and microscopic approach. *Estuarine, Coastal and Shelf Science* 64, 531-545.

Testau, J. L., Conand, F., 1983. *Estimation des surfaces des différentes zones du lagon de Nouvelle Calédonie*. Publ. ORSTOM Nouméa, 5p.

Thomas, Y., Courties, C., El Helwe, Y., Herbland, A., Lemonnier, H., 2010. Spatial and temporal extension of eutrophication associated with shrimp farm wastewater discharges in the New Caledonia lagoon. *Marine Pollution Bulletin* 61, 387-398.

Thomassin, B. A., Coudray, J., 1981. Presence of wide hardground areas on lagoonal bottoms of the coral reef complex of Nouméa (SW New Caledonia). *Proc. 4th Int. Coral Reef Symp.* 1, 511-522.

Torréton, J.-P., Rochelle-Newall, E., Jouon, A., Faure, V., Jacquet, S., Douillet, P., 2007. Correspondence between the distribution of hydrodynamic time parameters and the distribution of biological and chemical variables in a semi-enclosed coral reef lagoon. *Estuarine, Coastal and Shelf Science* 74, 667-677.

Torréton, J.-P., Rochelle-Newall, E., Pringault, O., Jacquet, S., Faure, V., and Briand, E., 2010. Variability of primary and bacterial production in a coral reef lagoon (New Caledonia). *Marine Pollution Bulletin* 61, 335-348.

Vincent, E., Lengaigne, M., Menkes, C. E., Jourdain, N. C., Marchesiello, P., Madec, G., 2009. Interannual variability of the South Pacific Convergence Zone and impications for tropical cyclone genesis. *Climate Dynamics*, DOI: 10.1007/s00382-009-0716-3.

Virly, S., 2006. Atlas des mangroves de Nouvelle-Calédonie In: «*Typologies et Biodiversité des mangroves de Nouvelle-Calédonie*» ZoNéCo program report, p 213.

Wantiez, L., 2008. Les récifs coralliens de Nouvelle-Calédonie en 2006 : état des lieux et réseau de suivi. *Rev. Ecol.* (Terre Vie) 63, 117-132.

Weinbauer, M. G., Kerros, M. E., Motegi, C., Wilhartitz, I. C., Rassoulzadegan, F., Torréton, J.-P., Mari, X., 2010. Bacterial community composition and potential controlling mechanisms along a trophic gradient in a barrier reef system. *Aquatic Microbial Ecology* 60, 15-28.

In: Lagoons
Editor: Mwinyikione Mwinyihija ISBN: 978-1-62808-092-6

Chapter 4

LAGOONS AS A WATERBIRD HABITAT: RESPONSE OF COMMUNITIES TO HUMAN IMPACT AND MANAGEMENT ACROSS SPACE AND TIME SCALES

Pablo Farinós, Francisco Robledano,* Carmen Perona and Antonio J. Soto
Department of Ecology and Hidrology, University of Murcia, Espinardo, Spain

ABSTRACT

Coastal lagoons (CLs) are transitional systems which harbor remarkable biological values and resources while performing key functional roles for ecological and social systems. Waterbirds are valuable top-level indicators of habitat conditions in CLs. Changes in the intensity of human influence (impact and/or management) cause dramatic changes in the composition, structure and conservation value of their waterbird assemblages, whatever the temporal or spatial scales at which they are examined. We use the results of two studies carried out in a representative Mediterranean lagoon complex of SE Spain (Mar Menor), to show the different physical scales and time spans over which waterbird communities change, using several parameters and indexes as signals of environmental change with a direct influence on biodiversity conservation.

The case study sites represent the extremes of a size gradient: on one hand, a small restored pond or sub-lagoon (Coterillo) with a few hectares of surface area, and on the opposite, the full littoral area of the Mar Menor CL (tens of square kilometres). They share the characteristics of being shallow hypersaline waters with a strong dependence on external influences, whether natural (atmospheric, hydrographic) or artificial (hydrological impacts, eutrophication, management). Particularly, the lack of proper management at the watershed scale (in the case of the main natural basin) or at local level (in the small artificial site) can cause a long-term habitat deterioration. These studies complement those previously made in the deepest part of the main lagoon basin.

Waterbird communities have been surveyed in both study sites through different monitoring programs. In the Coterillo site, formerly a salt pond connected to an industrial salina, wintering (monthly) and breeding (annual) waterbird census were made between

* Corresponding author: Francisco Robledano. E-mail: frobleda@um.es.

1999-2010. In the Mar Menor littoral waters, 15 sampling units were censused monthly between 2006-2010, covering two summer (April-September) and three wintering (October-March) periods. Abundance, richness, diversity and multispecific conservation indexes were used to describe waterbird community variation. Analysis of similarity (MDS) and indicator value (INDVAL) were then used to identify indicator species of ecologically meaningful spatial or temporal units (areas or phases) characterized by consistent impact- or management-related conditions.

The persistence of background threats (land use changes and watershed pollution), and the lack of continuous monitoring, particularly of impact variables, makes it difficult to learn from the results of waterbird studies and to take advantage of their indicator value for conservation purposes. Our results suggest that, besides offsetting the loss of natural habitats, the restoration of small secondary lagoons offers a powerful tool for managing impacts on the larger basin, and that waterbirds represent a key indicator of management success, both at the site and complex level, and regarding both biodiversity and ecosystem function.

1. Introduction

1.1. Mediterranean Lagoons: Values and Impacts

Coastal lagoons (CLs hereafter) are transitional waterbodies which harbor remarkable biological values and resources while performing key functional roles for ecological and social systems (Pérez-Ruzafa and Marcos, 2012). Depending on the bioclimatic and geographical setting, physical dynamics, historical trends of land reclamation and the stage of ecological succession in which they are found, CLs range along a wide gradient of size, depth, terrestrial influence, connectedness to the sea and internal heterogeneity (Pérez-Ruzafa *et al.*, 2011). They often appear as coastal wetland complexes in which several types of waterbodies and vegetated wetlands coexist. Human uses like fishing, hunting, navigation, fish rearing and salt harvesting have historically shaped the configuration of such wetland complexes, and are still conditioning their ecological functioning today. In the Mediterranean, most CLs are managed for fishing or fish culture, a type of resource use that can be developed with minor morphological changes (although often with critical effects on hydrological functioning). Surrounding land uses like agriculture and tourism have also historically modified the landscape setting of coastal plains and shorelines, becoming much more widespread and intensive in recent decades. Even in preserved CL systems (Nature reserves and Ramsar Sites), the indirect influence of such intensification results in water eutrophication and general loss of biological value (Aliaume *et al.*, 2007).

Of particular landscape impact has been, among the traditional wetland uses, the transformation of shallow basins and wetland sections for salt exploitation. Hyperhaline aquatic environments like exploited solar saltworks, salterns or salinas (Evangelopoulos et al., 2008) can be important alternative habitats for waterbirds and other biota (Oliva-Paterna *et al.*, 2006; Amat *et al.*, 2007; Ma *et al.* 2010). The abandonment of these artificial wetlands results in a dramatic change in land and water use, with a high risk of losing values of economic, cultural and environmental importance (Paracuellos *et al.*, 2002; Crisman *et al.*, 2009), including the role of treatment of nutrient-enriched waters (Crisman, 1999; Marques *et al.*, 2003).

The loss of hyperhaline biotopes, especially of solar saltworks, attained 74% in Spain, 63% in Italy and 55% in Portugal during the two decades preceding a review by Amat *et al.* (2007). Ecotourism and other activities, coupled with salt production, are compatible with the maintenance of most of the cultural, landscape and ecological values of traditional saltworks. In order to ensure such compatibility, rehabilitation or restoration of these sites requires careful planning and a continuous feedback of information from research and monitoring (Petanidou, 2000; Crisman *et al.*, 2009).

In recent decades a number of research papers have addressed the role of saltworks as alternative or complementary habitat for aquatic birds and their potential to compensate coastal habitat loss or supplement the remaining natural wetlands (e.g. Britton and Johnson, 1987; Velasquez, 1992; Masero, 2003; Sadoul *et al.*, 1998). In the Iberian Peninsula, most studies have been conducted in Atlantic coastal wetlands (Masero *et al.*, 2000; Múrias *et al.*, 2002) establishing conservation and management proposals (Velasquez, 1992; Masero, 2003; Sánchez *et al.*, 2006), while only a few have been developed in the Iberian Mediterranean coast, evaluating the consequences of their abandonment (Paracuellos *et al.*, 2002). More studies focus on the landscape-scale consequences of abandonment of large industrial saltworks and their effects on flagship species (e.g. Greater Flamingo; Béchet, 2009), than on the design, management and monitoring of small-scale restoration projects, notwithstanding that artisanal saltworks are not only in greater risk of abandonment or reclamation but can also be more easily restored. Water surfaces managed primarily for conservation are of great value and should be increased in number and diversity of bio-physical conditions (Múrias *et al.*, 2002; Tavares *et al.*, 2009).

The potential use of abandoned saltworks and other artificial wetlands as "green filters" or treatment systems for nutrient-enriched waters is a particularly relevant aspect in the management of CL complexes.

It brings together issues like wetland restoration, environmental services, and biodiversity. Waterbirds can benefit from these wetlands providing increased suitable habitat, and society can benefit from both the environmental services they perform (regulation of hydrological regimes, water purification, e.g. Maltby, 1986; Everard *et al.*, 2012), and from the multiple values and functions (recreational, educational, scientific, bioindicator) provided by their birds (Amat and Green, 2009; Wenny *et al.*, 2012).

1.2. Aquatic Birds as Bioindicators

Waterbirds are valuable top-level indicators of habitat conditions in CLs. Changes in the intensity of human influence (impact and/or management) cause dramatic changes in the composition, structure and conservation value of their waterbird assemblages, whatever the temporal or spatial scales at which they are examined. Waterbirds respond to changes imposed by a variety of stressors, constituting warning signals against undesirable changes (Amat and Green, 2009; Mallory *et al.*, 2007), and can be used as bioindicators both at suborganismic and at population-community-ecosystem level (Kushlan, 1993).

Among the biota of CLs and other transitional coastal waters, waterbirds are probably the more intensively and regularly monitored group, with the indispensable help of citizen science (Robledano and Farinós, 2010).

Their potential use as bioindicators has been widely discussed (Peakall and Boyd, 1987; Adamus, 1996; Green and Figuerola, 2003; Gregory *et al.*, 2003; Stolen *et al.*, 2004; Rönkä *et al.*, 2005; Amat and Green, 2009). Although recent research on bioindication in coastal waters has encouraged more intensive monitoring of waterbirds (Hubina, 2008; Rönkä *et al.*, 2008), in the particular case of CLs, birds have received little attention compared to other taxa (Robledano and Farinós, 2010), with few studies addressing waterbirds as integral components of the food webs (Acuna *et al.*, 1994; Žyydelis and Kontautas, 2008). In man-made or restored wetlands, aquatic birds are the preferred biological group on which to base the assessment of their ecological value and conservation status. The presence or abundance of waterbird species or guilds, and the changes in their relative abundance, have proven effective proxies of the biological integrity of wetlands and are considered good indicators of the ecological status of a site and the environmental changes that may occur in it (Bradford *et al.*, 1998).

1.3. Objectives

We use the results of two studies carried out in a representative Mediterranean lagoon complex of SE Spain (Mar Menor), to show the different physical scales and time spans over which waterbird communities change, using several parameters and indexes as signals of environmental change with a direct influence on biodiversity conservation.

On the basis of these studies we aim to accomplish the following objectives:

- To assess the utility of waterbird community monitoring as a tool for tracking ecological changes in CLs and their associated wetlands, in response to human influences at different scales, both spatial (site, lagoon basin, watershed) and temporal (short and mid- to long term)
- To evaluate the performance of different community descriptors and indexes as response indicators of ecological changes in these wetland complexes
- To integrate the results of waterbird monitoring programmes made at different spatial and temporal scales within CL complexes, into a general evaluation system of ecological integrity
- To devise a habitat restoration scheme based on the experience in waterbird monitoring of representative wetland types within the Mar Menor CL system, which maximizes biodiversity enhancement and environmental functionality of the restored and created habitats

The case study sites represent the extremes of a size gradient: on one hand, a small restored pond or sub-lagoon (Coterillo Pond = CP hereafter) with a few hectares of surface area, and on the opposite, the full littoral area of the Mar Menor CL (tens of square kilometres). They share the characteristics of being shallow hypersaline waters with a strong dependence on external influences, whether natural (atmospheric, hydrographic) or artificial (hydrological impacts, eutrophication, management). Particularly, the lack of proper management at the watershed scale (in the case of the main natural basin) or at local level (in the small artificial site) can cause a long-term habitat deterioration.

These studies complement those previously made in the deepest part of the main lagoon basin, in combination with which a general framework for habitat restoration and management can be set.

2. Study Area and Methods

2.1. General Characteristics of the Study Area

The Mar Menor Lagoon complex (Figure 1) is a 135 km^2 hypersaline coastal waterbody surrounded by a number of wetlands originated through several terrestrial and litoral processes leading to the isolation of small secondary basins, historically modified by man for different purposes (mainly salt production). Succession toward more terrestrial conditions due to sediment loading into these secondary basins, has led many of them to evolve into shallow wetlands or coastal steppes, where artificial saltpans have been created through dyking, channeling of freshwater flows and water pumping from the main lagoon.

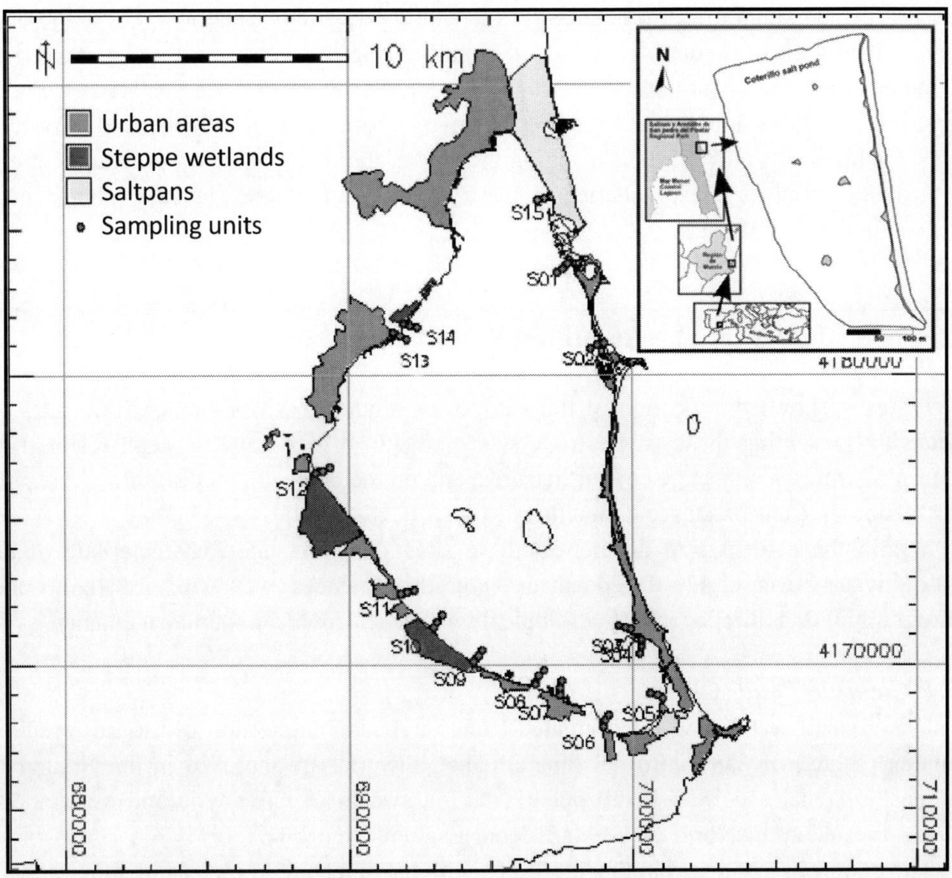

Figure 1. General map of the Mar Menor Lagoon and details of the study sites: Coterillo pond (upper right, including general geographic setting), and lagoon shoreline (main map) showing the 15 bird sampling units and the main associated wetlands: active saltpans (salinas) and saline steppe areas and other phreatophytic formations.

As a result the Mar Menor complex is now a mosaic of wetland habitats with an ample variety of depths (from salt steppes without a water table to the deepest sections of the lagoon with several meters) and sizes (from small ponds of a few hectares to the main lagoon of more than a hundred square kilometers. According to these morphological gradients, the range of environmental impacts and conservation issues is also extensive and conforms to the biophysical characteristics of each wetland or habitat type. The main lagoon basin has remained relatively unchanged in its basic features, although dramatic changes have been reported in several parameters and habitat characteristics. Among its main threats are the external influences (land use changes in the watershed, hydrological alterations), and the modification of littoral habitats due to urbanization, coastal works, and other sources of disturbance (Álvarez-Rogel *et al.*, 2006; Velasco *et al.*, 2006; Carreño *et al.*, 2008; Robledano *et al.*, 2010; Álvarez-Rogel *et al.*, 2011). Overfishing and eutrophication from sewage and agricultural pollution threaten the biotic communities of the lagoon, as well as the lixiviation of metal-contaminated waste originated in past mining activities (Marín-Guirao *et al.*, 2005; María-Cervantes *et al.*, 2009).

Associated wetlands around the lagoon have suffered from urban encroachment, tourism development and dumping of waste, sometimes invading physically their basins (particularly after the abandonment of salt production). Recreation pressure is usually higher in these areas due to easier access, although water-based activities also threaten even the remotest parts of the main lagoon. In recent decades most peripheral wetlands have been declared as protected areas, but this status does not always keep them safe from external influences like freshwater flows, contaminants and nutrients reaching them from the surrounding areas. Human pressure has also increased as the boundaries of these protected areas are gradually encroached by intensively developed areas.

2.2. Case Study Sites and Sampling Points

Figure 1 shows the location of the study sites around the Mar Menor. Two sites were selected, representing the extremes of a size gradient: on one hand, a small restored pond (Coterillo) with a few hectares of surface area, and on the opposite, the full littoral area of the Mar Menor CL (tens of square kilometres).

Within the evident size difference, these sites share the main characteristic of being shallow water areas, highly dependent on external influences, wether natural (atmospheric, hydrographic) or anthropogenic (hydrological impacts, eutrophication, management).

2.2.1. Coterillo Pond (CP)

It is a small restored wetland created from an abandoned section of the salt production complex (salina) of San Pedro del Pinatar. The saltworks are included in the Natural Park "Salinas y Arenales de San Pedro del Pinatar", a system of halophytic and psammophytic hábitats located in the North of the Mar Menor Lagoon (Figure 1).

The pond was excavated in the late 1950s with the objective of increasing the evaporating surface within the saltworks, abandoning its use in mid-1980. At the end of this decade, the pond was restored as part of a project for the conservation of *Aphanius iberus* (Valenciennes, 1846), funded by the EEC Regulation MEDSPA (Special Action Programme for the Mediterranean, 1986-1991).

Aphanius iberus is a Cyprinodontid fish endemic to the Iberian southeast, which usually does not exceed 6 cm in total length. Its diet is omnivorous, feeding mainly on crustaceans, insect larvae, detritus and algae (Oliva-Paterna *et al.*, 2006). The pond quickly became a valuable area for many waterbird species (including some included in Annex I of the Bird's Directive 79/409/EEC). In the following years the management objectives were gradually reoriented towards an interpretative and educational function.

Among other functions performed by the pond, we should highlight the "green filter" role exerted over the past decades. The water supply of the pond is a constant effluent coming from the experimental fishponds of a nearby research center (Centro de Recursos Marinos). This effluent carries both excretion products of fish raised in the fishponds and other organic debris, resulting in a general composition characterized by high concentrations of organic forms of P and N. Since its creation, the pond has undergone important changes in both morphological and physicochemical features, which could be related to the nature and composition of the effluent and the absence of any explicit management.

The study area consists of three basic units: the pond main basin (hereafter, the pond), a canal that runs longitudinally along its eastern shore, separated from it by a sand levee, and an adjacent salt marsh located to the north of the basin, created by subsurface diffusion of water from the main basin and therefore closely dependent on the pond in its water regime and water quality. The pond and canal are not fully independent since there are artificial connections between them, which add to the subsurface connection under the levee.

The pond surface is 8.2 hectares and it has an approximately trapezoidal shape (Figure 1). It contains four flat islands that are usually occupied for shelter and/or nesting by some bird species. The eastern canal is designed to help regulate the flow of water. Water enters the southern end of the canal and is discharged to the sea from the northern end through a pipe (which has not been permanently operational).

There are floodgates between the pond and the canal, but there has been no clear pattern of flow management. Moreover, the canal has suffered frequent breaks, maintaining a permanent connection with the main basin, although it has generally behaved as a semi-independent hydraulic compartment.

In the near future there are plans to reconnect the pond to the salt production network of the industrial saltworks, which implies the redirection of the effluent from the experimental fishponds to the sea, bypassing the pond system.

2.2.2. Mar Menor Shoreline (MMS)

To conduct census on which to assess the indicator value of waterbirds, it is important to choose sampling sectors with similar baseline conditions (physical features), but with differences in the composition of their communities large enough to reflect the impacts they support. These impacts depend on seasonal changes in human pressure, as well as on the spatial distribution of the terrestrial factors affecting littoral waters and adjacent terrestrial areas (e.g. urban development or intensive agriculture, and the impacts induced by them). Sectors directly influenced by artificial connections of the lagoon with the open sea, will differ from those which are not hidrologically disturbed, as well as those in more natural landscape contexts will do so from sites bordered by artificial structures (harbours, seaways or urban fronts). Therefore, we chose a total of 15 points, distributed along the entire coastline of the Mar Menor to reflect a representative sample of the shore, which will be detailed below.

The low lying nature of the lagoon shores, with ample extensions of shallow nearshore waters (gently sloping to the deepest sections, rarely under -7 m), allow the comparison of sampling units on the basis of land-use and impact features superimposed on the general configuration of this sedimentary coastal area.

Each unit (centered on one of these points) needs a configuration that guarantees coverage of a representative and equivalent area, both on nearshore and deeper water sections. The shore length of each unit extends to approximately 500 meters, from which we established four parallel bands whose external boundary is at 100, 250, 500 and 1000 meters from the shore. The outer edges of the bands were gradually more regular polygons to close as approximately rectangular sections, covering a total area around 50 ha. This allowed us to establish 60 census cells (4 bands x 15 points). The minor differences in size and shape between units and cells reflect differences in the morphology of the shore and the need to establish ground references to facilitate recognition of their limits.

Among the 15 points (=units), there are natural and urban or semi-urban sectors. Some include or lie close to a recreational or fishing harbor; others are located next to the mouth of one or several ephemeral watercourses (locally *ramblas*) or drainage canals. Moreover, some sampling units were fringed by halophytic vegetation, reedbeds or other natural communities, others were bordered by agricultural land or urbanizations, and some had combinations of both. Regarding water quality, clear water sections contrasted with areas affected by eutrophication to a different extent. Sampling points located east of the lagoon, have marine (Mediterranean) influence unlike those located on the western, continental side, whose hydrological and hydrochemical influences are land-based activities (e.g. fertilizer application).

2.3. Biological Data Gathering

Waterbird communities were surveyed in both study sites through monitoring programs covering different time spans and with different periodicity. In Coterillo Pond, bird census started in October 1999 and have been extended until May 2010, being carried out mostly by G. A. Ballesteros, head of the "Programa de Seguimiento Biológico en Humedales de la Región de Murcia" (Murcia Region's Wetlands Biological Monitoring Program), funded by the Autonomous Government. During this same period census of breeding pairs were conducted in the pond, including also the adjacent saltmarsh in the year 2000. The results of such census have been analyzed separately and in conjunction with those made in the pond (thereby constituting the pond+saltmarsh complex.

The methodology used in the census was the total direct count through comprehensive visual surveys in each of the water bodies, with suitable optical equipment (binoculars and spotting scope), from the banks and vantage points (Tellería, 1986), preferably in the early morning, trying to concentrate census effort in time, to improve the homogeneity of the data. The methodology used in the censuses traditionally applies equally to all groups included under the broad definition of waterfowl set by the Ramsar Convention (birds ecologically dependent on wetlands). For the study area, this refers essentially to open water bodies without dense associated emergent vegetation, making it possible to determine precisely the population of waterbirds.

The estimates of the number of pairs of waterbirds have been made by combining the direct counting of the breeders and young of the year, with the search and count of nests, eggs and chicks. For such estimates to be as accurate as possible, the criteria used as evidence of breeding have been those established by the European Bird Census Committee (EBCC, see e.g. Hagemeijer and Blair, 1997), adapted by Martí and Del Moral (2003). Besides birds, another biological feature was assessed in the pond: the general population and conservation status of *Aphanius iberus*. The aim was to determine relevant changes both in the density and in the distribution of this fish species within the pond and its peripheral canal, as a response indicator of the ecological changes that the system was undergoing. Through the characterization of population status we also tried to assess the pond's performance and adequacy as a habitat for the reintroduction of the species. We considered this endemic fish as an indicator with different ecological requirements than waterbirds, although related to them since it represents a potential food resource, especially for wading birds and terns.

Fish sampling was made through the placement of 13 Minnow traps over a 24 hour period, the traps being distributed uniformly along the canal and the west and east margins of the pond. This type of trap, devised to capture small fish, is used commonly for estimating population size on the basis of capture/recapture ratios, but in our case we only counted adults and juveniles *in situ*. It is a passive method in which the individuals captured are not removed (Bryant, 2000).

In the Mar Menor Shoreline, bird censuses were performed during three winters (October 2006 to March 2007, October 2008 to March 2008 and October 2009 to March 2010) and during two summers (April to September 2007 and April to September 2008). The surveys were carried out by P. Farinós almost entirely as a part of his Ph D.'s research fieldwork.

During these periods, direct counts of individuals of the wintering and summer species selected (all waterbirds except shorebirds, since the target assemblage was that using the littoral lagoon's waters) were accomplished in each of the sampling points. The counting was done with a spotting scope mounted on a tripod and binoculars. Fixed visual references were used on land and in water to estimate distances on which to determine the limits of the whole unit and its bands.

2.4. Recording of Environmental Variables

Coterillo Pond was surveyed for some morphological and limnological characteristics, although only three times since its creation (1995, 2007 and 2010, the two latter within the period covered by bird census). Additional qualitative observations were combined with the results of these surveys to make a general descriptions of the main macroscopic phases in the evolution of the pond.

Both the pond and canal were morphologically characterized through the measurement of their perimeter and depth. This was made with a GPS (Trimble GeoExplorer XT) for recording the perimeter and by setting seven transects in the pond and an equal number of ponts in the canal for measuring depth.

Along each transect three points (A, B and C) were marked to record depths, one in the center of the basin and the other two at 2,5 m of the East and West banks, respectively. Due to the shallowness of the pond and the prevailing low depth values, measurements were made manually with a rigid steel tape.

The physicochemical characterization of the pond and canal was made with a Multi 340L/SET probe, the sampling points corresponding to those used to measure depth but adding a new point in the pond, located in the vicinity of the point where the effluent was discharged. In each point we recorded the temperature (ºC), pH, conductivity (µS/cm) and salinity (g/l) of the water. To determine the concentration of the major nutrients (nitrate, nitrite, ammonium and phosphate) in the canal and pond, 15 samples were taken in high-density polyethylene bottles. The results of the laboratory analyses were expressed in parts per million (ppm) for nitrite, ammonium and phosphate, and in parts per billion (ppb) for nitrate. Nutrient concentrations give information on the productivity of the system and on the major biochemical processes occurring in it.

In the lagoon shoreline, the 15 units (centered on the census points) were characterized on the basis of the general landscape setting of the area (natural or artificial). A general description is given in Table 1. Bands within units represented a gradient of increasing depth extending offshore.

2.5. Community Descriptors and Statistical Analyses

In Coterillo Pond, bird censuses were analyzed separately for Winter (October-March) and Summer (April-September), two broadly defined phenological periods including, respectively, the true wintering and breeding periods, plus some transitional months in each case. As defined, these periods extended from the autumn passage and the building of the wintering community to its gradual departure, and from the spring passage and arrival of breeders to the post-nuptial period of dispersion of adults and young reared in the study area. "Winter" gathers together mainly birds having bred in more northern latitudes in passage or staying in the Mar Menor CL system in the cold months, while "Summer" does so with migrating birds having spent the winter in more southern areas plus those that settle in the study area as breeders.

With the aim of detecting changes in the conservation value of the bird community, we calculated two multispecific indexes adapted from Pons *et al.*, (2003), and previously used in other wetland habitats (Robledano *et al.*, 2010). In the SPEC index, species are classified according to the 'Species of European Concern' categories established by BirdLife International (2004). In the LRAE index, bird species were ranked regarding their threat status (extinction risk) as reflected by the Spanish BirdS Red Book (Madroño *et al.*, 2004), in which the species are evaluated against the IUCN Red List categories at the national level. Each category within these ranks was given a numerical value that was used to compute the index (Table 2).

The values are attributed in a geometrical progression of increasing threat status or conservation concern (Pons *et al.*, 2003). The value given to each species is multiplied by its abundance (logarithmically transformed), and the resulting values are summed across the community to generate a single index score for each year studied (Pons *et al.*, 2003; Paquet *et al.*, 2006). For breeding birds, total species richness and total number of breeding pairs were computed for each year. The total number of breeding pairs was split into three spatial divisions (pond, adjacent saltmarsh and the whole Regional Park to which the former belong), in order to explore the potential dependencies among them when interpreting numerical changes.

Table 1. Description of the 15 bird sampling units of the Mar Menor CL shoreline

Unit	Description
1. La Encañizada vieja	Located in the North part of La Manga (sandbar closing the lagoon), adjacent to an urbanization; a stretch of natural sandy coast with halophytic scrub and shallow flooded areas belonging to the *golas* (natural canals through the sandbar) managed traditionally for fishing with an ancient catching system (*encañizadas*) restored in this area but abandoned or destroyed elsewhere in the CL
2. Playa Estacio	Area located near the Tomás Maestre harbour in the northern part of La Manga. Sandy and gravel gently solping beach of predominantly urban nature, with blocks of flats adjacent to one of its ends. There is a remarkable presence of fishing nets in the adjacent part of the lagoon.
3. Isla del Ciervo (North)	Stretch of semi-urban steep rocky shore (western side of the volcanic hill Monte Blanco) and its adjacent lagoon area. It is located in the central part of La Manga, facing the north side of Isla del Ciervo and being limited by other nearby islands (Redonda and Sujeto).
4. Isla del Ciervo (South)	Adjacent to the previous unit and located in the old artificial isthmus (now removed) of the Isla del Ciervo, oriented towards its southern side. Gently sloping sandy beach close to an urban area and next to a harbour.
5. Canal de Marchamalo	Semi-urban sector located in the southernmost end of La Manga, adjacent to the Marchamalo coastal saltworks. It is a low sandy beach surrounded by halophytic scrub impacted by landfills and adjacent to residential buildings.
6. Playa Honda	Semi-natural sector next to a coastal urbanization. A stretch of sandy coast, bounded by an urban façade, although its general nature is that of a mixed urban-agricultural landscape. Intensive human use of the littoral waters in summer (mainly recreational activities).
7. Islas Menores.	Section of urban coast located between two harbours. Gently sloping sandy coast adjacent to a promenade bordered by a strip of low buildings on the terrestrial side. Affected by a continuous (baseline) stress due to the transit of boats from and to both harbours.
8. Punta Lengua de la Vaca (Desembocadura Rª Carrasquilla)	Natural stretch of stony coast (large sized material deposited by streams flowing into the lagoon) surrounded by agricultural landscape. The census point was located at the outlet of the Carrasquilla ephemeral watercourse.
9. Los Nietos	Sandy beach of fine particle size. The observation point was situated in the contact of the natural area (characterized with dense formations of helophytes), with the promenade belonging to a nearby urban sector.

Table 1. (Continued)

Unit	Description
10. Saladar de Lo Poyo	Natural sector within the wetland complex of Lo Poyo, a stretch of gently sloping sandy beach adjacent to large patches of psammophytic vegetation growing on sand dunes, and of halophytic scrub surrounding ancient saltworks. The whole area is influenced by two ephemeral watercourses flowing into its northern and southern ends, respectively (responsible for the disposal into the wetland of large amounts of mining waste with high concentrations of heavy metals).
11. Los Urrutias	Integrated into a stretch of semi-natural sandy beach, bounded by a harbour and a residential area, but keeping a dense patch of reedbeds of medium height. Noteworthy is the presence of buoys of maritime demarcation near the shore, usable for certain species as resting places.
12. Desembocadura Rambla Albujón.	Stretch of natural, gently sloping sandy beach with patches of reedbed and salt marsh. Its littoral waters show eutrophic conditions due to the inflow of nutrients through the channelized outlet of the Albujón ephemeral watercourse.
13. Los Narejos	Located within an urban stretch of coast with a promenade and a dense urban façade behind. Noteworthy is the presence of a pier on shore.
14. Playa de La Hita.	Stretch of natural, gently sloping beach into a dense beltof reedbeds; noteworthy is the proximity to disturbing elements of anthropic origin such as San Javier airport.
15. Salinas de San Pedro del Pinatar.	Sector of natural conditions within the Regional Park of Salinas de San Pedro del Pinatar. It is a stretch of gently sloping beach adjacent to several large saltpans of the industrial saltworks complex, which can be used by many species of waterbirds as a resting and refuge area.

Table 2. Numerical values attributed to bird species ranks within the SPEC and LRAE categories, used to compute the respective indexes of conservation value

Numerical value	SPEC	LRAE (Spanish Birds Red Book)	
4	SPEC 2	VU	Vulnerable
2	SPEC 3	NT, LC, DD	Near Threatened, Least Concern, Data Deficient
0	ON-SPEC	NE	Not evaluated

This was done through Pearson correlation analysis between Regional Park and pond, pond and saltmarsh, and Regional Park with the ensemble pond+saltmarsh. Finally, the waterbird community was analyzed on the basis of the distribution of its abundance among four functional groups or guilds, a concept widely used in community studies to refer to assemblages of species with ecological similarity not necessarily translated into taxonomic affinity (Wiens, 1989). Each guild can thus incorporate species from more than one family (and even from higher distinct taxonomic divisions).

Ducks (*Anatidae*): Represented in CP only by one family and virtually by a single species, *Anas acuta*, whose trophic amplitude (feeds mainly on plants but also insects, worms and occasionally mollusks; Gordon *et al.*, 1998), allows it to be considered as a generalist, i.e., capable to live in a variety of habitats and not specialized in a particular food source (Olmos and Boulhosa, 2000). In the context of CP succession, these species are globally regarded indicative of mature or stressed stages.

Wading birds: Birds generally large in size that feed in shallow waters, searching for food in the sediments or filtering water with specialized bills, or catching fish and other aquatic fauna. It includes three families (*Threskiornithidae, Phoenicopteridae* and the richest one, *Ardeidae*). Grey Heron *Ardea cinerea* (family *Ardeidae*) and Flamingo *Phoenicopterus roseus* (*Phoenicopteridae*) are considered specialists, i.e. species adapted to relatively narrow ecological conditions (Olmos and Boulhosa, 2000). In the context of the CP system, they are regarded as indicative of relatively undisturbed conditions.

Shorebirds: Birds that feed in tidal areas and other hábitats experiencing fluctuations that expose or make available productive sediments and shallow waters. They explore marshes, beaches, rocky coasts and wetlands (freshwater or salty), searching for invertebrates (insects, worms, small crustaceans). In CP several families (*Recurvirostridae, Burhinidae, Haematopodidae* and mostly *Charadriidae* and *Scolopacidae*) are represented. This guild is particularly sensitive to wetland degradation, with dramatic variations in their abundance and richness occurring along with changes in the ecological conditions of their habitats. This causes them to be generally considered as specialists (Hayman *et al.*, 1986). In the particular case of CP, these species are also considered indicative of undisturbed stages.

Gulls and terns (families *Laridae* and *Sternidae*): are omnivore species that often include in their diets, besides fish, refuse and carrion. Thus, they are less sensitive tan other guilds to degradation, pollution or human presence in their habitats, and can be considered opportunists or generalists depending on the species (Buzun and Mierauskas, 1987). In CP, they are globally regarded as indicators of mature or stressed stages (i.e. eutrophic, polluted conditions).

The change of waterbird abundance, Shannon Diversity, SPEC and LRAE indexes, were analyzed with the R Statistical Package (R Development Core Team 2007), using the non-parametric Kruskal-Wallis test and establishing the statistical significance at $p = 0.05$ in all tests. The Shannon index depends on the number of species (richness) and also on the uniformity of their contribution to total abundance (number of individuals of each species).

The main temporal phases within the survey period (between which relevant changes in the composition of the waterbird community occurred) were established through multivariate ordination techniques. These methods can be used to describe the degree of similarity (and thus to differentiate) among temporal or spatial subdivisions, in terms of the identity and relative abundance of the species (Lopes et al., 2005; Evangelopoulos et al., 2008; Zárate-Ovando et al., 2008; Sánchez et al., 2006). Data were normalized ($\log(x+1)$ transformation) and submitted to Multidimensional Scaling (MDS; Clarke, 1993) implemented through the Primer 6 software (Clarke and Gorley, 2006), to group years on the basis of their similarity and hence identify the main phases of change and their indicator species. This should allow to interpret phases as a reflection of stages in the succession of the pond ecosystem, with characteristic waterbird assemblages. The results obtained, both for winter and summer, did not allow to infer such major succession stages since at similarity levels above 60%, each year was individually recognized as a different group.

A subsequent Indicator Value Analysis (IndVal) based on the Monte Carlo Test, performed with the PC-Ord Software (McCune and Mefford, 1999), allowed to identify representative (indicator) species for any year or group of them (Dufrene and Legendre, 1997), i.e. for the temporal phases recognized by the MDS.

In the Mar Menor shoreline, like in the case of CP, two data sets were analyzed separately: winter and summer censuses. Three community variables, the mean annual abundance, the mean monthly abundance (between years) and the Shannon-Wiener index were calculated. In order to detect changes in the conservation value of the waterbird community, SPEC and LRAE indexes were also calculated.

The spatial and temporal variability of structural and conservation indexes was analyzed, using "season" and "year" as temporal factors to test for differences between winter and summer community and between years (within seasons), respectively. As in CP, the nonparametric Kruskal-Wallis test was used to determine the significance of these differences. "Basin" and "band" were also used as factors, to detect spatial variations in community structure and organization. "Basin" was used since in the Mar Menor lagoon two main sectors or *basins* with different hydrological characteristics are recognizez (Pérez-Ruzafa and Marcos, 2003), which may determine also different particular conditions for waterbirds. Factor "band" was considered because each species may exploit different areas of the depth gradient depending on their morphological characteristics. Again, statistical significance was set at $p = 0.05$ for all analyses.

Multivariate ordination methods were also applied to shoreline waterbird data, to describe the degree of similarity (or difference) between the temporal or spatial subdivisions (on the basis of previously described factors), in terms of relative abundance and species composition.

After data normalization through a log (x +1) transformation, a Multidimensional Scaling analysis (MDS) was performed (Clarke, 1993) with the software Primer 6 (Clarke and Gorley, 2006). Subsequently an Indicator Value analysis (IndVal) based on the Monte Carlo test was implemented through the PC-Ord software (McCune and Mefford, 1999). This analysis allowed to recognize indicator species with statistical significance for each class within each grouping factor (basin, band and classes inferred from calculated rank values of LRAE and SPEC) that is integrated into the analysis (Dufrene and Legendre, 1997). Qualitative classes of conservation value were inferred from rank values of conservation indexes and assigned to sampling units; then, units were grouped according to these classes and IndVal tests conducted to detect indicator species of each conservation class (separately for each season).

3. Results and Discussion

3.1. Coterillo Pond

Regarding morphological and limnological changes, the results of the three surveys available (1995, 2007 and 2010) are presented separately for the pond main basin and the canal, since despite their connections they are considered semi-independent hydrological units (Table 3). For the canal there are no data for 1995.

Table 3. Morphological and limnological changes detected in the surveys made in the pond (3 periods, upper section) and in the canal (2 periods, lower section)

POND (MAIN BASIN)					
	1995	*2000-2006*	*2007*	*2008-2009*	*2010*
Temperature (ºC)	25.36		23.25		25.90
pH	7.95		4.88		6.75
Average depth (cm)	27.93		23.67		35.89
Average salinity (g/l)	22.87		58.42		49.33
Nitrite (mg/l)	0.002		0.0125		0.00001
Nitrate (mg/l)	2.45		<0.2		0.00003
Ammonium(mg/l)	0.04		<0.06		0.00108
Phosphate (mg/l)	0.08		<0.2		0.00018

ADJACENT CANAL				
	1995-2006	*2007*	*2008-2009*	*2010*
Temperature (ºC)		23.53		25.57
pH		5.54		6.76
Average depth (cm)		24.83		37.33
Average salinity (g/l)		43.70		38.85
Nitrite (mg/l)		0.010		0.0120
Nitrate (mg/l)		<0.2		N.D.
Ammonium(mg/l)		<0.06		3.36275
Phosphate (mg/l)		<0.2		<0.002

Taking as a reference tistuation the state of CP in 1995, in 2007 the features indicative of a degraded or stressed state had worsened: lower depth, increase in salinity and excess primary production, as well as a decrease in pH. All this seems related to the evolution of the system towards an apparent eutrophic state following nutrient loading in the water supply, and in the absence of any explicit management of factors affecting the system productivity (water residence time, internal circulation, depth management, removal of decomposing algae…). The pond had evolved into a state of lower hydrological and ecological quality, far from the original objectives for which this artificial wetland was created. This trend was reversed in 2008 with the opening of the communication of CP to the sea, causing a *rejuvenation* of the system and a gradual return of the different physicochemical parameters of the water towards less stressed, baseline conditions. This new state (deeper water, more neutral pH and low salinity) could not be confirmed until the 2010 survey.

With regard to nutrients, the waste from the Marine Resources Center which is discharged with the effluent into the canal, is composed primarily of fish excreta, uneaten food and remains of other products used in fish culture. This waste, rapidly increasing nutrient concentrations (total phosphorus, ammonia, organic nitrogen) and organic matter content in the sediment and water column, decreases the concentration of dissolved oxygen and raises the general trophic status of the water body (Tartarotti *et al.*, 2004; González-Félix and Pérez-Velázquez, 2006; Urakawa *et al.*, 2006), wich results in the eutrophication of the system. In wetlands, the process of decomposition (mineralization) occurs at a significantly lower rate in the sediment, due to the prevalence of anaerobic conditions in most of its profile.

Consequently, a considerable accumulation of organic matter can cause mismatches in the system (Clymo, 1983). The continuous discharge of nutrients, mainly phosphate and nitrogen, stimulates the growth of algae and other photosynthetic aquatic life, accelerating eutrophication of the receiving environment, significantly reducing the concentration of dissolved oxygen and producing undesirable changes in aquatic populations (Knobelsdorf, 2005; Ruiz, 2008).

Regarding the population of *Aphanius iberus*, it should be noted that in this type of wetlands the fish community displays a very low species diversity, a fact that can be decisive for the stable presence of this endemic species. Usually only three species of fish (*Anguilla anguilla*, *Mugil cephalus* and *Atherina boyeri*) coexist with this small cyprinodotid. In 1995 and 2007 only *Mugilidae* were recorded in CP.

Thus, it is likely that the populations of *Aphanius* of CP benefit locally from reduced interspecific competition (Oliva-Paterna and Torralva, 2008). However, other abiotic factors may be determining the settlement of *Aphanius* populations, for example, the absence of hydrodynamism in these closed water systems (Smith and Able, 1994).

In the absence of a continuous monitoring of the population of *Aphanius iberus*, on the basis of the numerical estimates available (revealing a high density and an important number of pregnant females), the species can be attributed to an apparently good conservation status, as already observed by Farinós and Robledano (2007).

Two conclusions can be drawn from the population surveys made in the two periods studied (Table 4). First, at a general level, the proportion of females relative to males is quite low. In other studies (Fernández-Delgado *et al.*, 1988; Granado-Lorencio, 1996), some populations show a numerical dominance of females over males. This dominance may result from a differential response of sexes to reproductive stress and sexual maturation. The significant dominance of females in these fluctuating environments, could be interpreted as the basis of a blooming response allowing effective population recovery. This species presents a life strategy that enables it to live in environments with significant intra and inter-annual fluctuations in environmental parameters.

The general pattern of this strategy would be characterized by a low number of age classes, high growth rates and a short reproductive life, but with an extended breeding season (these aspects have not been confirmed but fit with the fluctuations known to occur in this ecosystem).

This is a typical strategy of an opportunistic species (Winemiller and Rose, 1992) adapted to extreme and unstable environments. Second, the distribution of *Aphanius iberus* in the system changed from 2007, when most of its population was concentrated in the canal, to be located preferently in the basin in 2010. This change may be due primarily to improved physical and chemical parameters, especially pH and probably dissolved oxygen.

Table 4. Population indexes (captures) of Aphanius iberus in 2007 and 2010

	POND		CANAL	
	2007	2010	2007	2010
Males	58	157	176	40
Females	249	691	492	158
Young	1	50	17	7
Total catch	308	898	685	205

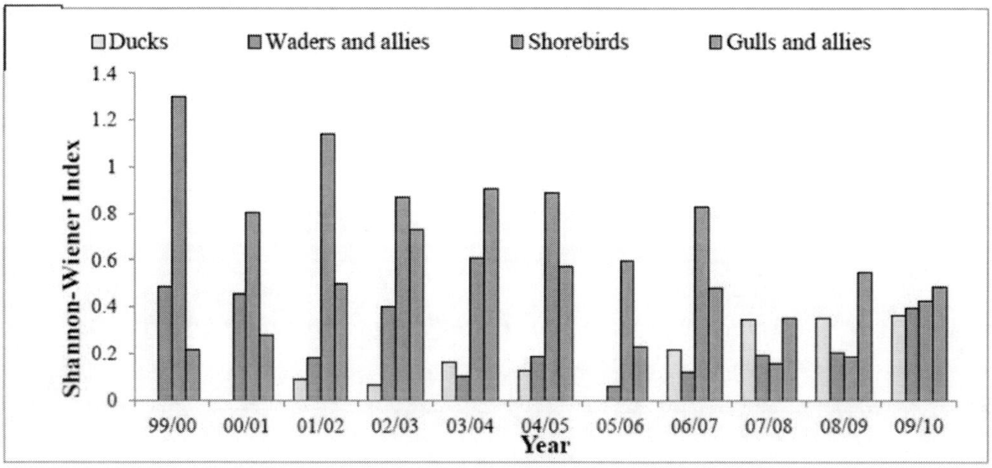

a

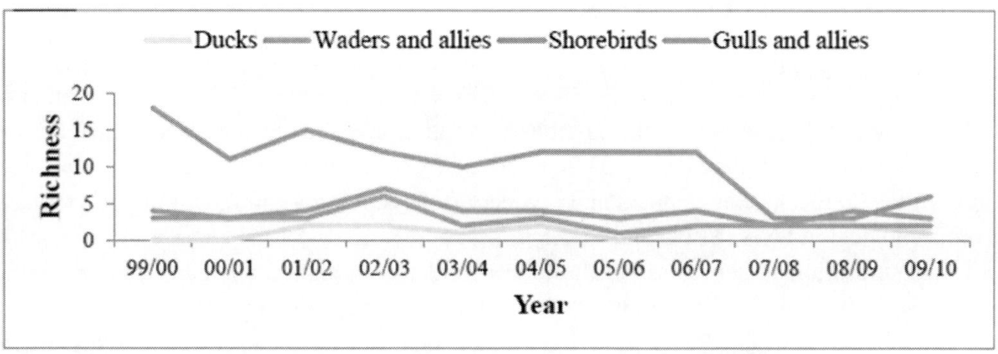

b

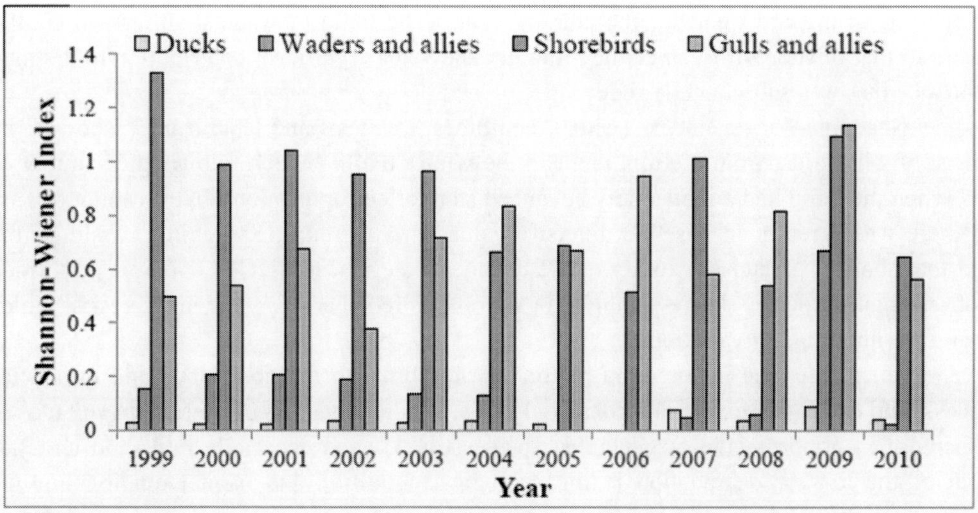

c

Figure 2. (Continued)

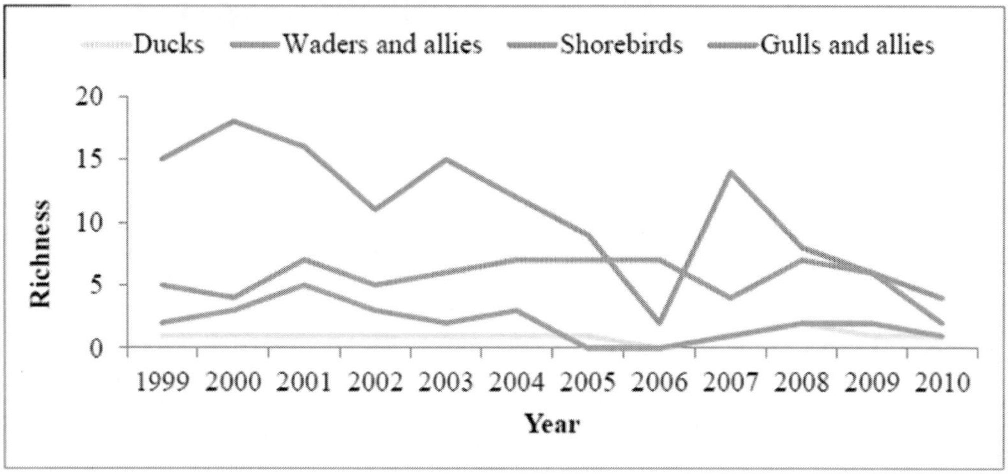

d

Figure 2. (a-d) From top to bottom : (a) Shannon-Wiener Index and (b) species richness within each waterbird guild in Coterillo Pond in winter; (c) Shannon-Wiener Index and (d) species richness of the same guilds in summer.

The long term variation of the composition, structure and conservation value of the waterbird community (Figures 2 and 3) is described by the changes in the relative abundance of bird guilds, in the species richness and diversity (total and intra-guild), and in the scores of the ornithological indexes (SPEC and LRAE).

The contribution of shorebirds to the total abundance remained relatively constant until the winter 05/06, when it markedly decreased possibly coinciding with the time when limnological conditions in CP started to deteriorate. As for ducks, they showed a reverse trend to that of shorebirds, since until the winter 05/06 they kept contribution to total abundance certainly modest (except in the winter of 2003). From 2005 it increased significantly, anticipating an upward trend for the coming years. The guild of waders, suffered a similar process to that of shorebirds, since they initially showed a significant contribution to the total abundance that was subsequently reduced.

The Shannon-Wiener index (which combines richness and abundance) showed an increase in all guilds, from the minimum in the winter 07/08 (which follows the summer of 2007 when the pond had a moderately advanced state of eutrophication), but it cannot be said that there was a return to baseline diversity values in all groups. Shorebirds would be the guild that showed greater sensitivity to the changes experienced by CP. This was confirmed by observing the species richness, which for shorebirds fell sharply, while in the rest of guilds remained without major variations.

The general decline in the contribution of shorebirds to the abundance and in species richness, and to a lesser extent in waders and allies, resulted in a loss of specialists, in favor of an increase in opportunistic or generalist species (ducks, and especially gulls and terns), a signal of the loss or degradation in the ecological condition and habitat quality. For the summer periods, the duck guild was not taken into account in the study, due to its much lower abundance in CP relative to winter.

The contribution of waders to summer abundance was considerably reduced compared with winter, but remained constant throughout the entire surveyed period.

The two markedly dominant guilds (shorebirds and gulls are and allies), which showed a reverse trend. Based on the characteristics and preferences described for each guild, one might expect that the contribution to the abundance of waders in 2007 was very low, which did not occur.

Two reasons can explain this fact. First, in 2007 the average depth was significantly lower than in the physico-chemical surveys of 1995 and 2010, resulting in a lower water table and therefore a larger surface of banks available for waders. Second, it could have been a certain delay in the response of birds to pond conditions, evidencing its effects in the winter 07/08 (and in the following summer of 2008) and not in the summer 2007 itself when the disturbance occurred.

The richness of shorebirds underwent drastic fluctuations. No data were available on limnological parameters for the summer of 2006, when very low numbers of shorebirds were recorded and when ducks and waders disappeared from CP. However, it is known that in May 2006, the communication with the sea was not open and the water level in the pond was very high, leaving little surface of exposed banks available for waders.

No more information is known about the state of the pond, but the response of shorebirds is consistent with the situation of other guilds, especially wading birds. In that same year, the abundance of gulls and terns was well above the peak of waders, as did the guild's species richness. The following year, the trend was completely reversed.

The frequent fluctuations in the waterbird community parameters, both in winter and summer, were in some cases difficult to explain due to the lack of a parallel monitoring of limnological and other environmental variables with similar continuity. In any case, the general trend observed in summer continued to be the gradual replacement of specialists by opportunists. Although gulls and terns kept a species richness around 4-7 along most of the study period, their contribution to total abundance increased every year, what could be a symptom of the system's deterioration.

Figure 3 shows the variation of three of the parameters used to characterize the waterbird community: SPEC, total abundance and total species richness. The latter showed a decreasing trend along the census period with significant differences among winters (Kruskal-Wallis test). This trend is related to the disappearance from CP of many members of the shorebird guild, the most negatively affected of all those represented. Among the species lost are all those of the family. In the last surveyed years a species of the *Sternidae* (*Sterna sandvicensis*) is also lost.

As for total abundance, it showed higher values in the last half of the survey period, until the winter 04/05, when the peak of abundance was recorded. Subsequently, it started to decrease until it reached a minimum in the winter 07/08. The trend of the SPEC index was concordant with those of total abundance and species richness.

Again, two neatly defined periods could be distinguished: until the winter 04/05 there was a rising trend, and then a decrease until 07/08.

At the end of the study period, the index rose again due mainly to four species, each belonging to a different guild, being incorporated to the community: a duck (*Anas acuta*), a wading bird (*Phoenicopterus roseus*), a shorebird (*Tringa totanus*) and a gull (*Larus genei*). Although the situation seemed to have improved with respect to the previous assessment made in 2007, the community was still far from the composition it had in 04/05, as a result of the loss of shorebird species.

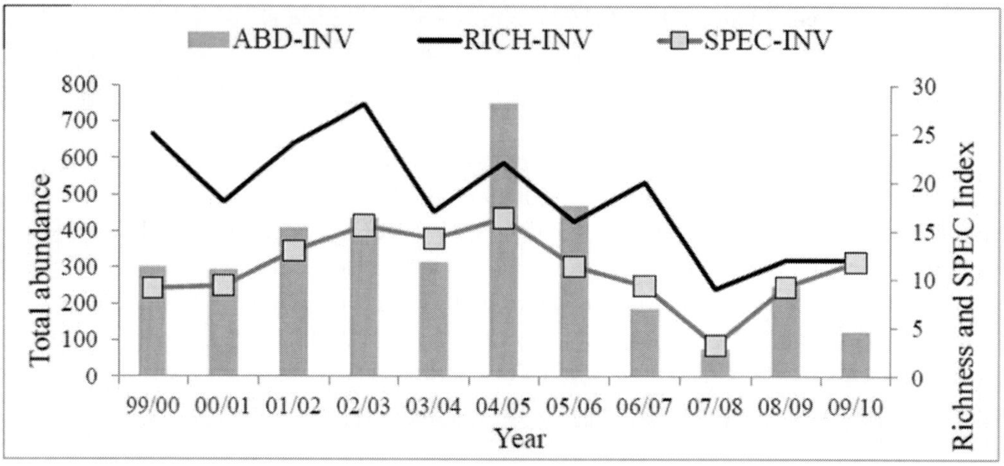

a

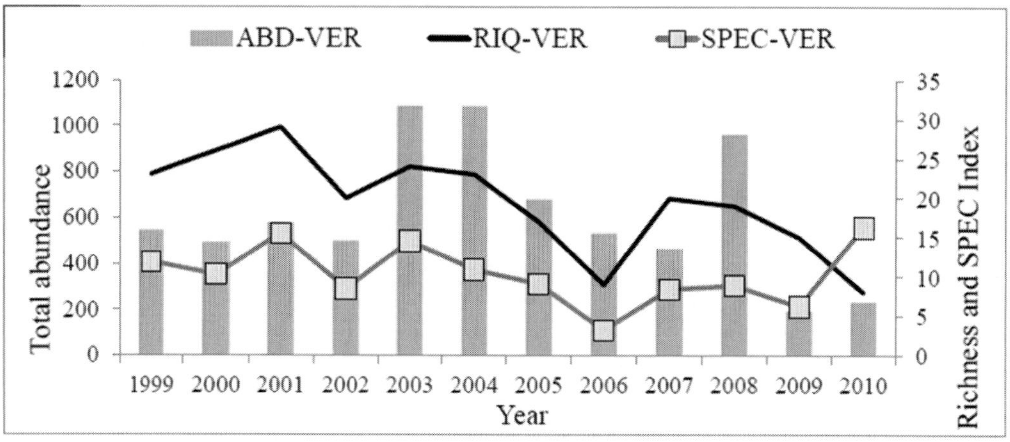

b

Figure 3. a-b From left to right: Total abundance (left ordinate axis), Total Richness and SPEC Index (right ordinate axis) in winter (=INV) and summer (=VER).

The ordination analysis on the basis of the census of the waterbird community, allowed recognizing three main temporal phases:

- 1999-2002: The community was dominated by shorebirds except in one winter (gulls and terns, 02/03). SPEC and LRAE indexes showed high values although the peak values occurred later.
- 2003-2007: Shorebirds end to be the dominant guild (in terms of their contribution to total abundance) in favor of gulls and ducks. Although the richness of the former guild continues to be the highest (except in the summer of 2006), it can be already said that shorebirds are receding and that the other two guilds alternate in the dominance of the community. Thus, it can be concluded that the specialist guild is being replaced by the more opportunistic or generalist ones.
- 2008-2010: Waders no longer dominate in abundance and their richnes decreases too, but since the total richness of the community also decreases, the result is a co-

dominance of shorebirds with wading birds, and larids, in 2008 and 2009, respectively.

Among the nesting families, the contribution of shorebirds and gulls and terns can be highlighted. This responds to several factors concurring in the Regional Park (and to some extent also in CP).

First, both represent an appropriate breeding habitat (particularly due to the presence of bare sand levees and islets, beaches, accumulations of sea drift debris, etc.). Second, they exhibit a high availability of food for these guilds, like benthic invertebrates and fish. *Sterna hirundo* and *Sterna albiforms* specialize precisely in the capture of small fish (Cramp, 1988), and both the Park and CP have abundant populations of *Aphanius iberus*.

Other *Sternidae* like the Gull-billed Tern (*Gelochelidon nilotica*) feed preferently on small reptiles and amphibians (Cramp, 1988), which are also abundant in the surroundings, especially the latter due to the development of a dense irrigation system in the nearby agricultural plain (Sebastián-González *et al.*, 2010).

One of the most abundant species, *Larus ridibunus*, can also be favored by the proximity of the sea where it uses to benefit from fish discarded by fishing wessels (Goutner, 1994).

In regard to shorebirds, represented by *Recurvirostra avosetta*, *Himantopus himantopus* and *Charadius alexandrinus*, their diet is based mainly on crustaceans, insects and worms. Data from Velasco *et al.* (1995) support the presence of several invertebrate prey (díptera, coleóptera, amphipoda, etc.) which make Coterillo Pond system (including the pond plus its adjacent saltmarsh) an attractive place for waterbird breeding in the context of the Regional Park to which it belongs.

Regarding percent contribution to total abundance, that of the families *Laridae* and *Sternidae* (gulls and terns) is the most remarkable, both in the Regional Park and in the Coterillo Pond+Adjacent Saltmarsh (CP+AS, hereafter) ensemble. The breeding community of the latter changed in a somewhat independent manner with respect to the general situation in the Regional Park, which depicts a response to the particular conditions prevailing in the pond and its tightly related adjacent saltmarsh (i.e., to the more fluctuant ecological regime of such environment, not dictated by the functioning of the industrial circuit of salt production as in most of the Regional Park).

In the ensemble CP+AS the terns seem to be replaced by more generalist species like *Larus melanocephalus o L. ridibundus* (virtually not represented in the early stages of the development of the pond ecosystem). This replacement is attributable to two factors: first, to the already mentioned deterioration of the initial ecological status of the CP, affecting negatively the specialist taxa.

And second, the settlement of a very high number of pairs of Black-headed Gull *Larus ridibundus* (190 in 2003, 257 in 2004, 202 in 2006) in a relatively small area (8.2 hectares, but much smaller if the water table is not accounted for), could have prevented the settlement of other breeding species. The opposite trend is seen in the Regional Park, with an increasing contribution of *Sternidae* over *Laridae*.

The species richness of the breeding waterbird community of CP represents around 40-60% of that of the whole Regional Park. The species richness of the AS is similar to that of CP during most of the survey period. As it can be seen in figure 4, only *Recurvirostra avosetta* changes in relation to its variation in the whole Regional Park, while five species change with a certain degree of coupling between the two parts of the ensemble CP+AS.

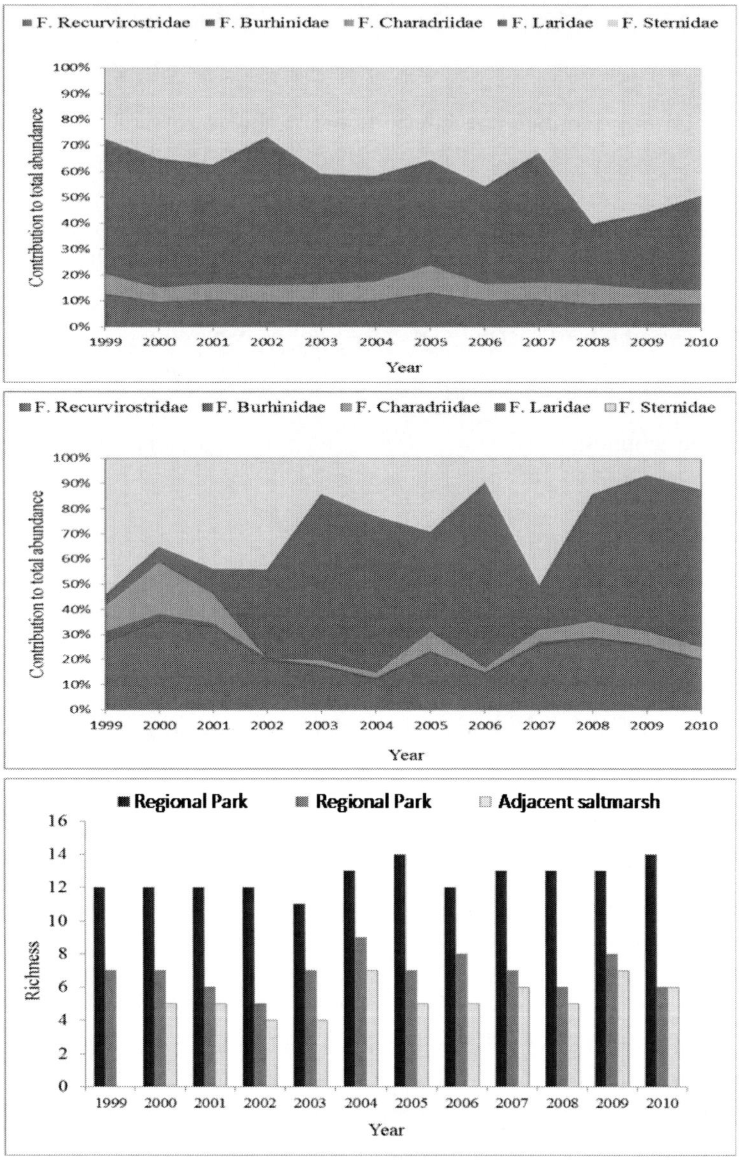

	REGIONAL PARK WITH COTERILLO POND	REGIONAL PARK WITH COTERILLO POND AND ADJACENT SALTMARSH	COTERILLO POND AND ADJACENT SALTMARSH
Himantopus himantopus	-0.12	-0.10	**0.50**
Recurvirostra avosetta	**0.54**	**0.54**	**0.82**
Larus ridibundus	0.02	0.03	**0.73**
Larus michahellis	-0.22	-0.22	**1.00**
Sterna albifrons	-0.43	-0.34	**0.78**

Figure 4. Percent contribution to total abundance of the different families of waterbirds breeding in the Regional Park, and in the ensemble CoterilloPond+Adjacent Saltmarsh (upper graphs). Annual change in the species richness of the breeding waterbird assemblage in each of these subdivisions (lower graph), and correlation coefficients among subdivisions for the breeding populations of particular breeding species (table).

From these observations, it was suggested that CP could probably function (in conjunction with other *satellite* wetlands, and obviously with the Regional Park) as a metapopulation-like system (Sebastián-González, 2011), at least for species behaving like *Recurvirostra avosetta.*

At a smaller scale the fraction of birds present in AS would represent a subpopulation of those settled in CP, both behaving as a coupled system (for the remaining species in the table of figure 4). The trends observed in the censuses covering the whole Park, however, are not reflected in the numbers of breeding pairs of these species recorded in CP. This fact, and the lack of correlation between the Regional Park and the ensemble CP+AS for most species, indicates that the breeding community of the latter responds to a greater extent to local ecological conditions and does not mirror overall trends or ecological changes which are appartent in the whole of the protected area system.

A summary of the results of indicator value analyses is shown in Table 5, which includes a qualitative assessment of the relative abundance and richness of species and guilds, as well as the definition of the three periods described according to the dominant strategy (degree of ecological specialization) among the waterbird community.

Regarding the winter periods, in that of 1990/00 *Phoenicopterus roseus* (the flagship species of the Regional Park) appeared as an indicator species. Its relative abundance along the study period was quite variable, occurring in five winters (although more than half of its total numbers occurred in that of 90/00). Flamingoes prefer brackish or saline, shallow water areas with depths usually under 60 cm, although it can exploit depths of 90-105 cm (Johnson, 1983) to feed on carotenoid-rich prey (McGraw *et al.*, 2001). Such preferences suggest that in these winters and especially in 99/00 water depth was not very high in the pond, although limnological surveys are lacking and other ecological interpretation is not allowed.

It can be highlighted that both the species richness and the Shannon diversity index showed a decrease from 2004 to 2007, probably due to the eutrophication process described by Farinós and Robledano (2007).

During that period *Phoenicopterus roseus* was not recorded, suggesting lowered food resources and/or worsened feeding conditions for the species. Velasco *et al.* (1995) recorded the presence of crustaceans of the orders *Mysidacea, Isopoda, Ansipoda* and *Amphipoda*, potential prey of flamingoes which tend to be sensitive to changes in pH and nutrient contents of the water. The eutrophication of CP could have reduced the density of these invertebrate groups, thus rendering the pond less attractive for flamingoes. However, since the response of flamingoes to eutrophication is not always negative, other factors can have been involved, e.g. the proliferation of filamentous algae, acting as a physical barrier against the access to prey living in the sediment or water column.

Sterna caspia was the indicator species of the winter 01/02 (it concentrated 100% of its abundance in that year).

Again the lack of limnological data precludes interpreting the relationship between its presence and the environmental conditions of the pond, although it could also have been occasional, since Paracuellos *et al.* (2007) described its phenological status as such in SE Spain. It is a cosmopolitan species wintering in Subtropical Africa. In Europe, it is usually found in the shores of the Baltic Sea, in the coast of the Black Sea and in all the perimeter of the Caspian Sea. It lives also in Kazajstan and in regions further East (Cabo and Camacho, 1981).

Table 5. Final summary of indicator species for years and temporal phases, derived from the analysis of total (summer and winter) census of waterbirds (ABD 100%: Total abundance of the species concentrated in that year; Guild >%ABD: Most abundant guild that year; Guild>RICH: Richest guild that year)

	95/99	99/00	00/01	01/02	02/03	03/04	04/05	05/06	06/07	07/08	08/09	09/10
Indicator species WIN		*Phoenicopterus roseus*		*Sterna caspia*		*Larus ridibundus*				*Larus michaellis*		
Indicator species SUM		*Pluvialis squaratola*		*Tringa erythropus*		*Larus ridibundus*		*Larus melanocephalus*				
Species with ABD 100% WIN	*Egretta alba* *Haematopus ostralegus* *Charadius dubius* *Vanellus vanellus* *Calidris alba*			*Anas crecca* *Calidris ferruginea* *Philomachus pugnax* *Numenius arquata* *Larus audouinii* *Sterna caspia* *Gelochelidon nilotica*			*Gallinago gallinago*			*Gallinula chlropus*		
Species with ABD 100% SUM			*Bubulcus ibis* *Ciconia ciconia*				*Larus audouinii* *Larus fuscus*			*Anas platyrrynchos* *Gallinula chlropus*		
Guild >%ABD WIN	SHBIRDS	SHBIRDS	SHBIRDS	GULLS	GULLS	GULLS	SHBIRDS	GULLS	DUCKS	GULLS	GULLS /DUCKS	
Guild >%ABD SUM	SHBIRDS	SHBIRDS	SHBIRDS	SHBIRDS	GULLS	GULLS	GULLS	GULLS	SHBIRDS	GULLS	GULLS	GULLS
Guild>RICH WIN	SHBIRDS	SHBIRDS	SHBIRDS	SHBIRDS	SHBIRDS	SHBIRDS	SHBIRDS	SHBIRDS	SHBIRDS	WADERS	SHBIRDS	SHBIRDS
Guild>RICH SUM	SHBIRDS	SHBIRDS	SHBIRDS	SHBIRDS	SHBIRDS	SHBIRDS	SHBIRDS	GULLS	SHBIRDS	SHBIRDS	SHBIRDS/ WADERS	SHBIRDS
Periods	A: Dominance of Specialists				B: Transition period to Generalists					C. Dominance of Generalists		

In the winter 03/04 the indicator species was *Larus ridibundus*. Besides being a local breeder, it is the most abundant of those present in CP (maximum of 314 individuals in the winter 02/03), being considered as a typical generalist species. It was recorded continuously in the pond since the start of the surveys in 1999.

According to the observations made by one of the data contributors (G.A. Ballesteros) in some years the low levels of the water table and the abundance of fish (*Mugilidae*) could favor the presence and feeding activity of this (and other) larids. Regarding the interactions with other species, a fact that can explain the decline of shorebirds in the last years of the survey period is the observation that in some wetlands, the numerical increase of the colonies of *Larus ridibundus* was responsible for a greater interspecific competition leading to the reduction of the populations of other waterbird guilds (Paterson, 1997).

Finally, *Larus michaellis* was the indicator species in the winter of 07/08. It is a species of wide geographical and ecological distribution, generalist in the choice of habitat and food and an opportunistic species with great trophic flexibility (Ramos *et al.,* 2009). At present, *Larus michaellis* colonizes not only islands, islets, cliffs and beaches se (Purroy 1997), but also periurbana reas and even cities due to the permanent availability of food in rubbish dumps (Bosch *et al.*, 1994, Burger and Gochfeld, 1996). Among the main biological effects of the expansion of *Larus michaellis* are its negative impact on other birds, especially through the displacement of endemic and endangered species (Swennen and Van de Meer, 1992) or through direct predation on many other (Oro *et al*., 2005). Moreover, their carrion feeding habits and their frequent use of rubbish dumps as feeding areas, increase their exposure to endoparasites and enhance their potential role as vector of infectious agents (Hervías, *in prep*.). Another species exclusive (100% of its abundance) in that same winter period was the Moorhen (*Galllinula chloropus*). Contrary to the majority of the species recorded, the Moorhen is a phytophagous species (Cramp and Simons, 1977), although the generalist and adaptable nature of this wide-ranging species precludes any attempt to draw conclusions from its occurrence in CP. In any case, its presence can be framed within the general succession process described by the change of the waterbird community.

As for those species exclusive (100% of their total abundance) in any other year, the winter of 1998/99 stands out with 5 species: *Egretta alba* and four shorebirds (*Haematopus ostralegus, Charadrius dubius, Vanellus vanellus* and *Calidris alba*). In the winter 01/02 four species are again found with this status (*Anas crecca* and the three shorebirds *Calidris ferruginea, Philomachus pugnax* and *Numenius arquata*). It can be seen that all except one of these species are wading birds or shorebirds, which supports the dominance of specialists over generalists in the first period of the ornithological survey.

Returning to the summer community, in 2000 the indicator species for this period was *Pluvialis squaratola*, whose diet is composed, as in most shorebirds, of crustaceans, mollusks, worms and insects (Turpie, 1995). Like other shorebirds of similar size it forages in waters less tan 5 cm deep (Connors *et al.,* 1993), which further indicates that in the summer concerned, the water level was low and consequently there were large tracts of exposed shores available for this guild to feed. In the summer of 2002 the indicator role corresponded to *Tringa erythropus*, another medium-sized shorebird (Myers, 1980). It is in the summer 2004 when for the first time a gull species (*Larus ridibundus*) appears as indicator, and in the summer of 2006 this condition corresponded to *Larus melanocephalus*, a very scarce breeder of which two pairs nested in CP that same year. In the Regional Park the higher numbers of breeding pairs were recorded between 2004 and 2007.

Given that, the presence of this gull in CP most probably resulted from the concurrence of the overall trend with suitable conditions in the pond area.

There were also species that concentrated all their abundance in any single summer: in 2001 two wading birds (*Bubulcus ibis* and *Ciconia ciconia*); in 2002 two gulls (*Larus audouinii* and *Larus fuscus*); and in 2008 a duck (*Anas platyrhynchos*) plus the Moorhen *Gallinula chloropus*. Among them, it should be highlighted the presence of Audouin's Gull (*Larus audouinii*). This gull has held one of its major breeding colonies in the nearby Isla Grossa (Oro, 1996), and 15 pairs were recorded in the Regional Park in 2007 (although not in CP). Its summer occurrence in CP leads in any case to a considerable increase in the value of the conservation indexes SPEC and LRAE. Its relative abundance depends largely on the presence and demography of *Larus michaellis* (Oro and Matínez Vilalta, 1994; Oro, 1996; Oro *et al.*, 1996; Martínez Abraín *et al.*, 1988). Despite being a larid, it can be categorized as specialist, due to its critical dependence on marine food resources.

As it is summarized in table 5, as a general rule, the species appearing as indicators (often being also exclusive) for years before 2004, belong to the "specialist type" (*Phoenicopterus roseus, Tringa erythropus, Pluvialis squatarola*). In this first period also acquires such status a tern (*Sterna caspia*), although it is an occasional visitor. Thereafter, all the indicator species identified are defined as generalists or opportunists, and belong to the same guild and family (*Laridae*): *Larus ridibundus*, *Larus melanocephalus* and at last, the most generalist one, *Larus michaellis*.

3.2. Shoreline Units

3.2.1. Spatial and Temporal Variations of Biological Indexes

In the lagoon shoreline, the changes in the indexes of abundance, species richness and diversity, and ornithological value of the waterbird community are assessed along shorter time scales, but the possibilities of spatial comparisons increase.

In winter, the Mar Menor community responds to internal structural and functional gradients, as the decreasing influence of the Mediterranean Sea into the lagoon -opposite the degree of confinement- (Pérez-Ruzafa *et al.*, 2011), given its regime of transitional waters, the hydrodynamic conditions, its physical structure and geomorphology, and several external influential factors such as human activities (Robledano and Farinós, 2010). The most abundant species in the wintering community are coot, grebes, cormorants and gulls (mainly Yellow-legged gull *Larus michahellis*) as shown in Figure 5. The interannual variations in the abundance of some species like coot or cormorant are not clearly related with previous trends described for these species on a long term scale (Robledano and Farinós, 2010), however the dominant role of certain species in the wintering community (grebes, cormorants or Yellow-legged gull) remains as when examined at that long term temporal scale. The interannual variations detected (as the decreasing trend of coot and cormorants –degradation indicators-) could be related with the short-term variations in the intensity of eutrophication agents affecting the CL in recent decades.

Differences in total abundance between the two main basins of the lagoon (northern and southern basin) were also detected (Figure 6). The southern basin, which includes the most natural areas (saline steppe, saltmarshes and reedbeds), displayed higher abundance values than the northern one (more anthropized).

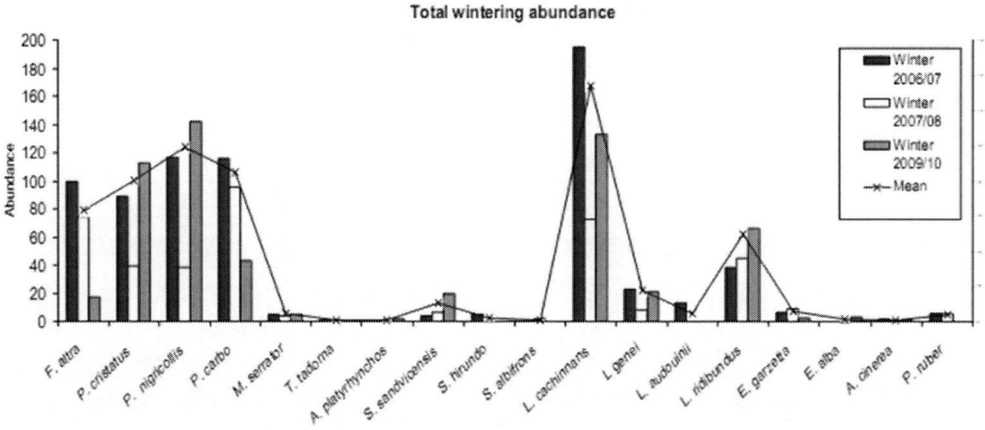

Figure 5. Total wintering abundance by species.

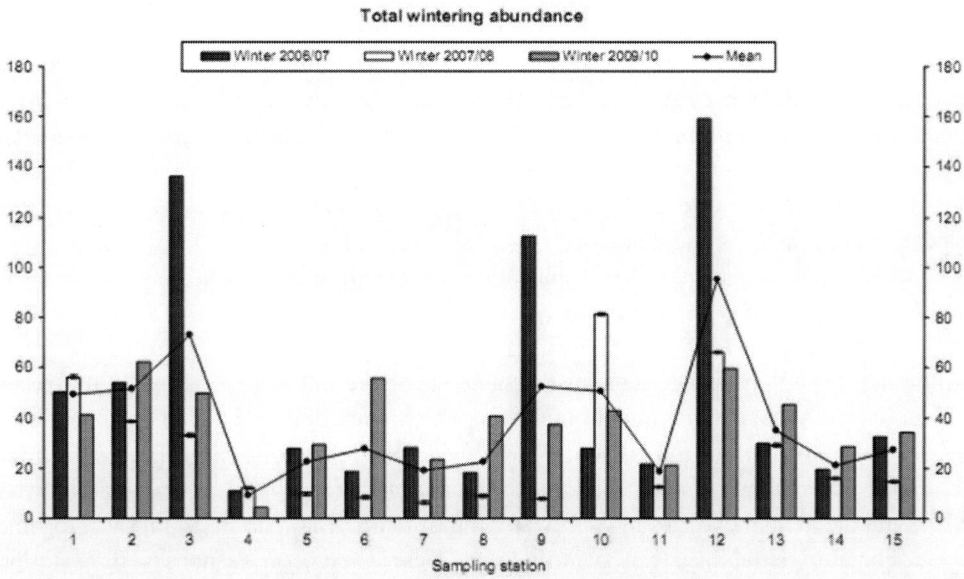

Figure 6. Total wintering abundance (per winter) in each sampling unit.

Furthermore, waterbird abundance differed significantly between bands, depending on trophic requirements and foraging strategies of different species (linked to the depth of the water column), or to the species-specific tolerance to human disturbance.

Comparing the present results with trends in wintering waterbirds recorded since the 70's, it is noted that Red-breasted merganser *Mergus serrator* maintains its gradual decline (Farinós and Robledano, 2010; Robledano *et al.*, 2011). This species, with a well-defined negative response to the trophic enrichment of the lagoon, remains relict to oligrotrophic areas, in particular, clear and exposed littoral waters close to the Regional Park of San Pedro del Pinatar which, as a secondary wetland associated to the lagoon (from which in fact originated), could be playing a shelter role for certain species (Taft and Haig, 2006). Conversely, grebes and cormorants display high abundances (although there is a gradual decline in the latter).

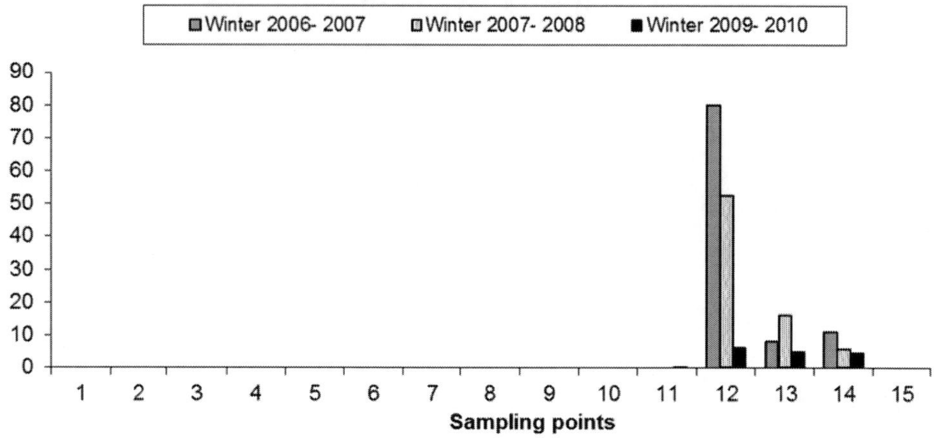

Figure 7. Distribution of *Fulica atra* among sampling units.

Great crested grebe *Podiceps cristatus*, Black-necked grebe *Podiceps nigricollis* and Great cormorant *Phalacrocorax carbo* are concentrated mainly in southern sectors with an intermediate level of eutrophication (related to an indirect influence of nutrient discharges through ephemeral channel outlets, probably due to their internal mobilization in the lagoon). As a result, the indicator role of these species, signalling moderate stages of eutrophication (Robledano *et al.*, 2011) is maintained.

The most obvious response to eutrophication has been the increasing trend of wintering Common coot *Fulica atra* in recent years, linked exclusively to sectors with direct discharge of nutrients (Robledano *et al.*, 2011).

Although its site preference remains the same in the present study (Figure 7), this recent period shows a gradual decline in coot numbers, which could indicate a slight setback in the process of eutrophication of the lagoon. Within the general state of incipient eutrophication, the complex interactions among biological compartments (phytoplankton, macrophytes, jellyfish) and the annual changes in wastewater inputs, agricultural drainage inflows, together with the climatic variability, can contribute to these short-term responses in waterbird numbers.

However, it is feasible that the negative trends shown by coot in other nearby regions (Robledano *et al.*, 2011) have been finally mirrored by the populations wintering in the Mar Menor CL.

On the other hand, the high abundance of Yellow-legged gull with respect to other *Laridae* could be explained by its eclectic and opportunistic behavior, and the probable competitive disadvantage of other species of gulls, allowing this expansive species to thrive in the increasingly disturbed lagoon environment.

Regarding species richness and the Shannon diversity index, southern sampling units -in general with more natural conditions- present the maximum values (Figure 8).

These units receive the indirect influence of nutrient discharges, resulting in an intermediate eutrophication stage. Higher community diversity is a characteristic feature of waterbird assemblages of moderately eutrophic waters (Fjeldså, 2004; Rönkä *et al.*, 2005; MacDonald, 2006).

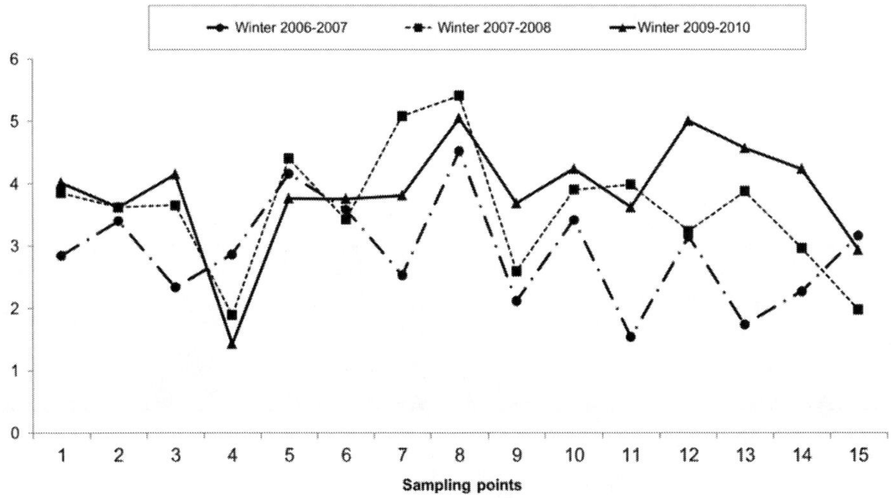

Figure 8. Variation of the Shannon-Wiener diversity index between years and sampling units in winter.

In addition to the higher productivity of their waters, the lower intensity of human activities in the units of the southern basin (Farinós and Robledano, 2010), and the species-specific tolerances to human disturbance (Burton, 2007) may be important factors that help explain these peak values of diversity and richness.

Differences in diversity and richness values were also detected between bands, since different species of waterbirds exploit the lagoon's littoral waters at different distances from the shoreline, according to their morphotype, feeding strategy or specific sensitivity to human presence. No differences were detected between basins.

In summer the response of the waterbird community to internal and external gradients is mediated by the increase (with respect to the winter season) of water-based human activities, and in general of stressing factors affecting the lagoon (Farinós and Robledano, 2010). As in winter, the total summer abundance differs significantly from the south to the north basin, higher abundance values being concentrated in the southern division (with more natural features). In addition, significant differences were also detected between bands according to the same features commented in winter (trophic requirements and foraging strategies).

The summer community (Figures 9-13) is dominated by *Laridae* and allies. The Yellow-legged gull is the most abundant species in summer, followed by Little tern *Sterna albifrons* and Common tern *Sterna hirundo*. However, except for the two species of terns mentioned, all species have lower abundances than in the wintering season. It is necessary to explain the occurrence in the summer dataset of some essentially wintering species (like coot or grebes). This is due to the fact that just two broadly defined phenological divisions were considered (winter and summer, see section 2.5). Therefore, these records come from the first transitional months included in the summer period. In any case, this differentiation of just two seasonal periods is consistent with the dynamics of human activities in the study area, whose benign climatic conditions allow typical 'summer' human activities (and the impacts derived from them) to take place along the six months of our *aestival* period (April to September).

Significant differences were found between bands in both richness and diversity, according to the features mentioned above. As in winter, no differences were found between basins for these two indexes.

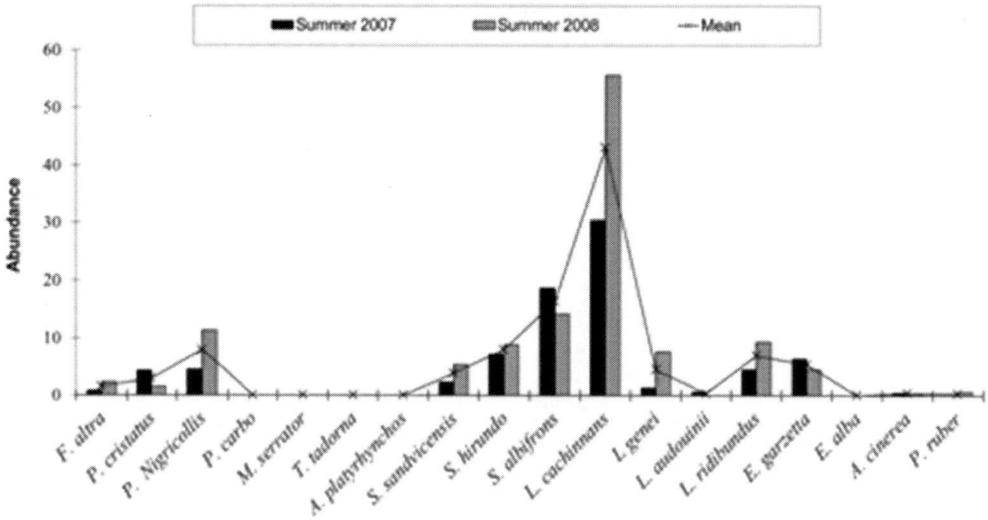

Figure 9. Total summer abundance by species.

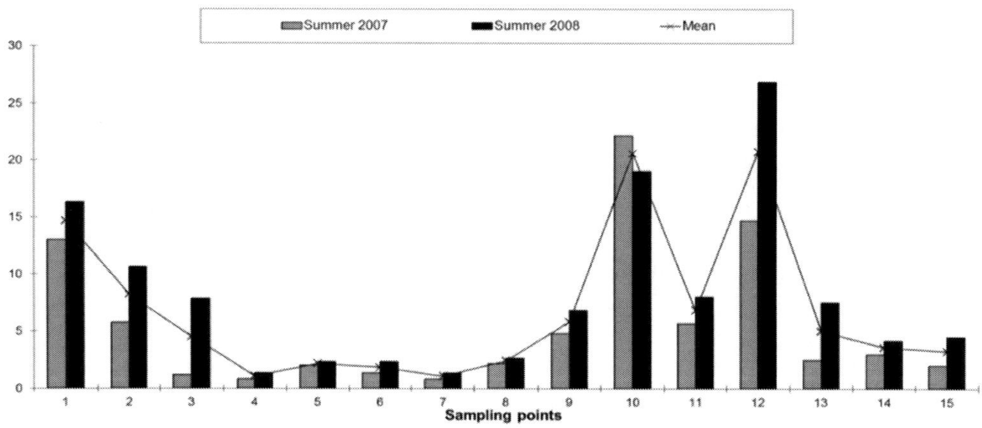

Figure 10. Total summer abundance in each sampling unit.

Is noteworthy that the highest mean values of diversity were recorded in units 1 and 2 (protected natural area and main channel of communication with the Mediterranean, respectively) and unit 12 (mouth of the main channelized watercourse discharging into the CL).

Such peak values were expectable in these areas since they represent scenarios favoring biological diversity, but owing to different causes: around the protected natural area, due to its shelter role (especially in summer when human disturbance increases in all the CL); in the mouth of the main discharge channel, due to the eutrophication effect that boosts biological productivity (and hence a temporary peak of bird diversity); and in the main communication with the Mediterranean, due to the increase of resources characteristic of lagoon-sea transitions, and to the diversifying effect of such areas on waterbird communities (sharing species from both ecosystem types).

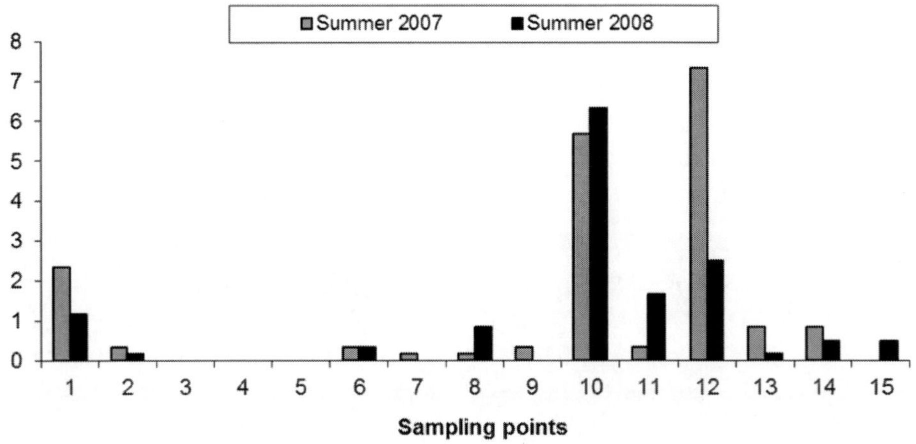

Figure 11. Abundance of Little tern *Sterna albifrons* in each sampling unit.

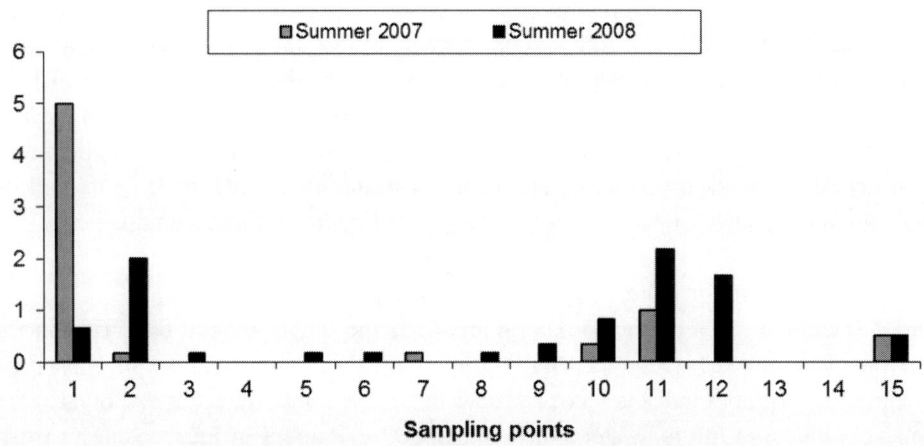

Figure 12. Abundance of Common Tern *Sterna hirundo* in each sampling unit.

3.2.2. Main Spatial and Temporal Gradients in Biological Indexes

After analyzing the temporal and spatial variation of biological indexes, it can be concluded that different temporal and spatial gradients are superimposed and affect the structuring of the waterbird community at different scales. First, a gradient of seasonal change in diversity and total abundance is found, with higher values in winter and lower in summer, related to the dominant phenology of the species recorded and to the intensity of human activities, respectively. This overall drop in diversity and abundance from winter to summer is a characteristic of southern coastal wetlands that receive an influx of wintering waterbirds, becoming refuges for some key species (Scott and Rose, 1996; Erwin, 1996). On the other hand, several overlapping spatial gradients are detected at different scales and both in winter and summer. There is a gradient of increasing abundance from the north to the south part of the basin, probably related to competition (intra and inter-specific) and to differences in the quantity and availability of food resources.

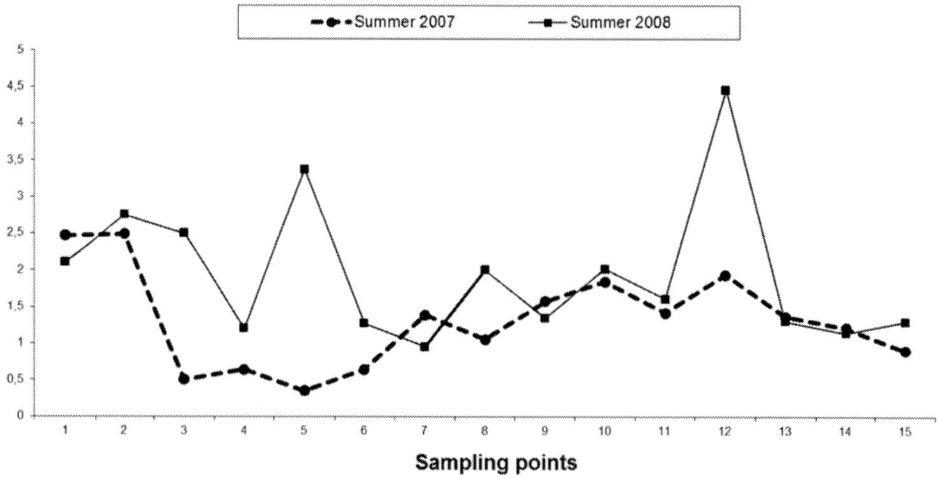

Figure 13. Shannon-Wiener diversity change between years and sampling units in summer.

There are also typical variations in abundance and diversity between bands, which are related to the morphotype and foraging strategies of the species involved. Finally, the variation in abundance and diversity among units can be explained on the basis of habitat quality gradients, occurring again in both winter and summer. These gradients summarize the different affinity of the species with the various types of lagoon shoreline, depending on their habitat structure and land-use scenario, interacting with the waterbirds' specific tolerances to human disturbances, since each unit represents a particular human impact landscape.

3.2.3. Spatial and Temporal Variations of Conservation Value Indexes

Unit 1 (belonging to a protected natural area) has the higher overall conservation value (considering both indexes) (Figures 14-15). Regarding the spatial gradient of conservation value within the CL, both indexes increase (both in summer and winter) in natural areas with different conditions: i) littoral waters suffering direct or indirect influence (due to internal mobilization) of agricultural drainages; ii) presence of other associated wetlands (probably offering alternative food resources or shelter), and iii) transitional areas that enhance biological diversity, like channels communicating the CL with the Mediterranean sea). The high conservation values attained by shoreline areas close to nutrient discharge points (and thus moderately enriched), seems related to the initial positive effect that the eutrophication process has on waterbird species richness and numerical abundance (Fuller, 1982).

With regard to the variation of LRAE index, it clearly shows a decline in the conservation value of the waterbird community from winter to summer, since wintering species tend to have a higher protection status. The presence of Flamingo *Phoenicopterus roseus*, Slender-billed Gull *Larus genei* or Sandwich tern *Sterna sandvicensis* as wintering species in the CL, certainly contributes to the higher winter scores of the conservation inexes, while the summer community hosts more generalist and widely distributed species. At the beginning of the summer season there is a significant decline in the LRAE index in almost all the CL (except for an increase in unit 10), related to the intensification of human activities (especially water-based ones).

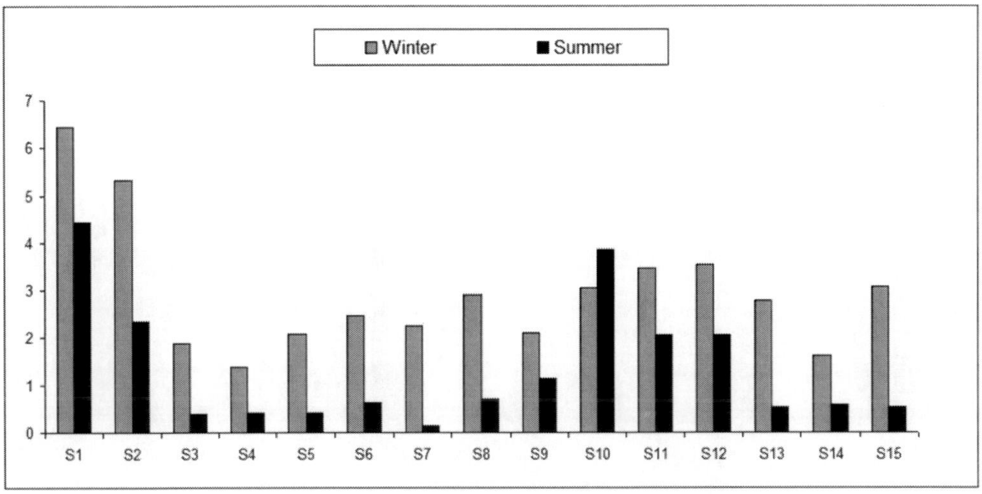

Figure 14. Mean values of the LRAE index in each sampling unit in both seasons. Winter values are means of 06/07, 07/08 and 09/10 data; summer values average 2007 and 2008 data.

In fact, it should be noted that unit 10 is one of the few areas remaining with a similar intensity of human activities (low level of recreational use in the adjacent land) in both seasons, which would confer it a shelter role in summer, when anthropogenic stress increases significantly in nearby areas.

With respect to seasonal variations in the index based on SPEC categories, again a decreasing trend occurs from winter to summer, also related to the higher protection status of many wintering species. It is noteworthy the remarkable increase in the conservation value of the community around unit 10 (also observed in the LRAE index, although less pronounced), for the reasons discussed above.

Moreover, a similar increase in the conservation value of the summer community was observed in unit 6 (near to a protected area) and in units 12 and 14 (natural areas influenced by agricultural drainages). In addition to the greater trophic offer expected in these sectors experiencing moderate eutrophication, their shared natural features (and sometimes proximity to protected areas), they could provide shelter for the summering species of higher protection status.

In short, it becomes evident that the winter community reaches overall greater conservation value, and that the higher conservation scores are also concentrated in natural or semi-natural areas, affected by various factors that can promote diversity and consequently improve conservation value (through the increase in species richness).

3.2.4. Similarity and Indicator Value Analyses

MDS analyses were based on different grouping factors: band (B1 to B4), basin (North vs South), and LRAE and SPEC categories (sampling units were grouped according to rank values for each index). Figure 16 shows how there is greater cohesion in the winter community than in the summer one (Stress = 0.17). This ordination confirms the existence of two distinct communities: wintering and summer (breeding).

With respect to the results of the analysis of winter data grouped by band, figure 17 shows how Band 1 is separated clearly from the other three (Stress = 0.18).

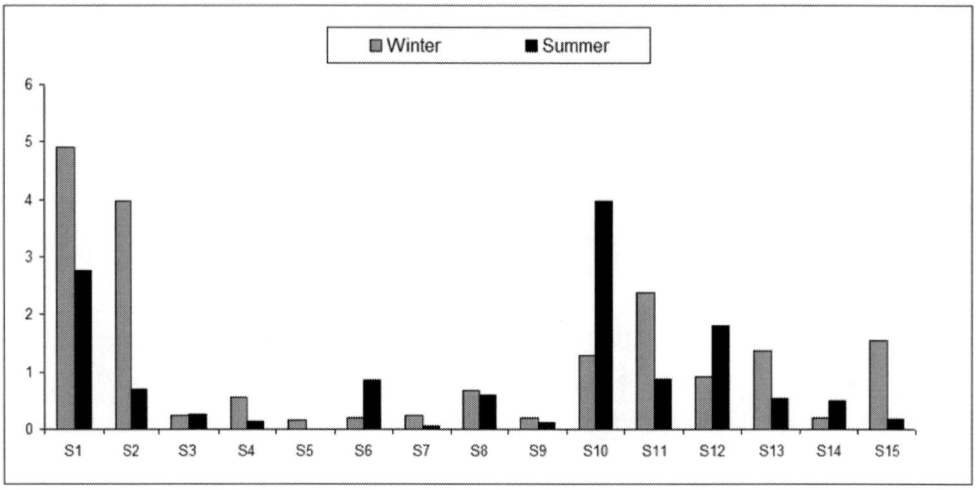

Figure 15. Mean values of the SPEC index in each sampling unit in both seasons. Winter values are means of 06/07, 07/08 and 09/10 data; summer values average 2007 and 2008 data.

This demonstrates that conditions of nearshore habitats determine a markedly different community composition in Band 1 (which attracts typical shallow water, wading and not-diving species).

Regarding the "basin" factor (Figure 18), it also emerged a grouping of units differentiating the northern from the southern one (Stress = 0.14). The southern basin appears as more cohesive (at least in waterbird community composition), probably related to the habitat characteristics shared by southern units. Also remarkable is the fact that the northern basin's units are separated into the more eutrophic (S12 to S14) and the more oligotrophic ones (S1, S2 and S15).

Furthermore, with respect to conservation indexes in winter (Figures 19-20), both display similar results: units with higher conservation values are grouped relatively homogeneously (Stress = 0.19). These are characterized by the presence of two species with high conservation value, Audouin's Gull and Flamingo.

Among common features displayed by these units stand out the closeness to a natural protected area, the marine influence due to the proximity to the communication with the sea (Estacio channel), and the oligotrophic nature of their waters.

However, in both cases (LRAE and SPEC) the stress value of the ordination diagram is 0.19, very close but under the recommended limit of 0.2 considered acceptable in a multidimensional scaling analysis (Clarke, 1993).

Regarding summer data, the same grouping factors were tested and similar results were obtained than in winter. Band 1 is clearly separated from the remaining ones (Figure 21), revealing a summer community linked to shoreline habitats and other community linked to the littoral water column.

Furthermore, the two basins are also separated in the ordination diagram (Figure 22) in a similar way to winter (Stress=0.12).

Finally, regarding conservation indexes, there is an apparent separation between units with a high or medium-to-high conservation value and units with lower conservation value, but the stress value (0.21) in both ordinations requires a cautious interpretation of the results. For that same reason, the ordination plots are not included.

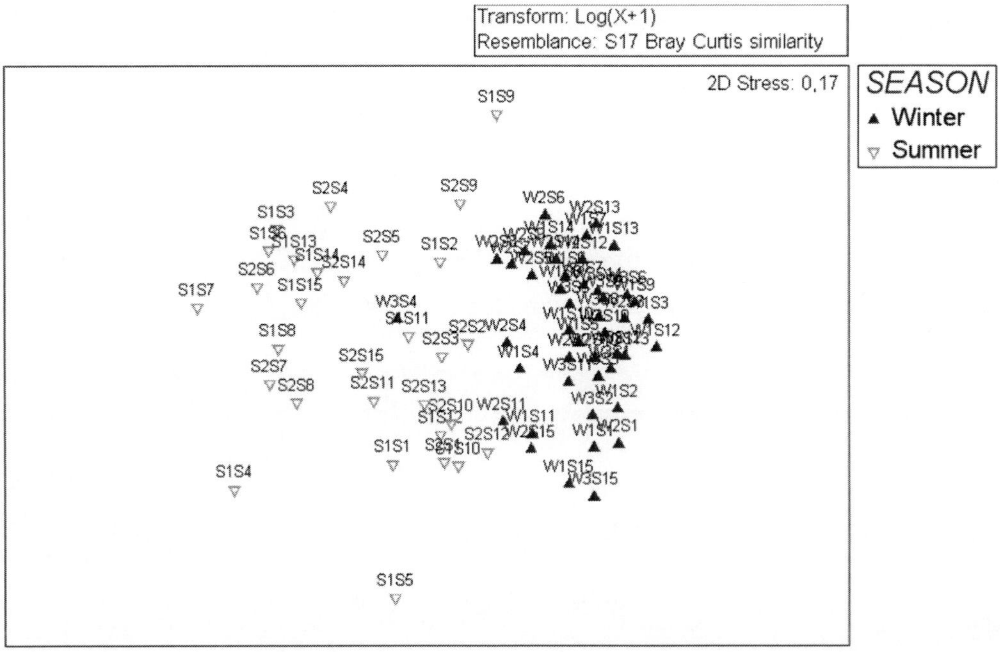

Figure 16. Multidimensional Scaling (MDS) differentiating samples on the basis of "season" as grouping factor.

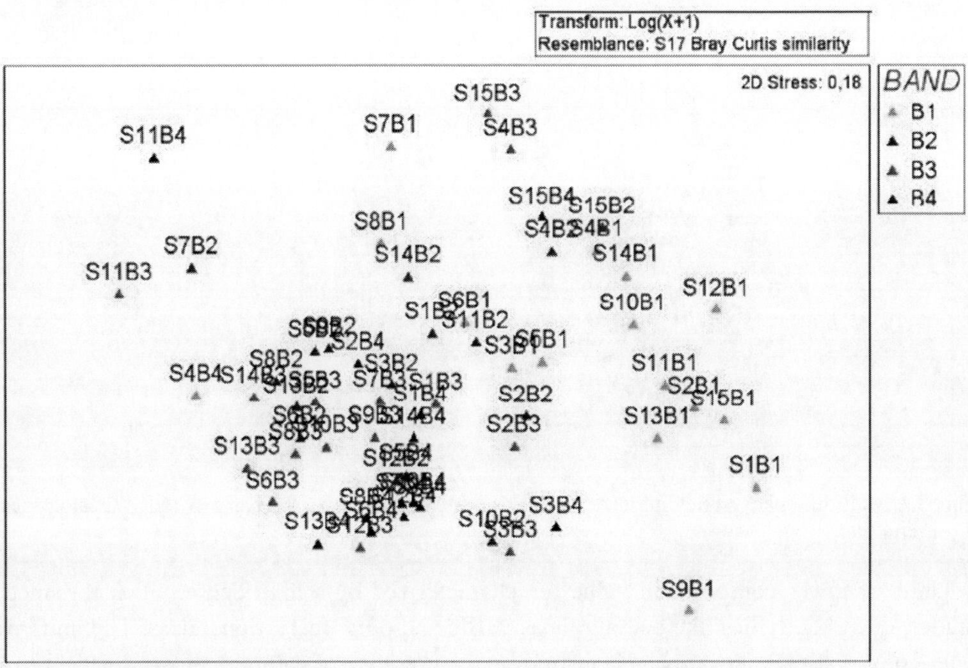

Figure 17. MDS differentiating winter samples on the basis of "band" as grouping factor.

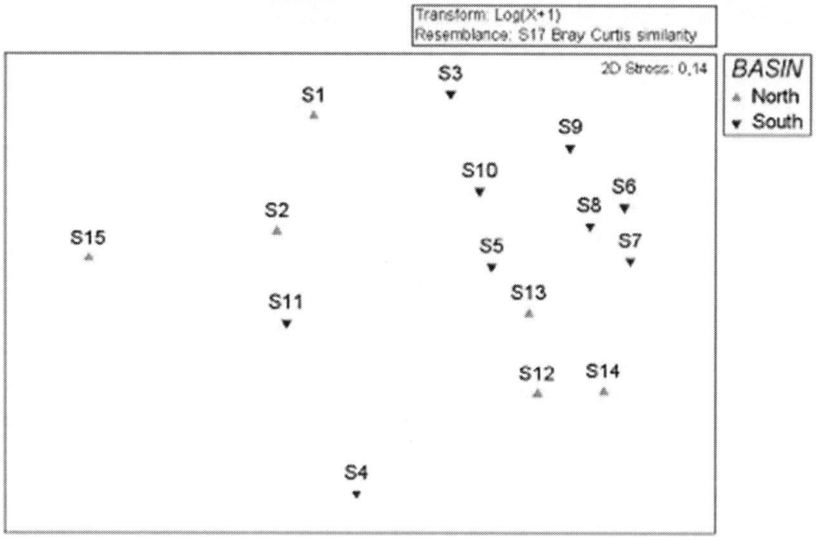

Figure 18. MDS differentiating winter samples on the basis of "basin" (North vs South) as grouping factor.

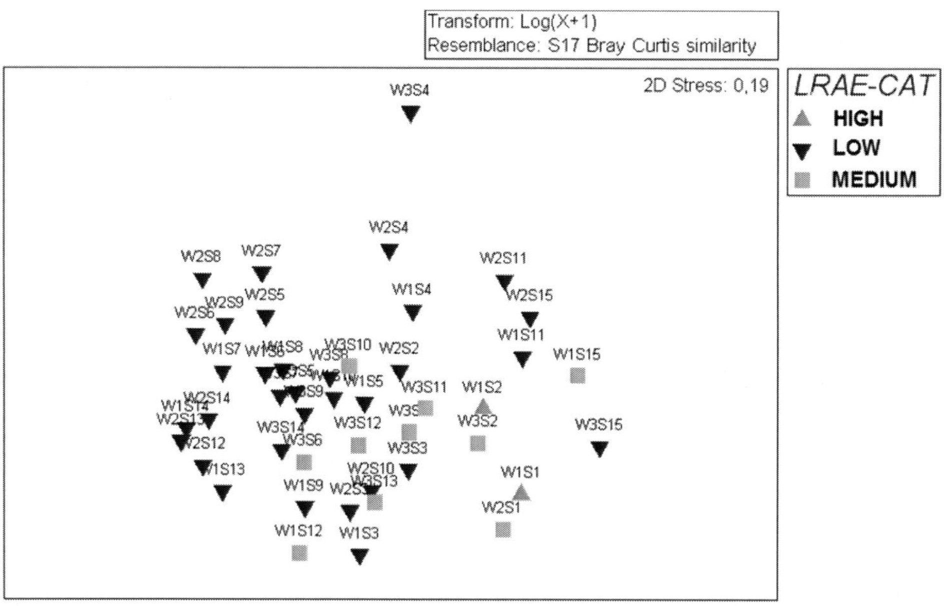

Figure 19. Multidimensional Scaling differentiating winter samples on the basis of "LRAE categories" as grouping factor.

Units of lower conservation value are characterized by a high degree of disturbance in summer, due to activities like bathing, boat traffic and other recreational uses. The units with higher conservation value are characterized by the presence of natural habitats adjacent to the shore, with minimal (or absent) representation of the above mentioned water-based activities.

After identifying groups of sampling units based on the factors studied (basin, band and SPEC and LRAE conservation indexes), "indicator" or representative species, based on the relative abundance of each group of units defined by MDS analyses, were identified trough subsequent IndVal analysis (Table 6). Significance level was set at p <0.05, although in certain specific cases, marginally significant results will be presented and discussed.

In winter, for "band" factor, Coot, Sandwich tern, Slender-billed Gull, Little Egret, Great Egret and Grey Heron were identified as indicators of the first band. Coot, a phytophagous species, restricts its activity to the first few meters of littoral waters, while herons and egrets are wading species and thus feed in shallow water. On the other hand, Black-necked grebe stands as indicator species of the fourth band (relatively deep water). This expert diver and primarily piscivorous species, although with greater flexibility and opportunistic character than other members of the same guild (Fjeldså, 2004), can efficiently exploit this habitat. With respect to the factor "basin", no indicator species were detected.

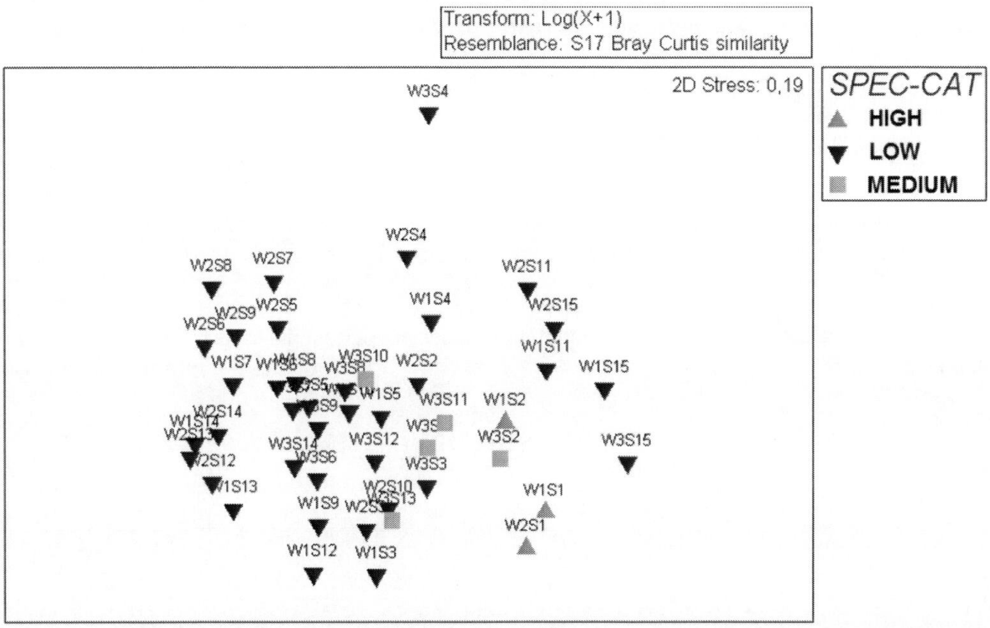

Figure 20. MDS differentiating winter samples on the basis of "SPEC categories" as grouping factor.

Table 6. Indicator species in winter and summer for different factor classes; significance values based on Monte Carlo Test

Factor	Classes	Winter indicator species (*p-value*)	Summer indicator species (*p-value*)
BAND	B1	*Fulica atra (0.03); Sterna sandvicensis (0.02); Larus genei (0.001): Larus ridibundus (0.001); Egretta garzetta (0.001); Egretta alba (0.012); Ardea cinerea (0.001)*	*Sterna sandvicensis (0.001); Sterna hirundo (0.001); Sterna albifrons (0.001); Larus michahellis (0.005); Larus ridibundus (0.001); Egretta garzetta (0.001)*
	B4	*Podiceps nigricollis (0.003); Podiceps cristatus (0.057)*	-
SPEC	High value	*Phoenicopterus roseus (0.002); Larus audouinii (0.067)*	-
	Medium value	*Sterna sandvicensis (0.036)*	-
LRAE	High value	*Sterna hirundo (0.038); Larus audouinii (0.047)*	-
	Medium value	*Egretta garzetta (0.069)*	-

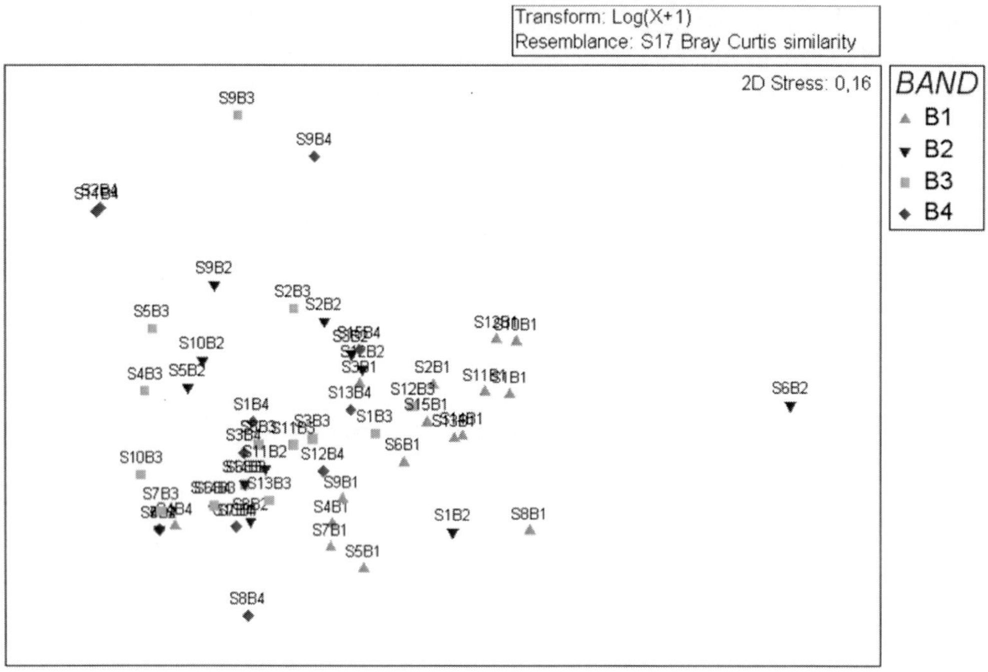

Figure 21. Multidimensional Scaling differentiating summer samples on the basis of "band" as grouping factor.

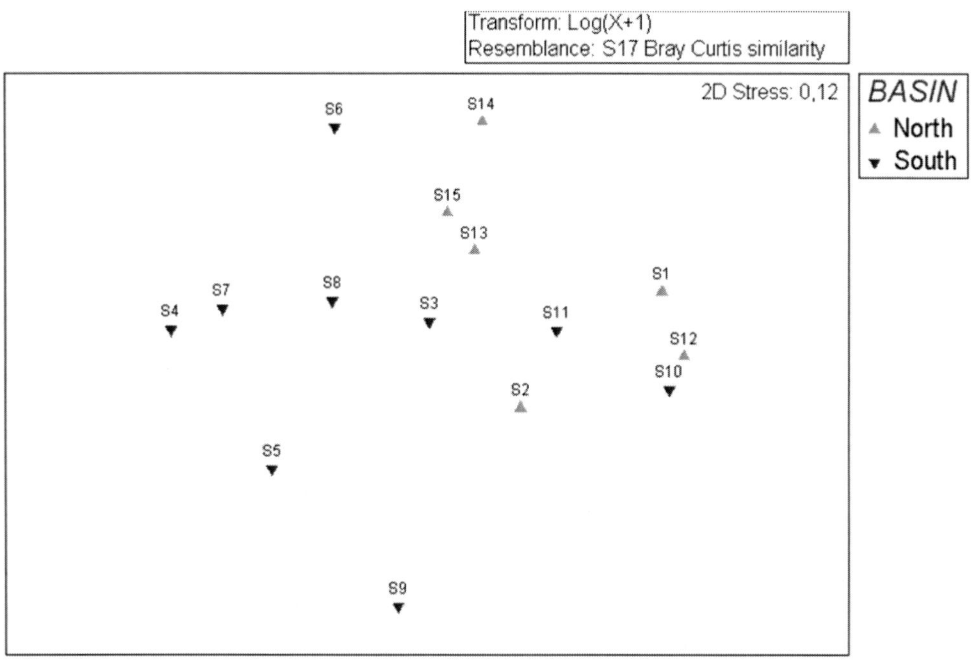

Figure 22. Multidimensional Scaling differentiating summer samples on the basis of "basin" as grouping factor.

In summer, Sandwich Tern, Common Tern, Little Tern, Yellow-legged Gull, Black-headed Gull, Little Egret and Grey Heron were identified as indicators of the first band. In the case of terns, this dependence on the first 100 m of water may be due to the specific requirements regarding the maximum depths they can exploit in plunge-diving. But it could also be related to the use of buoys delineating marine restrictions, as perchs or resting places. Finally, the indicator role of Yellow-legged Gull may be due primarily to the fact that with the decline of diversity in summer and the consequent temporary vacancy of certain ecological niches, this species takes over these empty niches, being more evenly distributed across the whole lagoon. The representativeness of the Black-headed Gull may be related to its wide range of diet (although still narrower than that of the Yellow-legged Gull), which also allows it to exploit several empty niches in summer. Neither indicator species were detected with respect to factor "basin"

In winter, regarding the SPEC conservation index, indicator species were detected for higher and medium categories. The indicator species of units with high conservation value were Flamingo and Audouin's Gull (although with marginal significance), present in more oligotrophic units, communicated with the Mediterranean sea and close to a natural protected area. The Sandwich Tern emerges as the indicator species of units with medium conservation value. This species is associated with maritime delineation lines of buoys (functional perches), located near the shoreline and being, a priori, the common element among the units with such medium category. There is probably some other common element to these areas which would help to explain the representative presence of Sandwich Tern, however it could not be detected at this level of environmental characterization of the units.

The analysis based on the winter value of the LRAE index, show Common Tern and Audouin´s Gull as indicator species of the maximum conservation category. These species were present in sampling units with overall high or medium conservation value. At least for the Audouin´s Gull, the reasons for this association are probably the same described for the SPEC index. It must also be highlighted the indicator role of Little Egret (with marginal significance) in medium conservation value areas. In summer, since there were not clear groups of units with respect to LRAE and SPEC conservation indexes, no Indicator Value analysis was performed to detect representative species of groupings based on their ranks.

4. Synthesis and Potential Application to Wetland Management and Restoration

In the Mediterranean, CLs experience various influences that compromise the maintenance of their ecological and economic values. The effects of such influences are evident at different spatial and temporal scales, and should be tackled with cost-effective mitigation measures. At the same time, monitoring and alert systems are needed to avoid undesirable changes going beyond irreversible thresholds. Ideally these surveillance methods should be easy to implement and able to integrate a wide range of processes and impacts. While land, air or water-based multiparametric sensors can gather large datasets from multiple variables that need further integration and interpretation, bioindicators have the advantage to provide by themselves an integrative response to a variety of stressors (Kushlan, 1993; Adamus, 1996; Amat and Green, 2009).

In the absence of regular limnological sampling schemes, the changes of the waterbird communities can be used to infer broad environmental changes with ecological and economic significance. Their monitoring can inform management and trigger adaptive responses to the benefit of the whole wetland ecosystem, particularly in artificial or heavily stressed sites.

The results of the two studies carried out in the Mar Menor CL complex (described in this chapter), show the multiple physical scales and time spans over which waterbird communities change in response to natural and anthropogenic influences. Numerical and distributional changes can be combined in several community descriptors (species richness, diversity, guild structure) and indexes (conservation value), which provide information on spatial differences in quality and on temporal changes in ecological integrity. These parameters and indexes, as well as the occurrence of indicator species, can be used as signals of environmental change relevant for habitat management and biodiversity conservation.

Regarding the objective of assessing the utility of waterbird monitoring as a tool for tracking ecological changes in CLs at different scales, the local-scale study (Coterillo Pond system) has shown how waterbird changes can be summarized in a few temporal phases along which the turnover of guilds and species indicate a transition from specialists to generalists (parallel to a deterioration of the ecological status of the site, due to the persistent input of untreated effluents from a fish farm). At the end of this process of ecosystem succession, following some rejuvenation triggered by water flushing through the wetland, conservation value indexes increase due to the appearance of new species of different families.

At an intermediate scale (CL shoreline), the spatial and temporal analysis of the ornithological descriptors and indexes studied indicated that winter and summer waterbird communities differed in their composition and structure, but also in their interannual variation. In winter, total abundance was similar troughouth the whole study period, the dominant species being those associated with increasingly eutrophic waters like grebes, cormorants and some phytophagous species. There are spatial variations at different scales, between bands –which determine species assemblages depending on the physical conditions affecting them-, between sampling points, and between the two main basins (North and South) of the Mar Menor CL. In the southern one, the units with higher waterbird abundance and diversity are located near the outlets of ephemeral watercourse channels, where nutrient loads from the watershed of the CL are discharged in a concentrated manner.

Moreover, although the north basin supports various disturbance agents (harbours, airport), several waterbirds with high conservation value are still present, being species more related to open and oligotrophic waters, and apparently also highly dependent on the shelter provided by the nearby natural protected areas (adjacent wetlands).

The summer community shows inter-annual variations in abundance, richness and value of the conservation indexes. Abundance seems more determined in this case by the degree of naturalness of the sectors, than by trophic conditions. Differences between north and south basin are more diffuse, and there seems to be a more homogeneous distribution of the waterbird community, probably also related with an equally uniform distribution of human disturbance intensity throughout the Mar Menor CL, and to the need of breeding species to forage closer to their respective nesting places.

The sectors with higher conservation value in both seasons, according to the two indexes computed (LRAE and SPEC), coincide in some features, i.e. natural habitat and landscape characteristics, and location away from urban centres, allowing them to host relevant species (regarding their threat degree and protection status).

But these features are lacking -and thus do not explain their conservation value- in other sectors with also high scores in these same indexes.

Regarding the second objective of evaluating the performance of different community descriptors and indexes as response indicators of ecological changes in CLs, it seems that monitoring schemes can benefit from the capacity of waterbirds to assess the implementation and effectiveness of management and mitigation measures (e.g. in relation to physico-chemical influences, changes in the landscape, in the intensity of land use, or in the degree of recreational pressure). Even so, part of the variability in waterbird community composition and structure remains unexplained, at least by the features used to define and characterize the ecological integrity (naturalness, hydrological alteration, openness) of our sampling units.

On the other hand, our third objective of integrating the results of waterbird monitoring programmes, run at different spatial and temporal scales within CL complexes, into a general evaluation system of their ecological integrity, depends critically on the availability of minimum financial resources to maintain a regular integrated survey of key biological (waterbirds, fish) and environmental variables (e.g. limnological conditions) in test sites like restored ponds and other man-made wetlands.

The last objective, devising a habitat restoration scheme based on the experience in waterbird monitoring of representative wetland types within the Mar Menor CL system, in order to maximize biodiversity and environmental functions (e.g. water purification) of the restored and created habitats, needs also a multi-scale spatial approach. The capacity of restored or newly created wetlands to to buffer the effects of increased freshwater inputs and excess nutrients (from agriculture, urban areas, aquaculture facilities) can also be exploited to improve biodiversity at a local (site) and landscape (complex) scales.

At the local scale small wetlands can supply a wide range of ecosystem services (water purification, recreation, wildlife habitat). At the watershed scale, a network of artificial wetlands can exert (with respect to the whole CL complex) a buffering function which otherwise would be performed by natural wetlands or by the main CL basin (Alvarez-Rogel *et al.*, 2011; Robledano *et al.*, 2011), but in that case at the expense of their biological and landscape values. Our experience in ornithological monitoring and conservation assessment of both natural wetlands (Martínez *et al.*, 2005; Robledano *et al.*, 2008; 2010; Farinós and Robledano, 2010; Robledano *et al.*, 2011), and artificial or restored ones (present chapter), reveals that conservation gains in unmanaged areas can be rapidly offset if effective monitoring and consequent adaptive management are not implemented, and wetlands are simply used as passive "green filters".

In any case, indirect evidence from ornithological monitoring suggests that the lack of management and monitoring, or the performance of opportunistic, short-lived versions of these, not supported by carefully planned field-based research, inevitably leads to a loss of value under an unpredictable, often unstable or low quality ecological status.

Thus, to balance the provision of environmental services and the conservation of ecological values, a two-stage design is needed, in which a part of the restored or created wetlands is used for the processing of the water supply (e.g. nutrient enriched effluents), while another part receives the treated water and can be managed to enhance biodiversity. Each section is engineered in a way that increases the performance of its main function, selecting the vegetation type, morphometric characteristics and flow regime best adapted to it.

Such wetlands usually represent anomalies in the prevailing environmental context (e.g. freshwater habitats within a saline landscape), which allows biodiversity to be enhanced both at the local and landscape scale.

Recently, at the request of the regional environmental authority, a consultant team (with participation of the University of Murcia), developed a proposal for the treatment of agricultural effluents that discharge into the Mar Menor CL from the adjacent coastal plain through the wetlands of its western shore (Esteve *et al.*, 2010). The proposal takes the form of a set of actions framed in the principles of ecological engineering (Marques *et al.*, 2003), combining actions at different scales. The set of actions were part of a dual and complementary approach of wetland re-creation and management:

1 First, the development of a Local Management System, applicable at a small spatial scale and focusing on the management of specific elements of the system (such as drainage canals or temporary ponds). At this level the main proposal was the creation of artificial wetland systems (functional units) composed by two ponds with different and well defined main functions: i) green filters, and ii) biotope recovery and increase of biodiversity.

2 Second, in a complementary way, the development of an Integrated Management System, applicable at a landscape scale. Basically, the proposal consisted in the joint management of the system of drainage canals, for concentrating and redirecting the water flows towards the functional units (local management system) that would perform the depuration role.

Once the combination of these two management systems was established as a basic wetland creation and management scheme, specific guidelines for the intervention in the area were set, in the sense of:

a Increasing water retention times to allow better depuration process of the organic content (managing drainage channels and artificial wetlands)

b Recovering saline steppe and saltmarsh biotopes and temporary water bodies associated to these habitats

c Increasing biodiversity, through the restoration of native vegetation and the recreation of biotopes, favoring the structural heterogeneity in the functional units (ponds)

d Controlling the expansion of reedbeds

e Establishing scientific and research criteria when planning the monitoring and evaluation schemes of the restoration actions and their effects, both on the present biota and on the quality of the waters draining into the lagoon.

As an example of the types of actions proposed, figure 23 outlines two types of pond which would integrate an ideal two-stage wetland system for mitigating the impact of agricultural drainages channeled to the Mar Menor lagoon through the saline steppe wetlands of its western shore. We advocate the use of these artificial wetlands in the treatment of agricultural drainage and other effluents impairing the lagoon ecosystems, based on the recognition of their potential to mitigate such impacts, while increasing habitat for various species.

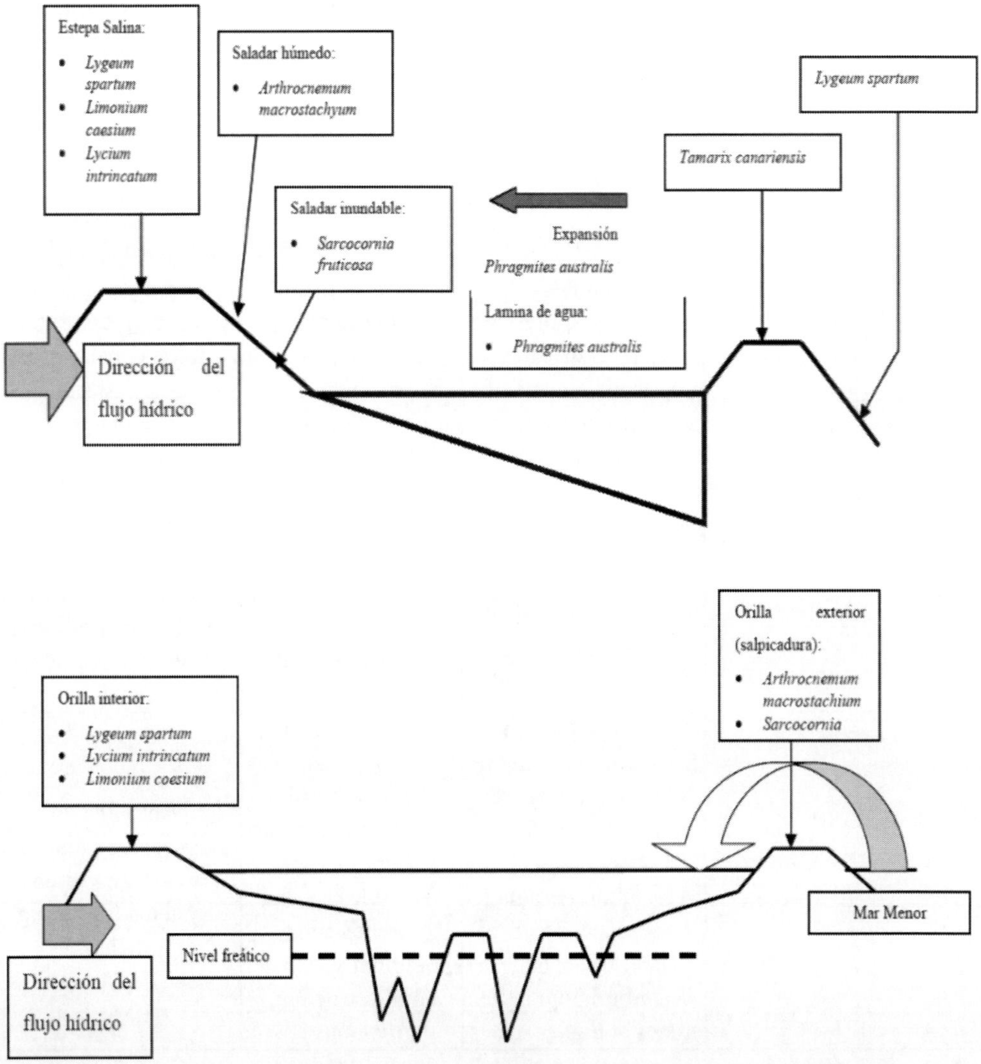

Figure 23. Schematic representation of the two types of artificial ponds which would integrate an ideal two-stage wetland system for mitigating the impact of agricultural drainages on the Mar Menor CL complex: on top "green filter" type (GF), and at the bottom "biodiversity enhancement" type (BE). On each, the arrow on the left side represents the direction of the water flow passing through the system. In GF, the narrower arrow represents the expected direction of expanding reeds *(Phragmites australis)*. In BE, the right arrow represents the lagoon's influence (splash), and the dashed line the phreatic water table (from Esteve et al., 2010).

Monitoring waterbird communities present in these wetlands -and in the natural ecosystems for which they act as buffers-, will be one of the main tools for evaluating their effectiveness. Experience with the conversion of ancient lagoon-based sewage treatment systems into endangered waterfowl refuges (e.g.: http://www.lifemalvasiamurcia.es/pdf/ propuesta_zepa_ lagunas%20moreras.pdf), illustrate the potential for developing suitable waterbird habitats as by-products of water quality management systems in heavily modified landscapes.

Table 7. Summary of spatial and temporal scales of waterbird community variation in response to impacts and other driving forces, and of management options available at each of these scales

Scales		Driving forces	Type of biological response	Management options
Space	Time			
Local	Short-term (seasonal)	Phenology Recreational pressure Habitat structure Seasonal course of activities, resource use and flows	Short-term and small scale numerical and distributional changes (band, season...)	Two-stage restored or created wetlands for addressing point-source impacts (green filter + habitat enhancement) Local restoration projects Routine management (water levels, recreational uses)
Intermediate (sampling sectors, internal gradients)	Mid-term (interannual)	Climate variability Reclamation projects Land use Recreational pressure	Changes at intermediate scales (among sectors/years, along spatial gradients)	Buffering systems of two-stage wetlands (to address diffuse impacts over medium-sized areas)
Whole CL complex	Long-term (decadal)	Climate change Socioeconomic trends Nutrient loading from agriculture and urban wastewater Agricultural drainage Landscape effects of land-use change Waterbird population trends at higher scales	Long-term habitat changes Noxious species blooms (microalgae, jellyfish) Waterbird species trends and community change (guild structure, diversity)	Networks of intercepting canals and two-stage wetland systems (buffering belts)

Finally, table 7 summarizes the different space and time scales over which the waterbird-based indicators change in response to different environmental drivers, the type of response shown and the management options available for addressing the effects of these drivers, particularly when related to anthropogenic influences on CLs.

ACKNOWLEDGMENTS

The authors are indebted to *Dirección General de Patrimonio Natural y Biodiversidad* of the Autonomous Community of Murcia Region, for allowing field work in Coterillo Pond and for supplying the results of ornithological monitoring programmes in that wetland and in other protected areas. The authors' gratitude goes specially to Gustavo A. Ballesteros for the help and advice provided in the gathering and exploitation of data from such programmes.

Part of the results included in this chapter belong to several academic projects made or directed by the authors, or draw on other research developed in the Department of Ecology and Hidrology of the University of Murcia, whose laboratory and field facilities were available to the authors.

Finally we thank all the team involved in the drafting of the study '*Adecuación biológica de los humedales incluidos en la ZEPA del Mar Menor*', particularly Emilio Diez de Revenga and Patricia Esteve of Ambiental SL.

REFERENCES

Acuna, R., Contreras, F. and Kerekes, J. (1994). Aquatic bird densities in two coastal lagoon systems in Chiapas State, Mexico, a preliminary assessment. *Hydrobiologia*, 279-280: 101-106.

Adamus, P. R. 1996. *Bioindicators for assessing ecological integrity of prairie wetlands*. EPA/600/R-96/082. US Environmental Protection Agency. Environmental Research Laboratory, Corvallis, Oregon.

Aliaume, C., Do Chi, T., Viaroli, P., and Zaldívar, J. M. 2007. Coastal lagoons of Southern Europe: recent changes and future scenarios. *Transitional Waters Monographs*, 1: 1-12.

Álvarez-Rogel, J., Jiménez-Cárceles, F. J. and Nicolás, C. E. 2006. Phosphorus and nitrogen content in the water of a coastal wetland in the Mar Menor Lagoon (SE Spain): Relationships with effluents from urban and agricultural areas. *Water, Air and Soil Pollution,* 173: 21-38

Álvarez-Rogel, J., Jiménez-Cárceles, F. J., Roca, M. J., and Ortiz, R. 2011. Changes in soils and vegetation in a Mediterranean coastal salt marsh impacted by human activities. *Estuarine, Coastal and Shelf Science* 73: 510–526.

Amat, F., Hontoria, F., Navarro, J. C., Vieira, N. and Mura, G. 2007. Biodiversity loss in the genus Artemia in the Western Mediterranean Region. *Limnetica*, 26: 387-404.

Amat, J. A. and Green, A. J. 2009. Waterbirds as Bioindicators of Environmental Conditions, In: Hurford, C., Schneider, M., Cowx, I. (Eds.), *Conservation Monitoring in Freshwater Habitats: A Practical Guide and Case Studies*. Springer, Netherlands, pp. 47-60.

BirdLife International. 2004. Birds in Europe. *Population Estimates, Trends and Conservation Status. Birdlife Conservation* Series, 12. Cambridge.

Bosch, M., Oro, D. and Ruiz, X. 1994. Dependence of yellow-legged gulls (*Larus cachinnans*) on food from human activity in two Western Mediterranean colonies. *Avocetta* 18: 135-326 139.

Bradford, D. F., Franson, S. E., Neale, A. C., Heggem, D. T., Miller, G. R., and Canterbury, G. E., 1998. Bird Species Assemblages as Indicators of Biological Integrity in Great Basin Rangeland. *Environmental Monitoring and Assessment*, 49: 1-22.

Britton R. H. and A. R. Johnson, 1987. An ecological account of a Mediterranean salina: The Salin de Giraud, Camargue (S. France). *Biological Conservation*, 42: 185-230.

Bryant, M. D. 2000. Estimating Fish Populations by Removal Methods with Minnow Traps in Southeast Alaska Streams. *North American Journal of Fisheries Management*, 20:923-930.

Burger, J. and Gochfeld, M. 1996. Family *Laridae* (gulls). *Handbook of the birds of the world.* Lynx Editions, Barcelona 3: 572-623.

Burton, N. H. K. 2007. Landscape approaches to studying the effects of disturbance on waterbirds. *Ibis* 149:95-101.

Buzun, V. A. and Mierauskas, R. 1987. Behaviour of the Mediterranean Gull (*Larus melanocephalus*) Tem in the period of colony formation and at early stages of breeding]. *Byull. Mosk. Obshch. Ispyt. Prir. Otd Biol.*, 92 (3): 27-36.

Carreño, M. F., Esteve, M. A., Martinez, J., Palazón, J. A. and Pardo, M. T. 2008. Habitat changes in coastal wetlands associated to hydrological changes in the watershed. *Estuarine, Coastal and Shelf Science* 77: 475-483

Cabo, J. M. and Camacho, I. 1981. *Aves acuáticas de la Mar Chica de melilla, julio 1977, junio 1978 y verano 1979.* Mediterránea, núm. 5.

Clarke, K. R. and Gorley, R. N. 2006. PRIMER v6: *User Manual/Tutorial. PRIMER-E, Plymouth.*

Clarke, K. R. 1993. Non-parametric multivariate analyses of changes in community structure. *Australian Journal of Ecology* 18: 117-143.

Clymo, R. S. 1983. Peat. In: Gore, A. J. (ed), *Mires: Swamp, Bog, Fen and Moor.* Elsevier, New York, pp. 159-224.

Connors, P. G., McCarrery, B. J. and Maron, J. L. 1993. Speciation in golden-plovers (*Pluvialis dominica* and *P. Fulva*): Evidence from the breeding grounds. *Auk* 110 : 9-20.

Cramp, S. and Simmons, K. E. L. (Eds) 1977. *The Birds of the Western Paleartic, Vol. I.* Oxford University Press. Oxford.

Cramp, S. (Ed.), 1988. *Hand-book of the birds of Europe, the Middle East and North Africa. The birds of the Western Paeartic, Vol. V.* Oxford university Press. Oxford.

Crisman, T. L. 1999. Conservation of Mediterranean coastal saline ecosystems: the private sector role in maintaining ecological function. *Proceedings of the Post Conference Symposium "Saltworks: Preserving Saline Coastal Ecosystems"*, Global NEST, Samos. Pp: 39-47. http://www.gnest.org/Conferences/Saltworks_post/039-047.pdf

Crisman, T. L., Takavakoglou, V., Alexandridis, T., Antonopoulos V. and Zalidis, G. 2009. Rehabilitation of abandoned saltworks to maximize conservation, ecotourism and water treatment potential. *Global NEST Journal*, 11: 24-31.

Dufrene, M. and Legendre, P. 1997. Species Assemblages and Indicator Species: The Need for a Flexible Asymmetrical Approach. *Ecological Monographs* 67: 345-366.

Edwards, A. L. *"The Correlation Coefficient."* Ch. 4 in An Introduction to Linear Regression and Correlation. 1976. San Francisco, CA: W. H. Freeman: 33-46.

Evangelopoulos, A., Koutsoubas, D., Basset, A., Pinna, M., Dimitriadis, C., Sangiorgio, F., Barbone, E., Maidanou, M., Koulouri, P. and Dounas, C. 2008. Spatial and seasonal variability of the macrobenthic fauna in Mediterranean solar saltworks ecosystems. *Aquatic Conservation: Marine and Freshwater Ecosystems* 18: 118 - 134.

Everard, M., Harrington, R. and McInnes, R. J. 2012. Facilitating implementation of landscape-scale water management: The integrated constructed wetland concept. *Ecosystem Services*, 2, 27-37.

Esteve, P., Farinós, P., Diez de Revenga, E. and Molina, J. 2010. *Adecuación biológica de los humedales incluidos en la ZEPA del Mar Menor. Estudio Técnico sobre Tratamiento de Drenajes Agrícolas en el Entorno del Mar Menor.* Dirección General de Patrimonio Natural y Biodiversidad, Región de Murcia.

Farinós, P. and Robledano, F. 2007. *Evaluación del estado ecológico de la Charca de experimentación e interpretación "El Coterillo" en el Parque Regional de Salinas y Arenales de San Pedro del Pinatar.* Consejería de Desarrollo Sotenible y Ordenación del Territorio. Comunidad Autónoma de Murcia (Informe inédito).

Farinós, P. and Robledano, F. 2010. Structure and distribution of the waterbird community in the Mar Menor coastal lagoon (SE Spain) and relationships with environmental gradients. *Waterbirds*, 33(4): 479-493.

Fernández-Delgado, C., Hernando, J., Herrera, M. and Bellido, M. 1988. Age, growth and reproduction of *Aphanius iberus* (Cuv. and Val., 1846) in the lower reaches of the Guadalquivir river (south-west Spain). *Freshwater Biology* 20: 227-234.

Fjeldså, J. 2004. *The Grebes.* Oxford University Press, Oxford.

Fuller, R. J. *1982: Bird Habitats in Britain.* Poyser, Calton.

González-Félix, M. L. and Pérez-Velázquez, M. 2006. Un Panorama de los Presupuestos de Nitrógeno para Cultivo de Camarón. Avances en Nutrición Acuícola VIII. *VIII Simposium Internacional de Nutrición Acuícola.* 15-17

Gordon, D. H., Gray, B. T. and Kaminski, R. M. 1998. Dabbling duck-habitat associations during winter in coastal South Carolina. *Journal of Wildlife Management* 62:569–580.

Goutner, V. 1994. The Diet of Mediterranean Gull (*Larus melanocephalus*) Chicks at Fledging. *J. Orn.* 135: 193-201.

Granado-Lorencio, C. (Ed). 1997. *Conservación, Recuperación y Gestión de la Ictiofauna Continental Ibérica.* Publicaciones de la Estación de Ecología Acuática. EMASESA. Sevilla.

Green, A. J. and Figuerola, J. 2003. Aves acuáticas como bioindicadores en los humedales, 47-60, In: Paracuellos, M. (Ed.). *Ecología, manejo y conservación de los humedales.* Instituto de Estudios Almerienses (Diputación de Almería), Almería.

Gregory, R. D., Noble, D., Field, R., Marchant, J., Raven, M. and Gibbons, D. W. 2003. Using birds as indicators of biodiversity. *Ornis Hungarica*, 12-13: 11-24.

Hagemeijer, E. J. and Blair, J. M. (Eds.).1997. *The EBBC Atlas of European Breeding Birds: Their Distribution and Abundance.* T and AD Poyser. Londres.

Hayman, P., Marchant, J. and Prater, T. 1986. *Shorebirds: an identification guide to the waders of the world.* Houghton Mifflin Company, Boston, Mass.

Hervías, E., Mayo, I., Díaz, J., Chitimia, L. J., Ortiz, C., Martínez-Carrasco, M. R. and Ruiz de Ybáñez, M. R. (in prep.). Parasites of Yellow-legged Gull (*Larus michahellis michahellis*) from south west mediterranean coast, Spain.

Hubina, T. 2008. *Development of a GIS to estimate the effect of abiotic and biotic factors on the abundance of waterbirds in the Grado-Marano Lagoon.* Ph.D. Thesis, Università degli Studi di Trieste.

Johnson, A. 1983. *Etho-écologie du flamant rose (Phoenicopterus ruber roseus Pailas) en Camargue et dans l'Ouest. Paléatlique.* Thése Université Paul Sabatier, Toulouse.

Knobelsdorf, M. 2005. *Eliminación biológica de nutrientes en un ARU de baja carga orgánica mediante el proceso VIP.* Tesis doctoral. Univ. Politécnica de Catalunya. España.

Kushlan, J. A., 1993. Colonial Waterbirds as Boindicators of Environmental Change. *Colonial Waterbirds*, 16: 223-251.

Lopes, R. J., Múrias, T., Cabral J. A. and Marques, J. C. 2005. A Ten Year Study of Variation, Trends and Seasonality of a Shorebird Community in the Mondego Estuary, Portugal. *Waterbirds* 28: 8-18.

Ma, Z., Cai, Y., Li, B. and Chen, J. 2010. Managing Wetland Habitats for Waterbirds: An International Perspective. *Wetlands*, 30: 15–27.

MacDonald, M. A. 2006. The indirect effects of increased nutrient inputs on birds in the UK: a review. In: *RSPB Research Report 21. Royal Society for the Protection of Birds*, The Lodge, Sandy, Bedfordshire.

Madroño, A., González C. and Atienza, J. C. 2004. *Libro Rojo de las Aves de España. Dirección General para la Biodiversidad-SEO/Birdlife*. Madrid.

Mallory, M. L., Robinson, S. A., Hebert, C. E., and Forbes, M. R. 2007. Seabirds as indicators of aquatic ecosystem conditions: A case for gathering multiple proxies of seabird health. *Marine Pollution Bulletin*, 60: 7-12

Maltby, E., 1986. *Waterlogged wealth*. Earthscan, London.

María-Cervantes, A., Jiménez-Cárceles, F. J. and Álvarez-Rogel, J. 2009. As, Cd, Cu, Mn, Pb, and Zn Contents in Sediments and Mollusks (*Hexaplex trunculus* and *Tapes decussatus*) from Coastal Zones of a Mediterranean Lagoon (Mar Menor, SE Spain) Affected by Mining Wastes. *Water, air and Soil Pollution,* 200:289-304

Marín-Guirao, L., Marín Atucha, A., Lloret Barba, J., Martínez López, E., and García Fernández, A. J. 2005. Effects of mining wastes on a seagrass ecosystem: metal accumulation and bioavailability, seagrass dynamics and associated community structure. *Marine Environmental Research* 60: 317–337.

Marques, J. C., Nielsen, S. N., Pardal, M. A. and Jørgensen, S. E. 2003. Impact of eutrophication and river management within a framework of ecosystem theories. *Ecological Modelling*, 166: 147-168.

Martí, R. and Del Moral, J. C. (Eds.) 2003. *Atlas de las aves reproductoras de España. Dirección General de Conservación de la Naturaleza-Sociedad Española de Ornitología.* Madrid.

Martínez-Abraín, A. and Dolz, R. 1988. Avance de los primeros resultados del primer censo exhaustivo de aves marinas nidificantes en el litroral de la Comunidad Valenciana. En: C. López-Jurado (Ed.): *Aves Marinas, GIAM Formentera*, 99 119-126.

Martínez, J., Esteve, M. A., Robledano, F., Pardo M. T. and Carreño, M. F. 2005. Aquatic birds as bioindicators of trophic changes and ecosystem deterioration in the Mar Menor lagoon (SE, Spain). *Hydrobiologia*, 550: 221-235.

Masero, J. A. 2003. Assessing alternative anthropogenic habitats for conserving waterbirds: Salinas as buffer areas against the impact of natural habitat loss for shorebirds. *Biodiversity and Conservation*, 12: 1157-1173.

Masero, J. A., Pérez-Hurtado, A., Castro, M., Arroyo, G. M. 2000. Complementary use of intertidal mudflats and adjacent salinas by foraging waders. *Ardea* 88:177–191.

McCune, B. and Mefford, M. J. 1999. *Multivariate Analysis of Ecological Data. Versión 4.20.* MjM Software, Gleneden Beach, Oregon, US.

McGraw, J. K., Hill, G. E., Stradi, R. and Parker, R. S. 2001. The influence of carotenoid acquisition and utilization on the maintenance of species-typical plumage pigmentation in male American goldfinches (*Carduelis tristis*) and northern cardinals (*Cardinalis cardinalis*). *Physiol Biochem Zool.*, 74(6): 843-852.

Múrias, T., Cabral, J. A., Lopes, R., Marques, J. A. and Goss-Custard, J. 2002. Use of traditional salines by waders in the Mondego estuary (Portugal): A conservation perspective. *Ardeola*, 49: 223-240.

Myers, J. P., William, S. L. and Pitelka, F. A. 1980. An experimental analysls of prey availability for sanderlings (Aves: Scolopacidae) feeding on sandy beach crustaceans. *Can. J. Zool.* 58: 1564-1574.

Oliva-Paterna, F. J., Torralva M. and Fernández-Delgado, C. 2006. Threatened Fishes of the World: Aphanius iberus (Cuvier and Valenciennes, 1846) (Cyprinodontidae). *Environmental Biology of Fishes* 75: 307-309.

Oliva-Paterna, F. J. and Torralva, M. 2008. *El fartet en la Región de Murcia: biología y conservación*. Dirección General del Medio Natural. Consejería de Desarrollo Sostenible y Ordenación del Territorio. Murcia.

Olmos, F. and Boulhosa, R.L.P. 2000. A meeting of opportunists: birds and other visitors to *Mabea fistulifera (Euphorbiaceae)* inflorescenses. *Arajuba* 8: 93-98.

Oro, D. and Martínez Vilalta, A. 1994. Factors affecting kleptoparasitism and predation rates of Audouin's Gull *Larus audouinii* by the Yellow-legged Gull *Larus cachinnans* at the Ebro Delta. *Colonial Waterbirds*, 17: 35-41.

Oro, D. 1996. Are migrating Peregrine Falcons Falco peregrinus a threat to breeding Audouin's Gull *Larus audouinii* at the Ebro Delta. *Colonial Waterbirds*, 19: 270-272.

Oro, D., De León, A., Minguez E. and Furness, R.W. 2005. Estimating predation on breeding European storm-petrels (Hydrobates pelagicus) by yellow-legged gulls (*Larus michahellis*). *J. Zool., Lond.* 265:421–429.

Oro, D., Jover, L. and Ruiz, X. 1996. Influence of trawling activity on the breeding ecology of a threatened seabird, Audouin's Gull *Larus audouinii*. Marine *Ecology Progress Series*, 139: 19-29.

Paquet, J. Y., Vandevyvre, X., Delahaye L. and Rondeux, J. 2006. Bird assemblages in a mixed Woodland-farmland landscape: The conservation value of silviculture-dependant open areas in plantation forest. *Forest Ecology and Management*, 227: 59-70.

Paracuellos, M., Castro, H., Nevado, J. C., Ona, J. A., Matamala, J. J., Garcia, L. and Salas, G. 2002. Repercussions of the abandonment of Mediterranean saltpans on waterbird communities. *Waterbirds*, 25: 492-498.

Paracuellos, M., Fernandez-Cardenete, J. R. and Robledano, F. 2007. Ecología y Conservación de los Humedales Litorales y sus aves en el Sudeste Ibérico. En: Barea-Azcón, J., Travesí, R., Moleón, M., Ballesteros, E., Luzón, J. M., and Tierno, J. M. (Eds.). *Biodiversidad y conservación de fauna y flora en ambientes mediterráneos.* Sociedad Granatense de Historia Natural, Granada: 483-541.

Paterson, A. M. 1997. *Las Aves Marinas de España y Portugal*. Lynx Edicions. Barcelona.

Peakall, D. B. and Boyd, H. 1987. Birds as bio-indicators of environmental conditions, 113-118, In: Diamond, A. W. and Filion, J. (Eds.). *The Value of Birds*. ICBP Technical Publication 6, Cambridge.

Pérez-Ruzafa, A. and Marcos, C. 2003. *Los recursos naturales de la Región de Murcia. Un análisis interdisciplinar*. Cap. 4.3.3. Servicio de publicaciones de la Universidad de Murcia, 1ª edición.

Pérez-Ruzafa, A. and Marcos, C. 2012. Fisheries in coastal lagoons: An assumed but poorly researched aspect of the ecology and functioning of coastal lagoons. *Estuarine, Coastal and Shelf Science*, 110: 15-31.

Pérez-Ruzafa, A., Marcos, C., Pérez-Ruzafa, I. M. and Pérez-Marcos, M. 2011. Coastal lagoons: "transitional ecosystems" between transitional and coastal waters. *Journal for Coastal Conservation*, 15 (3): 369-392.

Petanidou, T. 2000. The postmodern saline landscape in Greece and the European: salinas for salt or what? *Proceedings of the Post Conference Symposium "Saltworks: Preserving Saline Coastal Ecosystems"*, Global NEST, Samos. Pp: 67-80.

Pons, P., Lambert, B., Rigolot E. and Prodon, R. 2003. The effects of grassland management using fire on habitat occupancy and conservation of birds in a mosaic landscape. *Biodiversity and Conservation*, 12: 1843-1860.

Purroy, F. J., (Coord) 1997. *Atlas de Aves de España (1975-1995). SEO/BirdLife*. Lynx Edicions. Barcelona.

R Development Core Team. 2007. R: *A language and environment for statistical computing.* R Foundation for Statistical Computing, Vienna. http: //www.R-project.org.

Ramos, R., Ramírez, F., Sanpera, C., Jover, L. and Ruiz X. 2009. Diet of Yellow-legged Gull (*Larus michahellis*) chicks along the Spanish Western Mediterranean coast: the relevance of refuse dumps. *Journal of Ornithology,* 150: 265- 272.

Robledano Aymerich, F., Pagán Abellán, I. and Calvo Sendín, J. F. 2008. Waterbirds and nutrient enrichment in Mar Menor Lagoon, a shallow coastal lake in southeast Spain. *Lakes and Reservoirs: Research and Management*, 13: 37-49.

Robledano, F., Esteve, M. A., Farinós, P., Carreño M. F. and Martínez-Fernández, J. 2010. Terrestrial birds as indicators of agricultural-induced changes and associated loss in conservation value of Mediterranean wetlands. *Ecological Indicators*, 10: 274-286.

Robledano, F., Esteve, M. A., Martínez-Fernández, J. and Farinós, P. 2011. Determinants of wintering waterbird changes in a Mediterranean coastal lagoon affected by eutrophication. *Ecological Indicators*, 11: 395-406

Robledano, F. and Farinós, P. 2010. Waterbirds as bioindicador in cooastal lagoons: background, potencial value and recent research in Mediterranean areas. In: *lagoons: biology, management and environmental impact:* Adam G. Friedman. Nova science Publishers, Inc. ISBN: 978-1-6171-738-6.

Rönkä, M. T. H., Saari, C. L. V., Lehikoinen, E. A., Suomela, J. and Häkkilä, K., 2005. Environmental changes and population trends of breeding waterfowl in northern Baltic Sea. *Ann. Zool. Fenn.* 42, 587–602.

Rönkä, M., Tolvanen, H., Lehikoinen, E., von Numers, M., and Rautkari, M. 2008. Breeding habitat preferences of 15 bird species on south-western Finnish archipelago coast: Applicability of digital spatial data archives to habitat assessment. *Biological Conservation*, 141: 402-416.

Ruiz, O. 2008. *Nitrifying and desnitrifying bacterial communities in the sediment and rhizosphere of free water surface constructed wetland.* Tesis doctoral. Univ. de Girona-España.

Sadoul N., Walmsley J. G. and Charpentier, B. 1998. *Salinas and Nature Conservation.* Conservation of Mediterranean Wetlands, 9. Tour du Valat, Arles.

Sánchez, M. I., Green, A. J. and Alejandre, R. 2006. Shorebird Predation Affects Density, Biomass, and Size Distribution of Benthic Chironomids in Salt Pans: An Exclosure Experiment. *Journal of the North American Benthological Society* 25: 9-18.

Sebastián-González, E., Sánchez-Zapata, J. A., and Botella, F. 2010. Agricultural ponds as alternative habitat for waterbirds: Spatial and temporal patterns of abundance and management strategies. *European Journal of Wildlife Research,* 56: 11-20.

Sebastián-González, E. 2011. *Patrones, procesos y conservación de comunidades: Aves acuáticas en una red de humedales artificiales*. Ph. D. Thesis, Universidad Miguel Hernández, Elche (Spain).

Smith, K. J. and Able, K. W. 1994. Salt-marsh tide pools as winter refuges for the mummichog, *Fundulus heteroclitus*, in New Jersey. *Estuaries* 17, 226–234.

Stolen, E. D., Breininger, D. R. and Frederick, P. C. 2004. Using waterbirds as indicators in estuarine systems: successes and perils, 409-422 In: Bortone, S. A. (Ed.). *Estuarine Indicators*. CRC Press, Boca Raton, Florida.

Swennen, C. and Van de Meer, J. 1992. Variation in egg size of Common Eiders. *Ardea* 80: 363-373.

Taft, O. W. and Haig, S. M.. 2006. Importance of wetland landscape structure to shorebirds wintering in an agricultural valley. *Landscape Ecology* 21: 169-184.

Tartarotti, B., Baffico, G., Temporetti R. and Zagarese, H. 2004. Mycosporine-like amino acids in planktonic organisms living under different UV exposure conditions in Patagonian lakes. *Journal of Plankton Research*, 26: 753-762.

Tavares, P. C., Alves, D. and Shapouri, M. 2009. Structural Changes in Macroinvertebrate Communities Associated with Reduction in the Management of Coastal Saltpans. *Journal of Marine Biology*, doi:10.1155/2009/629310.

Tellería, J. L. 1986. *Manual para el Censo de Vertebrados Terrestres*. 278 pp. Ed. Raíces. Madrid.

Turpie, J. K. 1995. Non-breeding territoriality: causes and consequences of seasonal and individual variation in grey plover *Pluvialis squatarola* behavior.

Urakawa H., Murata, Sh., Fujiwara, T., Kuroiwa, D., Maki, H., Kawabata, S., Hiwatari, T., Ando, H., Kawai, T., Watanabe, M. and Kohata, K. 2006. Characterization and quantification of mmoniaoxidizing bacteria in eutrophic coastal marine sediments using polyphasic molecular approaches and immunofluorescence staining. *Environmental Microbiology*. 5: 787–803.

Velasco, J., Millán, A., Miñano, J., Martínez, B., Perán, A., Sánchez, I. and Guerrero, C. 1995. *Producción de quironómidos en una laguna experimental en el Parque Regional de las Salinas y Arenales de San Pedro del Pinatar. Ensayo para una propuesta de manejo y gestión para su utilización como zona de alimentación de aves limícolas.* Preliminary Report. Consejería de Medio Ambiente, Región de Murcia.

Velasco, J., Lloret, J., Millán, A., Marín, A., Barahona, J., Abellán, P., Sánchez-Fernández, D. 2006 Nutrient And Particulate Inputs Into The Mar Menor Lagoon (Se Spain) From An Intensive Agricultural Watershed. *Water, Air, and Soil Pollution* 176:37-56.

Velasquez, C. R. 1992. Managing Artificial Saltpans as a Waterbird Habitat: Species' Responses to Water Level Manipulation. *Colonial Waterbirds*, 15: 43-55.

Wenny, D. G., Devault, T. L., Johnson, M. D., Kelly, D., Sekercioglu, C. H., Tomback, D. F. and Whelan, C. J. 2011. The Need to Quantify Ecosystem Services Provided By Birds. *The Auk*, 128: 1-14.

Wiens, J. A. 1989. *The Ecology of Bird Communities. Vol. 1.* Foundations and Patterns. Cambridge University Press, UK.

Winemiller, K. O. and Rose, R. A. 1992. Patterns of life-history diversification in North American fishes: implications for population regulation. *Canadian Journal Fisheries and Aquatic Science* 49: 2196-2218.

Zárate-Ovando, B., Palacios, E. and Reyes-Bonilla, H. 2008. Estructura de la comunidad y asociación de las aves acuáticas con la heterogeneidad espacial del complejo lagunar Bahía Magdalena-Almejas, Baja California Sur, México. *Revista de Biología Tropical*, 56: 371-389.

Žyydelis, R. and Kontautas, A. 2008. Piscivorous birds as top predators and fishery competitors in the lagoon ecosystem. *Hydrobiologia*, 611: 45-54.

In: Lagoons
Editor: Mwinyikione Mwinyihija

ISBN: 978-1-62808-092-6

Chapter 5

MULTI-BIOMARKER APPROACH AS AN INTEGRATIVE TOOL FOR THE EVALUATION OF HUMAN IMPACTS IN THE LAGOON OF VENICE, ITALY

Vanessa Moschino, Nicoletta Nesto and Luisa Da Ros*
Institute of Marine Sciences, ISMAR, National Research Council, CNR, Venice, Italy

ABSTRACT

The Lagoon of Venice is a complex shallow coastal system and one of the largest Mediterranean transitional environments, including a variety of aquatic and terrestrial habitats such as marshes, tidal flats, islands and canals. Since the Middle Ages, various anthropogenic interventions have influenced the Lagoon structure, and human activities must be considered the main extrinsic driving force in the evolutionary history of the Lagoon environment and its ecosystems. There is no other lagoon in the world where the mingling of human activities and natural ecology has been so enduring and profound. For all these reasons, the Lagoon of Venice and its ecosystem represents a unique and perfect natural laboratory to analyze and document changes in biological communities in response to anthropogenic forcing. Over the last two decades, several monitoring and research programs have been carried out by local authorities and research institutions to assess the environmental status of the Lagoon, particularly concerns being posed on its chemical contamination. Industrial activities strongly increased in the area after World War II and a peak of pollution was reached during the 1960s and 1970s due to the growth of local industries, such as oil refineries and several chemical production plants, and urbanization. Various classes of anthropogenic chemical contaminants, at first mainly trace metals and more recently also persistent organic pollutants, have been discharged into the Lagoon from various sources and have been accumulated in sediments. Several research studies performed to assess the environmental quality of the Lagoon in relation to chemical pollutant impacts have employed suites of biomarkers to integrate tentatively biological responses detected at various levels in different organisms, and thus achieve a realistic account of the ecological quality of this complex habitat. Some species have been privileged as biomonitors for the Lagoon environment, among which the mussel *Mytilus galloprovincialis*, the Manila clam *Ruditapes philippinarum* and the grass goby

* Contact: luisa.daros@ve.ismar.cnr.it.

Zosterisessor ophiocephalus are the organisms most frequently used for analyzing the effects of anthropogenic stressors throughout the biomarker approach.

1. Introduction

The Lagoon of Venice ecosystem represents a unique and perfect natural laboratory to analyze and document changes in biological communities in response to both natural and anthropogenic forcings. Coastal lagoons are widely recognized as the most productive environments thus particularly valuable not only for their commercial and recreational worth, but also for their ecological relevance (Kennish and Paerl, 2010). As a matter of fact, they are also particularly vulnerable to human activities, which generate anthropogenic stressors of various nature as water quality problems, contamination, and habitat alteration, which are all factors potentially capable of altering the structure, function and productivity of the lagoon ecosystems. Assessing the value of lagoon ecosystems requires the involvement of various scientific expertise such as climatology, hydrology, ecology, fisheries assessment and management, as well as ecotoxicology, which is the focal point in helping to elucidate the possible impact/effects of pollution on the biotic compartment (Brehmer et al., 2011). In this context, pollution is considered a critical issue in transitional environments due to the combination of its effects with those of the variable abiotic factors, which force estuarine organisms to severe restrictive conditions. Some authors suggested that estuarine species, living near their tolerance limits, are probable more vulnerable to additional stress. On the contrary, other authors stated that species colonizing transitional environments might be adapted to tolerate pollution stress due to their pre-existing high tolerance to the highly variable abiotic factors (Chapman and Wang, 2001). The lagoon of Venice offers exciting possibilities in the framework of the ecotoxicological studies, in particular for assessing the biological effects derived from human impact since there is no other lagoon in the world where the mingling of human activities and natural ecology has been so enduring and profound. Various point and non-point sources of pollutant inputs converged on its basin, sometimes localized in specific areas, mainly deriving from the industries of the Porto Marghera district, the sewage and urban runoff from the cities of Venice and Chioggia, and the run-off from the drainage basins which are intensively urbanized and cultivated. Other sources are peculiar to Venice, such as contamination from boat engines and waste products from the glasswork industry (Guerzoni et al., 2007; Giusti and Zhang, 2002).

This chapter provides an up-to-date review of ecotoxicological research focused on the evaluation of the effects of anthropogenic stressors throughout the biomarker approach in various biomonitor species. Several research studies performed to assess the environmental quality of the Venice lagoon in relation to chemical pollutant impacts have employed suites of biomarkers to integrate tentatively biological responses at various levels in different organisms, aiming to achieve a realistic account of the ecological quality of their specific habitats. This approach was first applied in the 1990s on *Mytilus galloprovincialis* and *Zosterisessor ophiocephalus* as a bioindicator species for evaluating the impact of micro-organic pollutants at the biochemical level (Livingstone et al., 1995), then extended to other species, e.g. *Ruditapes* (= *Tapes*) *philippinarum* and *Carcinus aestuarii*, at various levels of biological organization, from genetic to physiological-organismic. (Table 1).

Table 1. Studies on biomarkers performed in the Lagoon of Venice in mussel, *Mytilus galloprovincialis*, clam, *Ruditapes (Tapes) philippinarum*, grass goby, *Zosterisessor ophiocephalus*, and crab, *Carcinus estuarii* (see text for explanation of abbreviations)

	Monitoring approach	Biological Level	Biomarkers	References
Mussel	Passive	Biochemical	SOD; CAT; P450; GST; Baf Met	Livingstone et al., 1995
	Passive	Cellular/tissue	NRR	Lowe et al., 1995
	Passive	Physiological	SFG	Widdows et al., 1997
	Passive	Cellular/tissue	NL	Lowe and Da Ros, 2000
	Passive - Active	Cellular/tissue	NRR; LMS	Da Ros et al., 2002
	Passive - Active	Biochemical; Physiological	SOD; ADH; NADPHcred; SOS	Nasci et al., 2002
	Passive	Biochemical; Cellular	ADH; CAT; NRR; SOS	Nesto et al., 2004
	Passive	Genetic	DNA ad.; MN	Venier and Zampieron, 2005
	Passive	Genetic, Biochemical; Physiological	MN; MT; SOS	Pampanin et al., 2005a
	Passive	Physiological	SOS	Pampanin et al., 2005b
	Active	Biochemical	TOSC;CAT; MDA;MT	Pampanin et al., 2005c
	Passive	Cellular/tissue; Physiological	AM; SOS	Marin et al., 2006
	Passive	Biochemical; cellular/tissue; physiological	MDA; MT; LMS; NL; LF SOS	Nesto et al., 2007
	Active	Cellular/tissue; physiological	NRR; NL; LF; L/C; SOS	Moschino et al., 2011
Clam	Passive	Passive	LMS; SFG	Da Ros et al., 1998
	Active	Biochemical; Cellular/tissue	Baf Met; NADPHcred; SOD; CAT; LMS	Nasci et al., 2000
	Passive - Active	Cellular/ tissue; Physiological	LMS; SFG; SOS	Marin et al., 2001
	Passive	Biochemical	MT	Irato et al., 2003
	Passive	Cellular/tissue	LMS	Da Ros and Nesto, 2005
	Passive	Biochemical	AChE	Matozzo et al., 2005
	Passive	Cellular/tissue; Physiological	AM; SOS	Marin et al., 2006
	Passive	Biochemical; Physiological	Vg-like; SOS	Munari et al., 2010
	Passive	Biochemical; Cellular/tissue	AChE; Vg-like prot; Hem; HLA; SOS	Matozzo et al., 2010

Table 1. (Continued)

	Monitoring approach	Biological Level	Biomarkers	References
	Passive	Cellular/tissue; physiological; behavioural	NL; LF; SOS; Reb	Moschino et al., 2010
	Passive	Cellular/tissue; physiological; behavioural	NL; LF; L/C; SOS; Reb	Moschino et al., 2011
	Passive	Biochemical; Cellular; Physiol; Behav;	MT; LF; SOS; Reb	Moschino et al., 2012
	Passive	Biochemical	AChE;SOD; CAT; SOS	Matozzo et al., 2012
Grass goby	Passive	Biochemical	SOD; CAT; EROD	Livingstone et al., 1995
	Passive	Cellular/tissue	NL	Lowe and Da Ros, 2000
	Passive	Biochemical/Physiological	EROD; FA	Franco et al., 2002
	Passive	Genetic	DNA ad	Venier and Zampieron, 2005
	Passive	Biochemical; Physiological	MDA; MT; EROD; PAH met; LSI; GSI	Nesto et al., 2007
	Passive	Biochemical; Cellular/tissue	MDA; ACR; HNE; NT; MMCs	Pascoli et al., 2011
Crab	Passive	Biochemical	MFO	Fossi et al., 1998
	Passive	Biochemical	AChE; EROD; Vg-like	Locatello et al., 2009
	Passive	Biochemical	Vg-like prot	Munari et al., 2010
	Passive	Biochemical	AChE; CYP4SO; Vg-like	Ricciardi et al., 2010

2. THE LAGOON OF VENICE

The Lagoon of Venice is the largest wetland in the Mediterranean Basin and one of the largest in Europe, covering a surface area of around 550 square kilometers (Figure 1). It is located along the northeastern coastline of the Adriatic Sea, extending from the Sile River at its north side to the Brenta River at the south border. Classified as a transition water body, according to the criteria of the Water Framework Directive recently adopted by the European Union (WFD, 2000/60/EC), the Lagoon is basically a shallow water system with an average depth of 1 m, crossed by a rather complicated net of canals (i.e. dredged channels). In fact,its land isonly around 8%, including Venice city center itself and several minor towns similarly located on a number of small islands interconnected by bridges. Most of the surface area (around 80%) consist of mud flats, tidal shallows and salt marshes, while the rest (about 11%) is permanently covered by open water. The Lagoon is connected to the Adriatic Sea by three inlets: Lido, Malamocco and Chioggia, opening between its mainland marine border at the south and two elongated coastal islands, which are a sort of coastal strip between the Lagoon and the sea. Sited at the northern end of the largely enclosed Adriatic sea, the Lagoon is subjected to high variations in water level, the most extreme being the spring tides known as "high waters", which several times a year flood much of Venice.

The Lagoon was shaped about six to seven thousand years ago, when the marine transgression following the Ice Age flooded the upper Adriatic coastal plain. It is the larger surviving lagoon (other smaller lagoons are located northeastward), of a vast estuarine system that in Roman times extended from Trieste, along the Northeast coast of Italy, toward the South, up to the city of Ravenna. Starting from the fifth/sixth century, this aquatic area,being sufficiently isolated from the mainland but not too far from it, had provided naturally sheltered ambient conditions suitable for the development of human settlements and for raising their economic expansion over the centuries, particularly during the XVIth - XVIIth centuries. Effectively, the Lagoon was inhabited since ancient times, but it was only during and after the fall of the Western Roman Empire that mainland inhabitants, running away from the barbarian invaders, settled in this sheltered ambient contributing to lay the first groundwork for the foundation of the city of Venice. Today, besides Venice city center, other main cities inside the Lagoon are Chioggia (at the southern inlet), Murano and Burano (in the center of the Lagoon), and Lido and Pellestrina on the two coastal islands. However, most of the inhabitants of the Venice municipality, as well as its economic core, its airport and its harbor, stand on the north-western border of the lagoon, around the mainland towns of Mestre and Marghera. At the northern end of the lagoon, there are two famous seaside resorts, Jesolo and Cavallino-Treporti. Nowadays, the Lagoon's natural habitats, although under threats by various anthropogenic pressures, still provides support for various lucrative human activities linked to leisure interests, sometimes conflicting with the commercial/industrial ones, such as the seaport, fishing and aquaculture.

The sediments of tributary rivers balanced the natural sinking over centuries, but the Lagoon would be doomed to disappear and become alternatively land or sea if it were not for human intervention. Effectively, starting from fifteenth centuries, various hydraulic works were implemented, the earliest ones to prevent the lagoon from turning into a marsh, the latest nowadays, to contrast the current trend toward the complete disappearance of the Lagoon for flooding, due to concurrent human and natural factors, e.g. pumping of aquifers since the nineteenth century which has amplified natural subsidence, and the increased frequency of "high waters" mainly due to sea-level rising. Originally, many of the Lagoon's islands were marshy, but gradual hydraulic interventions for their drainage made them fit for human setting.

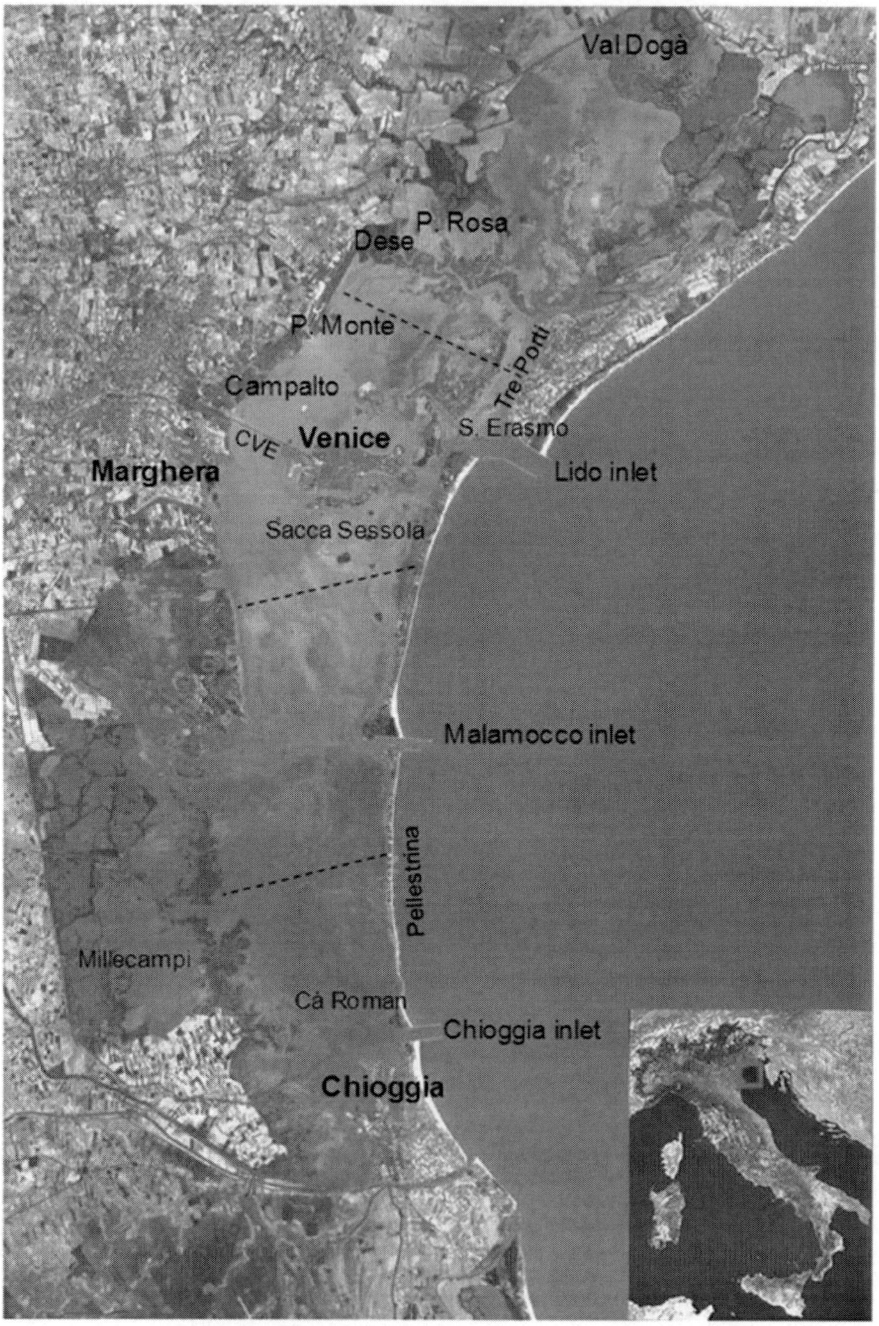

Figure 1. The Lagoon of Venice with the most cited sampling locations. Broken lines indicate the four basins.

Lagoon hydrodynamic conditions are driven by real forcing (wind, tide and water level at the inlets) and by evaluating their general spatial patterns, a clear-cut partitioning of the Lagoon into three sub-basins has long been accepted: northern (where the town of Venice is located), central and southern. However, considering the relatively major influence of the efflux/influx at the inlets, it has recently been suggested that a four-sector separation would give a more reliable compartmentalization (Tagliapietra and Ghirardini, 2006; Solidoro et al., 2010). An even more ecologically sound partitioning was suggested recently by Tagliapietra and Ghirardini (2006), with the aim to facilitate comparison both of parts of the same Lagoon and between different lagoons in the framework of ecological status evaluation,. In this context, they suggested that the Lagoon should be considered set off into various sub-basins, to be regarded as consistent water body units, whenever management actions focused on habitat protection are to be undertaken.

3. Pollution Impacts in the Lagoon of Venice

The Lagoon is classified as a transitional water body, i.e.,following the definition provided by the WFD, a body of surface waters placed between land and sea, which may be affected by both freshwater inputs from tributary rivers or from their nearby mouths and coastal waters. Thus, the Lagoon waters have partly saline characteristics as a result of their proximity to the sea; on the other hand they are also significantly influenced by freshwater inputs.

Overall, the freshwater inputs into the Lagoon from the tributary discharge, draining about 185,000 hectares of land of which 75% is used for intensive farming generating heavy loads of nitrates and partly phosphates, are evaluated in about 40 m^3 s^{-1} (Collavini et al., 2005). An estimated 7,000 metric tons of nitrates are released into the lagoon each year, of which 53% comes from agriculture and livestock and 31% are of urban origin. Farming and civil sources, respectively, account for 46% and 39% of the 800 metric tons of phosphates that enter the lagoon each year. Natural rivers and drainage channels, which flow into the lagoon, bring in dissolved nutrients,sediments, suspended materials and pollutants, particularly pesticides. The situation concerning agricultural effluents is quite problematic despite the numerous regulations that set severe quantitative and qualitative restrictions on materials and pollutants that can be released into the lagoon. Due to the too numerous and diffuse water sources entering the Lagoon from the mainland, it is difficult to monitor and control the discharge of agricultural pollutants.

However, there are still other worrysome sources of incoming pollution as main streams, i.e. industrial runoff from the port and the industrial zone of Porto Marghera, which is basically a petrochemical plant located on the northwest edge of the Lagoon, and urban effluents from cities located within the Lagoon itself, i.e. Venice and Chioggia, and other villages sited on populated islands, such as Burano and Murano. Since the 1950s Porto Marghera has been one of the main hubs of the Italian chemical and oil industry, where about 1,000 companies with up to 15,000 employees, have been producing a range of basic and intermediate chemicals (among others, ethylene, propylene, ammonium sulfate, caprolactam, chlorine, sodium hydroxide), afterwards transported by pipe and road to various destinations in Italy and Europe and by sea to rest of the world. Much of the industrial zone, presently declining mainly due to several economic contingencies, grewuntil the 1970s and production levels remained high until the 1980s, mostly throughout the development of chemical plants, steel works, and an oil refinery. Over those years, lacking any legal regulation, industrial wastes had been discharged quite massively into the Lagoon and partly into uncontrolled dumps located on the Lagoon borders. This situation is now quite improved, and illegal waste discharges declined markedly since the 1980s,especially with the introduction of tough, legally enforced environmental regulations. It is now estimated that 80% of the industrial waste from the mainland is treated in purification plants. The peak inputs during the 1970s and 1980s have been documented by a number of studies carried out in the past twenty years (among many others: Pavoni et al., 1987, Frignani et al. 1997), which all have highlighted the presence of a wide range of contaminants in the Lagoon, particularly in the sediments. Attention was mainly focused on heavy metals such as Hg, Ni, Cd, Cr and Pb, but more recently it has been made clear by several investigations on the Lagoon sediments that persistent non-biodegradable organic-micro pollutants must also be considered as a major heritage of the past industrial activity in Porto Marghera in the 1950s to 1970s (Guerzoni and Raccanelli, 2004). These pollutants, commonly known as POPs -Persistent Organic Pollutants, include dioxins, polychlorinated biphenyls -PCBs, polyaromatic hydrocarbons - PAHs, hexachlorobenzene - HCB. Moreover, it has been recently estimated by public environmental agencies that there are a number of abandoned dumps in the lagoon containing around 5×10^6 m^3 of pollutant materials. It has also been repeatedly declared in public debates by local administrators that pollution derived from industry discharged between the 1930s and 1996s, have triggered a significant decrease in water transparency, a considerable decline in the number of plant and bird species, and a very important reduction of the amount of the marine eelgrass that consolidates the bed of the lagoon.

It is worth noticing that, although the Lagoon waters are at present virtually free from regular discharge of industrial waste, pollutants are released from sediments, which have accumulated over the years, and still retainsignificant levels of heavy metals and non-biodegradable chlorinated aromatic compounds (Bellucci et al., 2000; Frignani et al., 2005). Among the most toxic and highly persistent ones, there is a group of 210 congeners of polychlorodibenzodioxins and polychlorodibenzofurans (PCDD/F), commonly referred as "dioxins", the level of which has been monitored over recent years by several authors (Raccanelli et al., 2009; Guerzoni et al., 2007) and are widely distributed in the Lagoon sediments and the city's canals. Dioxins, like other POPs, may enter the aquatic food web and ultimately affect human beings as final consumers of bio-accumulating marine organisms such as fish and edible mollusks. Effectively, up to now, most investigations on locally cultivated mussels/clams have shown that the average level of contamination in these molluscs is to be considered neither particularly alarming, nor substantially different from other medium polluted coastal areas in the world (Micheletti et al., 2004). It is now clear from the numerous research investigations and other various monitoring studies carried out by local public environmental agencies in the last twenty years that the industrial pollution, originally resulting from wastewater discharges of Porto Marghera, is at present mainly caused by sediment resuspension. Therefore, sediments have to be considered as the main secondary source, a sort of "pollutant repository" from where pollutants seep out or spread toward ambient water.

It has also been shown that Venice, particularly its natural channels, is contaminated with a mix of dioxins from urban sources such as an urban waste incinerator located on the Lagoon edge, and boat-engine emissions. Effectively, there are other sources of pollutants potentially affecting the Lagoon waters, but minor when compared to industrial and agriculture ones: the discharge of urban sewage of the various Lagoon city centers directly into the city canals from houses, after the only preliminary mechanical treatment (from private dwellings) or after both mechanical and biological treatments (from hotels, shops, restaurants); the atmosphere inputs, attributed to the traffic and to the industrial waste treatment plant located in the proximity of the north-east Lagoon edge (Guerzoni and Raccanelli, 2004).

It has also been observed that much of the persistent or low biodegradable pollutants conveyed in the Lagoon waters from these different sources are mainly diluted by tides which contribute to exporting resident waters and bringing in marine waters. To this end, it is worth noticing that residence time for Lagoon waters ranges from a few days, in the area close to the inlets, up to about 2 months in the northernmost inner areas.

Chemical monitoring of water, sediments and biota aimed at evaluating the environmental quality of the Lagoon started in the 1960s, with particular focus on the central part of the basin. Effectively, this part of the Lagoon is generally considered the most impacted, being affected both by the industrial area on its northern edge, by the presence of Venice city center in the middle, and by urban, tourist and commercial motor-boat traffic.

Despite the large number of international, national and regional research projects that have been performed over the last 20 years with the aim of evaluating the quality status of this environment (see, for instance, the results of the 1990–1992 UNESCO Project "The Venice Lagoon Ecosystem", Lasserre and Marzollo 2000), the condition of the Lagoon still gives rise to controversy, and an overall accorded picture of its ecological health status has not been yet achieved. It is however, to be highlighted that only in the last twenty years, in particular starting from the above mentioned UNESCO Project, the biomonitoring approach based on biomarkers has begun to be extensively adopted.

4. THE MOST USED BIOMARKERS IN LAGOON MONITORING STUDIES: AN OVERVIEW

The biomarker approach is widely known to play a fundamental role in the framework of ecosystem health assessment. The analyses of biomarkers have been developed to meet the requirements for more sensitive and highly significant indicators of sublethal effects, allowing the detection of early biological changes, which may therefore result in long-term physiological disturbances. Biomarkers are defined as measurable internal markers of changes detected at molecular, cellular or physiological levels, which offer great potential for understanding environmentally mediated impairments that enable cause-and-effect relationships to be established early. Since the 1980s biomarkers have been developed for providing early warning signals of adverse exposure and/or deleterious effects on biological systems due to contaminants and to natural stressors. Initially, single responses have been studied in single organisms, but soon the importance was highlighted that measuring several biomarkers at the same time at least in one species or, even better, in more than one species, gives rise to the concept of the so-called multi-biomarker approach. This multiparametric approach, which provides for the use of various complementary biomarkers at different levels of biological organization, should allow not only assessment of the effects on suitable indicator organisms (also known as *sentinel* organisms) of different stressors affecting the marine environments (diagnostic purpose), but also to permit predictionsof animal and population health status (prognostic purpose). Biomarkers are commonly categorized in "exposure" and "effects", the former are related to responses induced by specific groups of contaminants mainly detected at molecular and cellular levels (i.e. enzymatic induction/inhibition, production of particular metabolites), which do not necessary lead to affect the general health status, the latter are related to toxicological/pathological effects in the organisms, not necessarily linked to specific contaminants.

Numerous studies carried out in marine and coastal waters both in Europe and elsewhere have provided evidence that an appropriate suite of biomarkers of effects and exposure may offer new insights into the causality of adverse effects at higher biological levels and may also be successfully employed to evaluate sub-lethal and chronic/cumulative long-term impacts. The biomonitoring of coastal waters and estuaries may be approached throughout different experimental designs: passive biomonitoring, when organisms are collected from natural populations at sites where they already exists; active biomonitoring, when organisms are collected from one unpolluted/unstressed population and then translocated for a fixed period at the various sites to be monitored (de Kock and Kramer, 1994).

The biomarker approach is particularly useful in a complex environment such as the Lagoon, where often-contrasting interests,economic, social and political, are to be taken seriously into account and balanced by evaluating their environmental sustainability, and by foreseeing their cascade-effects on the biological resources. In this context, it is clear that use of early signs of biological impairment, in particular when key physiological functions are affected, can offer valuable indications beyond the degree of actual exposure, as they may be useful as a predictive tool (Moore and Widdows, 1998). Thus, this approach, in both passive and active procedures, has started to be applied in biomonitoring the Lagoon from the 1990s; since then, various biomarkers have been evaluated and different species have been tested, to reach a better comprehension of the biological effects of pollutants within the Lagoon environment.

Genotoxicity Biomarkers

Genotoxicity is referred to effects caused by carcinogenic and mutagenic contaminants, which could potentially lead to structural DNA damage, to genomic instability and altered gene expression. These compounds are considered theoretically the most dangerous substances occurring in ecosystems because they can exert their toxic effects beyond exposed individuals, being potentially active through several generations. Several methods have been developed and proposed for evaluating DNA alterations, which include single and double strand breakages, modified bases, DNA–DNA crosslinks, and DNA–protein crosslinks (Viarengo et al., 2007). A set of assays to evaluate DNA effects should include methods for DNA adducts and to analyze DNA strand breaks and synthesis, e g. comet assay, or single-cell gel electrophoresis, developed for the detection of double strand DNA breakage. In the studies performed in the Lagoon environment, DNA damage and chromosomal alterations have been assessed through the evaluation of DNA adducts and micronucleus frequency in bivalves (*M. galloprovincialis*) and DNA adducts in fish (*Z. ophiocephalus*) by Venier and Zampieron (2005). DNA adducts are caused by the covalent binding of reactive xenobiotics to nucleophilic DNA positions and provide direct evidence of the genotoxic effects induced by different classes of pollutants. The adduct levels estimate the real amount of molecules covalently bound to DNA which depends on contaminant uptake and metabolism, chemical adduct structure, DNA repair and cell replication or cell death (Venier and Zampieron, 2005). DNA adduct formation detected with the ^{32}P-postlabelling assay have been successfully used for monitoring environmental exposure to aromatic-like genotoxic compounds and for understanding cancer development (Philips and Castegnaro, 1999).

The micronucleus assay is a well known cytogenetic technique first developed in mammalian species and then proven to be suitable for application to the aquatic environment, in both fish and invertebrate species. Micronuclei arise from chromosome or chromosomal fragments, which are not incorporated into daughter nuclei during cell division and mitosis (Bolognesi and Degan, 2001). The presence of micronuclei is an indicator of chromatin breakage, which may be caused by clastogens or spindle dysfunctions, ultimately caused by toxic compounds (Viarengo et al., 2007)

Haemocytes and gill cells are the target tissues most frequently considered for the micronuclei determination in bivalves, whereas in fish they are found in various cell types such as erythrocytes and gill, kidney, hepatic and fin cells (Bolognesi and Hayashi, 2011).

Biochemical Biomarkers

Oxydative Stress Biomarkers

In living organisms various physical, chemical and biological agents may produce reactive oxygen species (ROS) which cause oxidative stress effects in cellular structure and functioning, i.e. enzymatic modifications, protein, lipid and carbohydrate peroxidation, DNA lesions and ion homeostasis alterations. To cope with this problem all eukaryotic cells are provided with complex defense systems whose aim is to inhibit/reduce the oxyradical formation. In particular, exposure to pollutants may alter the oxidative status of marine organisms, which may increase the intracellular formation of ROS and modify the efficiency of antioxidant enzymatic system (Regoli, 2000). Antioxidant enzymes could be induced by increasing production of ROS as a protection mechanism against oxidative stress or inhibited when deficiency of the system occurs, suggesting oxidative stress toxicity (Winston and Di Giulio, 1991; Cossu et al., 1997). Oxidant – mediated effects, commonly used as biomarkers, include both increased activities of antioxidant enzymes tending to inhibit oxyradical formation such as superoxide dismutase (SOD), catalase (CAT) and accumulation of non–enzymatic compounds such as malondialdehyde (MDA). MDA is a metabolite indicating generic oxidative stress caused by the excess of membrane lipid peroxidation. MDA content may be induced at extremely variable levels in relation to species, type and burden of contaminants, and exposure times. Moreover, it has been demonstrated that MDA reacts with DNA to form adducts (Marnett, 1999; Hugget et al., 1992; Gérard-Monnier et al.,1998). SODs are a group of metalloenzymes that catalyze the transformation of superoxide anions ($O_2^{-}\cdot$) into hydrogen peroxide (H_2O_2), which is subsequently metabolized by CAT into molecular oxygen and water. These two enzymes play an important antioxidant role but their use as a biomarker, in environmental risk assessment, are still controversial because both induction and inhibition may be observed after exposure to environmental pollutants (van der Oost et al., 2003).

Another antioxidant enzyme is Glutathione-S-transferases (GST), which is involved in the cellular defense against oxidative stress by conjugating activated xenobiotic metabolites to electrophilic compounds (glutathione) through broad-specific transferases (Habig et al., 1974). In the Lagoon of Venice, the activity of this enzyme was determined in mussels according to methods reported in Fitzpatrick et al. (1995).

In the Lagoon of Venice the antioxidant enzymes SOD and CAT were measured spectrophotometrically in fish livers according to the methods described by Livingstone et al. (1992 a,b). Sub-lethal exposures to some metals, such as Ni and Cu, have been observed to cause a rapid and high accumulation of MDA both in *M. galloprovincialis* and *Hediste diversicolor*, whereas organic contaminants, such as B(a)P did not determine significant variations, nor after 24 or 48 h exposure. The quantification of MDA were also performed in fish livers, according to Gérard-Monnier et al. (1998), and in plasma using the thiobarbituric acid reactive substance assay (TBARS) (Yoshida et al., 2005). Also the total oxyradical scavenging capacity (TOSC) has been used in the Lagoon to measure the mussels' biological resistance to different oxyradicals (Pampanin et al., 2005c), following the methods reported in Winston et al. (1998) and Regoli and Winston (1999). This biomarker is useful to predict adverse oxidative stress effects on the health status of these organisms (Regoli, 2000).

Mixed-Function-Oxydase-System

The activities of Ethoxyresorufin O-deethylase (EROD) and benzo[a]pyrene hydroxylase (BPH), NADPH-cytocrome C reductase (NADPHcred) and total Cytocrome P450 (P450) are considered a very sensitive catalytic probe for detecting the response of the Mixed-Function Oxidase system (MFO) to the presence of organic pollutants in fish and mussels (Livingstone et al., 1985; Gkosøyr and Forlin, 1992; Van der Oost et al., 2003). This microsomal enzymatic system, whose catalyst center is constituted by the cytochrome P450, is involved in phase I of the detoxifying metabolism of organic contaminants, in particular, it catalyzes the transformations of lipophilic xenobiotics in more water-soluble compounds in order to facilitate their excretion (Bucheli and Fent, 1995). Hydrophobic organic pollutants may induce the activity of the MFO enzymatic system in many species, including mussels and fish, and the exposure to specific classes of contaminants may induce specific cyt P450 isoforms in marine organisms. In the Lagoon of Venice these responses were repeatedly studied in mussels (Livingstone et al., 1995; Nasci et al., 2002), fish (Livingstone et al., 1995; Franco et al., 2002; Nesto et al., 2007) and crabs (Locatello et al., 2009; Fossi et al., 1998), and only once in clams (Nasci et al., 2000). EROD activity was determined in fish according to the methods reported in Burk and Mayer (1974) in a number of Lagoon studies (Livingstone et al., 1995, Franco et al, 2002, Nesto et al., 2007)

Acetylcholinesterase Activity

Cholinesterases (ChEs) are considered ubiquitous enzymes whose physiological function is to remove acetylcholine from synaptic clefts, making the nerve cell suitable to receive a new signal, therefore being essential for the normal functioning of the sensory and neuromuscular systems (Murphy, 1986). Many pesticides (organophosphates and carbamates) are reported to be AChE inhibitors, ultimately determining an accumulation of acetylcholine in the synaptic cleft. The accumulation of this neurotransmitter assists in maintaining the depolarized postsynaptic membrane, ultimately causing a continual stimulation of the muscle fibers, which can lead to paralysis and then to death (Fulton and Key, 2001). The inhibition of this enzyme has been proposed and used as a biomarker of exposure for the pesticide in many species of aquatic organisms (Fulton and Key, 2001; Roex et al., 2003; Corsi et al., 2003) not only vertebrates but also invertebrates, such as bivalves and crabs (Talesa et al., 2001;). In vertebrates, AChE activity has been proven to be extremely sensitive to neurotoxic pesticides; however in *Mytilus* sp. and generally in molluscs, this biomarker can be employed for exposure to high concentrations of pesticides. Recently, it was pointed out that AChE activity in marine mussels is significantly reduced by pesticides only at high concentrations (Galloway et al., 2002; Rickwood and Galloway, 2004). In the Lagoon of Venice, AChE has been evaluated in the bivalve *R. philippinarum* (Matozzo et al., 2005, 2010, 2012;as well as in crab *C. aestuarii* (Locatello et al., 2009; Ricciardi et al., 2010).

Vitellogenin-Like Proteins

Vitellogenins, the precursors of vitellins, are glyco-lipophosphoproteins that are taken up by the ovary as an egg yolk energy source for embryo development in oviparous organisms (Wallace 1985). In fish, vitellogenins are synthesised in response to endogenous estrogens, released into the bloodstream and then stored in developing oocytes. Whereas their levels increase in sexually mature females, they are lower in female juveniles and generally undetectable in males. In invertebrates, specifically in molluscs, the synthesis of vitellogenin(Vg)-like proteins is regulated by neuropeptides and estrogens in combination, and they may be influenced by exogenous sources of estrogens as already evidenced in vertebrates (Li et al., 1998; Puinean et al., 2006). Vg-like protein levels in immature females and males is considered a useful biomarker of exposure to xenoestrogenic compounds and a wide range of organic xenobiotic compounds, common in the aquatic environment, have been shown to activate the production of Vg-like proteins (see for review Matozzo et al., 2008). These xenoestrogen substances, such as pesticides, PCBs, household products and pharmaceuticals are able to disrupt the endocrine system in vertebrates as well as invertebrates. Although endocrine disruption in invertebrates has received less attention compared with fish, partly because the knowledge regarding invertebrate endocrinology is more limited (de Fur et al., 1999), it is commonly recognized that they are sensitive to endocrine disruptors at environmentally relevant concentrations.

In the Lagoon environment the levels of vitellogenin-like proteins have been detected both in bivalves *R. philippinarum* (Munari et al., 2010; Matozzo et al., 2010) and crabs *C. estuarii* (Locatello et al., 2009; Munari et al., 2010) through the indirect alkali-labile phosphate assay (ALP) proposed by Blaise et al. (1999) and Gagnè et al. (2001).

Metallothioneins

Metallothioneins are low-molecular weight, cysteine-rich proteins, which have been found in many vertebrates including numerous species of fish, as well as in aquatic invertebrates such as molluscs and crustaceans. The thiol group (-SH), which are characterized the cysteine residue allows true MTs and MT-like proteins to bind metals (Amiard et al., 2006). The biological functions of metallothioneins are still controversial. Given their metal-binding capability, it is generally recognized that MTs play a role in the homeostasis of essential metals (Cu, Zn) storing the necessary amount to fulfill enzymatic and other metabolic demands (Viarengo and Nott, 1993). Moreover, MTs are involved in metal detoxification, acting as chelating agents for intracellular excesses of nonessential metals, such as Ag, Cd and Hg (Amiard et al., 2006). In addition to their dual role in the homeostasis and detoxification of metals, MTs may play a role also as oxidative stress protectors, as suggested by some authors who have shown that the cellular levels of MTs can be increased by various oxidant chemicals, i.e. paraquat, diquat, diamide (Amiard et al., 2006; Viarengo et al., 2000). However, MTs induction is generally considered as a biochemical biomarker of exposure to metals due to their metabolic function. Many methods have been proposed in literature for determining MTs concentrations (Ryvolova et al., 2011), the data presented in this chapter referred to the spectrophotometric method of Viarengo et al. (1997), the most used in the Lagoon to assess MT levels on mussels (Pampanin et al., 2005a, c; Nesto et al., 2007) and clams (Moschino et al., 2012).

PAHs Bile Metabolites

The levels of PAHs bile metabolites in fish are indicative of a recent exposure to such pollutants. Once inside the body, PAHs are metabolized by an efficient enzyme system and the metabolites are discharged into the bile, which is periodically emptied of its contents. Since PAH exposure cannot be reliably determined by measuring fish tissue levels for their high rate of metabolism (Varanasi et al., 1989), the evaluation of PAH metabolites as fluorescent aromatic compounds (Aas, et al., 2000) is preferred as a valuable biomarker of environmental risk assessment concerning PAH-contaminated sites (van der Oost et al., 2003). In the Lagoon this biomarker has been evaluated in *Z. ophiocephalus* using the method of Aas et al. (2000) by Nesto et al., (2007).

Cellular and Tissue Biomarkers

Lysosomal Biomarkers

The lysosomal system of many marine species, not only mollusks and fish but also annelids, and crustaceans, is known to be particularly sensitive to environmental perturbations and, for this reason, its alterations are widely used as indicators of physiological stress. Lysosomes are membrane-bound organelles, containing various hydrolytic enzymes, present in almost all cells of eukaryotic organisms, which are involved in various cellular processes including digestion, defense, and reproduction (Moore, 1988; Viarengo et al., 2007). In particular, lysosomes of hepatic cells in fish and digestive cells in molluscs are strongly implied in the detoxification of metabolisms, being involved in the sequestration and accumulation of a wide range of chemicals, such as metal ions, PAH, pharmaceuticals (Moore et al., 2006). Lysosomal alterations are considered valuable biomarkers of effects; essential measurements in the framework of environmental risk assessment and are generally classified into composition changes, membrane destabilization, and variations in morphometric parameters (size and number) (see, for instance, Moore, 1994; Marigomez et al., 1996 for *M. edulis*; Ringwood et al., 1998 for *Crassostres virginica*). At present, measurements of the various lysosomal parameters are widely applied, both in freshwater and marine organisms for monitoring coastal and estuarine environments, especially in temperate and sub-tropical areas (see, for instance, Cajaraville et al., 2000; Domouhtsidou and Dimitriadis, 2001; Moore et al., 2006; Guerlet et al., 2007).

Composition changes of lysosomes are mainly referred to accumulations of lipofuscins and neutral lipids, which are generally detected through histochemical techniques (Schmorl reaction by Pearse, 1972 for lipofuscins and following the method of Bancroft, 1967 for neutral lipid). Lipofuscins are cell pigments, which, due to their undergradable nature, tend to accumulate in the lysosome vacuolar system, particularly of fish hepatocytes and digestive gland cells of mollusks, thus increasing with the organism's age. Lipofuscin granules are mainly composed of oxidatively modified proteins and lipid degradation products, along with carbohydrates and metals (Viarengo and Nott, 1993). It is generally accepted that lipofuscins may contribute to metal detoxification, being either retained as residual bodies or extruded into the extra-cellular space (Moore 1988).

Neutral lipids are the predominant lipid class in bivalves and their abnormal accumulation in digestive cell lysosomes have been indicated as a general stress response to organic pollutant exposure (Marigomez and Baybay-Villacorta 2003).

Enhanced catabolic activity within lysosomes can provoke enlargement, damageto membrane integrity, and, as final effect, leakage of acidic hydrolases in the cytosol, eventually leading to cell death (Koehler et al., 2002). The stability of the lysosomal membranes, basically evaluated as variations in permeability, may be monitored in marine mussels according to two methods, first that of Moore (1976), which cytochemically measures *in vitro*the activity of N-acetyl-β-hexosaminidase in digestive cell lysosomes (LMS), and secondly the *in vivo* methodology of Lowe et al. (1995) based on retention of the cationic probe neutral red within the lysosomal compartment of haemocytes over time (NRR assay). Results achieved by these two techniques are directly correlated, both being sensitive to general stress and similarly suitable as early warning tools in biomonitoring programs, providing at the same time potentially prognostic signalings for animal health (Moore et al., 2006; Viarengo et al., 2007).

Lysosomal enlargement is widely accepted as a general stress biomarker (Cajaraville et al., 1995), responding to a variety of stressors including organic and metallic pollutants and changes in several natural environmental variables. The measurement of lysosomal size is determined using N-acetyl-β-hexosaminidase as a marker enzyme and then through an automated image analysis system (Cajaraville et al., 1991).

In the Lagoon of Venice lysosomal alterations on both the haemolymph and digestive gland cells have been measured in *M. galloprovincialis* and *R. philippinarum*in, in a number of studies(see Table 1 for the details).

Autometallography

Autometallography (AM) is a non-specific histochemical procedure for detecting metal ions in biological sections (Danscher, 1984, 1991; Danscher and Montagnese, 1994; Danscher et al., 1987). It applies silver amplification of metal ions, which are detected as black silver deposits (BSDs), basically on the principles of photography, under light or electron microscope (Danscher, 1981). This method is based on a microscopic approach and has been proposed as a technically simpler and more cost-effective tool to assess metal contents in biological samples, mostly from mollusks. Moreover, it gives greater insight into the mechanisms of metal toxicity, by providing information about the cellular and subcellular localization of metals and elucidating the cellular response to toxic metal uptake (Herwig et al., 1989; Soto et al., 2002). The potential usefulness of AM in the Lagoon of Venice has been evaluated only in one investigation that considered, in particular, mussels and clams (Marin et al., 2006).

Physiological Biomarkers

Scope for Growth

Scope for Growth is the result of a variety of vital functions reflecting the balance between processes of energy acquisition, such as feeding and digestion, and energy expenditure, such as respiration and excretion, and provides a direct measure of the energy available for production (growth and reproduction) (Smaal and Widdows, 1994). SFG is a biomarker at the individual/whole organism level with a high ecological relevance. This measure ranges from maximum positive values in optimal environmental conditions, when the organism is acquiring energy from the food ingested, decreasing to negative values during severally stressed periods. Seasonal variations in the physiological energy budget of bivalves are usually related to intrinsic factors like body size and reproductive conditions, and to extrinsic factors like temperature and food availability (Bayne and Newell, 1983; Widdows, 1985). Some studies have been performed on *M. galloprovincialis* and *R. philippinarum* from the Lagoon of Venice (Widdows et al., 1997; Da Ros et al., 1998; Marin et al., 2001). SFG was determined according to the method proposed by Widdows et al. (1985), by evaluating the clearance rate, respiration rate and the assimilation efficiency that were then used to calculate SFG.

Survival in Air Test

The "Survival in air" or "stress on stress" (SOS) test evaluates a biological response at the organismal level that exploits the natural ability of bivalves to survive periods of aerial exposure (Eertman et al., 1993). This is a rapid, simple and low-budget method suitable for examining whether environmental parameters may have perturbing effects on the fitness of individuals (de Zwaan and Eertman, 1996). The usefulness and ease of this test have made it attractive for both field and laboratory conditions (Veldhuizen-Tsoerkan et al., 1991; Viarengo et al., 1995; Marin et al., 2001), and it has been also widely deployed in Lagoon studies (see Table 1).

Somatic Indices

The liver somatic index (LSI), which indicates in fish the percentage of liver weight with respect to body weight, may provide an indication of the liver metabolic activity (Slooff et al., 1983). This parameter has been proposed as a general index of stress, since the presence of pollutants may determine an increase of the liver metabolic activity through the activation of enzymes involved in the detoxification processes. The final stress effect is detectable as increases of volume and consequently the weight of the liver itself. Gonad somatic index (GSI), expressed as a percentage of the gonad weight with respect to the body weight, is a gross index providing information on the stage of gonad development on the different phases of the reproductive cycle of the species. This index, not a real biomarker but rather a supporting parameter, is considered sensitive to variations of some environmental chemical-physical parameters as well as to the presence of contaminants (Corsi et al., 2003). Both LSI and GSI were determined in field and laboratory studies on *Z. ophiocephalus* from the Venice Lagoon (Nasci et al., 1991; Nesto et al., 2007).

Fluctuating Asymmetry

Fluctuating asymmetry (FA) is the morphological deviation from perfect bilateral symmetry in fish, which may cause physiological injuries, and ultimately widespread morbidity and/or changes in the community structure. It has been recommended by a number of studies as a sensitive biomarker for early warning signals of environmental disturbances (see the review of Allenbach, 2011).

A large number of meristic (counted) characters and morphometric (measured) characters were used to determine the FA index (Allenbach, 2011). The number of pectoral fin rays, the maximum length of pectoral fin rays, the preorbital and postorbital and the area, perimeter, and maximum-minimum radius of otoliths were determined in *Z. ophiocephalus* of the Venice Lagoon (Franco et al., 2002).

Behavioral Biomarkers

Reburrowing Test

The ecological relevance of burrowing behavior in infaunal species, such as clams, becomes apparent when we consider their need to avoid predation (Pearson et al., 1981). On this basis, a rapid bioassay has been developed and successfully used in toxicity sediment tests (Phelps et al., 1985; Phelps, 1989). It gives an indication of the indirect mortality resulting from increased exposure to predators (Phelps, 1989) and can provide a reliable and cost-effective short-term response (minutes to hours) to the introduction of chemical or environmental perturbation.

5. Biomarker Results in the Most Used Sentinel Organisms: What They Say on the Pollutant Effects

Mytilus galloprovincialis

Native to the Mediterranean coast, the Black and Adriatic Seas, the Mediterranean mussel, *M. galloprovincialis,* predominantly inhabits the natural and/or artificial coherent substrate along the coastline. This species, due to is high tolerance to a wide range of environmental parameters, is widespread in marine as well as in lagoon environments. In the North Adriatic region, natural beds of *M. galloprovincialis* are found not only along the coast, but also in some coralligenous reefs, locally known as "tongue", at a depth ranging between 18 – 30 m, and distant 6 – 24 km from the coast (Ponti et al., 2011). This species is also widespread within the lagoon of Venice, both near the inlets and in its inner parts, colonizing wharfs, "bricole" (i.e. a man-made marine structure consisting of three or more wood piles that delimitated the canals), and the banks of the "fondamente" and "rive" (i.e. the walls delimiting the city channels). This species has been widely used as sentinel species in the Venice lagoon environment both through the passive and active biomonitoring approach.

The first study performed in the Venice lagoon involving the use of *M. galloprovincialis* as a bioindicator was carried out in the framework of the UNESCO-MURST Venice Lagoon Ecosystem Project aiming at evaluating the pollution state of the lagoon environment

(Lasserre and Marzollo, 2000). SOD, CAT, NADPH-cytochrome c reductase, glutathione S-transferase activities and P450 were evaluated on digestive glands of mussels seasonally collected in 1992 at 2 lagoon sites, one near the industrial area located in the central basin, Canale Vittorio Emanuele (CVE) and one in the northern basin (Livingstone et al. 1995; Livingstone and Nasci, 2000). Despite contaminant body burdens (aliphatic hydrocarbons, PAHs and PCBs) were higher in mussels from the industrial area, biochemical biomarker values were generally similar comparing individuals collected at the industrial site with those of the northern basin (SOD: 658 and 653 u SOD g^{-1} ww, respectively; CAT: 1.8 and 1.6 mmol g^{-1} ww, respectively; P450: 65.9 and 63.9; GTS: 1.79 and 1.98 pmol min^{-1} g^{-1} ww, respectively; Baf metabolism: 3.02 and 4.05 pmol min^{-1} mg^{-1} protein, respectively). During 1999, another survey was performed analyzing various biochemical determinations, i.e. SOD, CAT, ADH and NADPHcred, in both transplanted and indigenous mussels from Marghera (industrial pollution) and CVE (urban and industrial pollution) in the central basin, Treporti (anthropic and motorboat traffic pollution) in the northern basin, Salute and Chioggia (urban pollution) and a reference site (clean farming area at Pellestrina), all in the central basin except Chioggia located at South. Biochemical biomarkers varied among sites and between indigenous and transplanted animals. (Nasci et al., 2002). Comparing these results with those obtained in the previous research, an increase in CAT and SOD activities was evident at CVE: approximately 8 mmol g^{-1} ww vs 1.6 mmol g^{-1} ww, respectively, for CAT and more than 1000 u SOD g^{-1} ww vs 653 u SOD g^{-1} ww, respectively for SOD. CAT and ADH were subsequently evaluated by Nesto et al. (2004) in organisms seasonally sampled during the years 2001–2002 in four sites of the Lagoon distributed among the central (impacted by industrial and urban pollution and a reference) and northern basins (S. Erasmo used as reference as well). CAT varied from 0.8 to 3.4 mmol g^{-1} ww in the reference sites and from 0.7 to 2.3 mmol g^{-1} ww in the most polluted site with the most relevant differences between polluted and reference samples identified in spring, when natural and endogenous factors appear to be less effective.

CAT and MDA together with TOSC and MT were evaluated in mussels transplanted in several lagoon sites differently impacted and spread over the four basins (Palude della Rosa, Campalto, Marghera, Sacca Sessola, the reference site Ca` Roman, and Chioggia) during the spring 2003 (Pampanin et al., 2005c). Inhibition in CAT activity was observed in all sites in comparison to the reference sites, confirming the presence of an oxidative pressure also detected with TOSC analysis. However, MT content in mussels revealed no significant differences in all sites, ranging from 29 to 69.5 μg g^{-1} of tissue, and was below the values reported in other studies (Viarengo et al., 1997; Petrovic et al., 2001). MTs were also evaluated by Pampanin et al. (2005a) in the canals of the Venice historic center and by Nesto et al. (2007) in mussels seasonally collected at Sacca Sessola and Campalto-S. Giuliano (central basin) in 2005-2006. In the former study, a significant induction of MTs was generally observed in the canal sites, particularly in spring (values ranging from about 100 μg g^{-1} of tissue in the reference site to up to 200 μg g^{-1} of tissue in the most affected canal), whereas in the latter, MT levels and temporal trends were similar at the two sites, showing decreasing values in summer. Their seasonal patterns, clearly overlapping most of the metal ones, may confirm that metals play a major role in their induction.

Genotoxicity biomarkers (micronucleus frequency and DNA adducts) were first evaluated by Venier and Zampieron (2005) in mussels collected during the summers of 1998 and 1999 from 10 sites distributed in the whole lagoon including an unpolluted site (near Lido inlet), and sites influenced by agricultural runoff and industrial wastes (near Porto Marghera) as well as two urban sites near Chioggia. Micronucleus frequency was determined in both heamolymph and gills, whereas DNA adducts were in gills only. DNA adducts were not detected in mussels from the unpolluted site, neither in urban ones, whereas a typical diagonal radioactive zone (DRZ), produced by overlapped adduct spots and indicative of a mixture of bulky adducts, was evident in two sites located along the industrial channels, where the highest adduct levels were recorded following butanol enrichment (3.710 and 1.216 DNA adducts/10^8 total nucleotides) and significantly different if compared to the level observed in the reference site (0.430 DNA adducts/10^8 total nucleotides). Frequently, micronucleus was significantly higher in mussels collected at the sites located along the industrial channels (5.3 ‰ in gill cells and 12.3 ‰ in haemocytes, respectively), likely caused by a mixture of environmental contaminants, which characterized these sites. Micronucleus frequency was also investigated on heamolymph of mussels collected from three sites located in the canals of the Venice historic center and one reference area (Pampanin et al., 2005a). Chromosomal alterations were generally significantly higher in mussels collected in the historical center (about 3 ‰) with respect to those from the reference sites, showing a significant correlation with DDD, DDT and PCBs. Comparing the results of the two studies, genetic damages and consequently the risk of further alterations at higher biological levels were higher in individuals from the industrial zone compared to those from the Venice city center.

Lysosomal biomarkers were measured in the haemolymph (NRR assay) and digestive glands (LMS, LF, NL and L/V) in both indigenous and transplanted mussels (Figure 2). These parameters were first evaluated by Lowe et al. (1995) and Lowe and Fossato (2000) in mussels sampled in six sites of the lagoon during two surveys performed in 1992 and 1993 respectively, highlighting the high potential of these responses in studying pollutant effects. In particular, NRR times ranged from 138 min and 120 min at sites where mussel farms were located in the central and southern Basins, decreasing to 0 and 33 min in the most impacted sites. Da Ros et al. (2002) comparing the results of NNR assay and LMS in six differently impacted sites (Chioggia in south basin, Treporti in north basin, and Venice city center, Canale Vittorio Emanuele-CVE, Marghera in the central one) stressed how lysosomal response measured in haemocytes is a more valuable biomarker of anthropogenic stress for both passive and active biomonitoring in marine coastal environments. Subsequently, Nesto et al. (2004) highlighted that the NRR assay did not appear to be very sensitive in discriminating sites as well as to seasonal variations, mean retention times being very similar among sites and seasons (from 75 min to 30 min), and low in the reference site, at least when compared to those found in the previous studies. Similar results were also obtained in transplanted mussels by Moschino et al. (2011), when no significant differences were detected among the ten monitored sites,not even in June (retention times ranging from 36 to 48 min) or October (retention times from 36 to 72 min). These results seem to indicate a general standardization of this response in recent years with a particular decrease of the retention times in mussels from sites historically considered as reference, probably reflecting uniform general redistribution of pollutants which now characterize the lagoon environment.

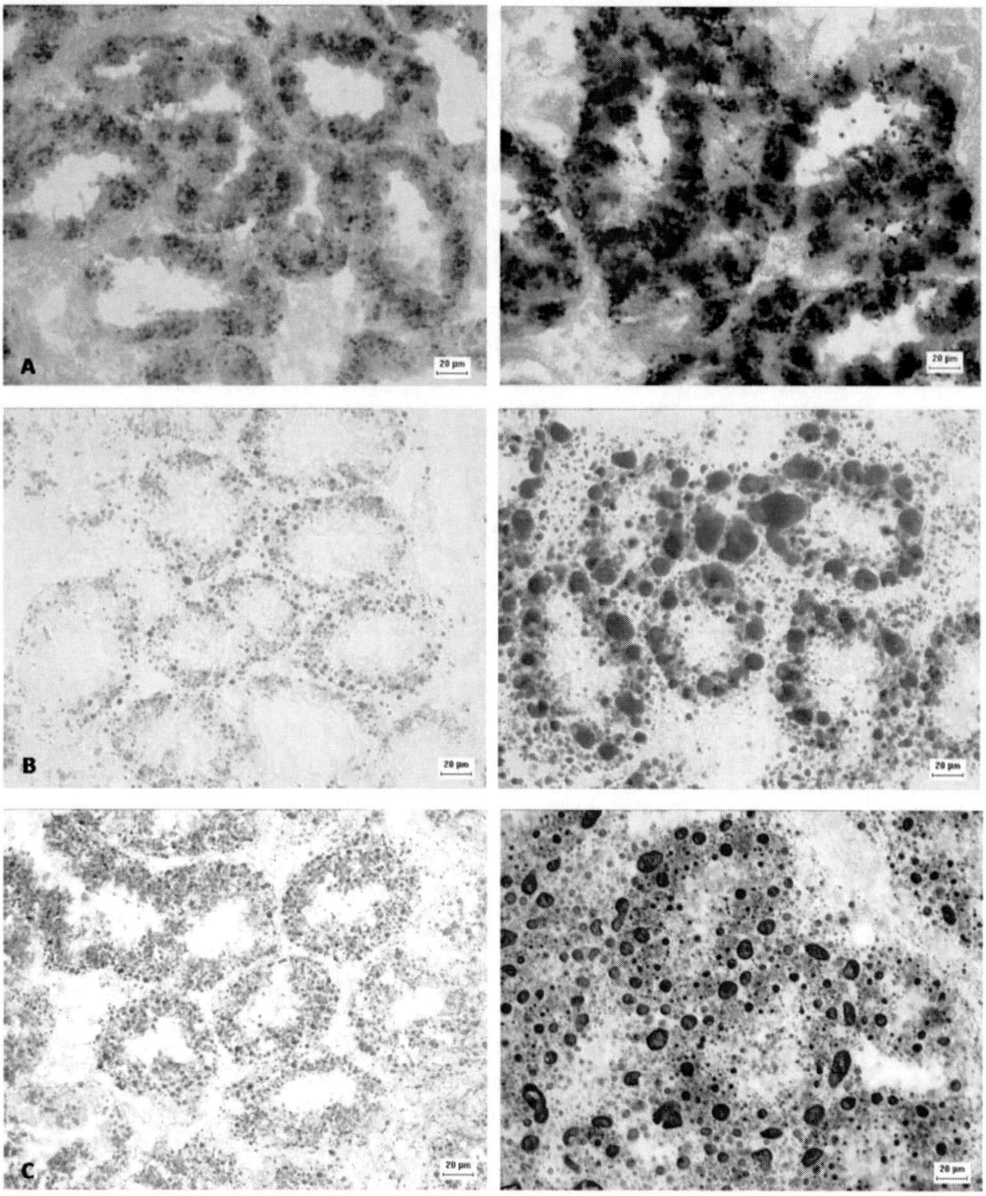

Figure 2. Lysosomal biomarkers: accumulation of lipofuscins (A), neutral lipids (B) and lysosome enlargement (C; lysosomes are marked with N-acetyl-ß-hexosaminidase) in *M. galloprovincialis* from reference (left) and polluted (right) sites in the Lagoon of Venice (methods are described in the text).

Higher differences among sites were found when considering other lysosomal alterations such as NL accumulation and L/V, which were,however, responsive only in spring (Moschino et al., 2011).

Authometallography was assessed only once in mussels from 2 lagoon sites in central (industrial inpact, Porto Marghera) and south (urban impact, Chioggia), evidencing promising results: BSD amountswere significantly correlated with total metal concentrations in sediments, consequently discriminating sites on the basis of their metal pollution levels (Marin et al., 2006).

When considering physiological biomarkers, SFG in *M. galloprovincialis* was evaluated in the only study of Widdows and Nasci (1997), which shows that mussels sampled from the most contaminated sites (CVE and Venice city center) exhibited a significant reduction in clearance rate and scope for growth compared to those living in the outer regions of the Venice Lagoon. The low SFG values (i.e. < 5 J g^{-1} h^{-1}) were indicative of severely stressed mussels and poor water quality, probably due to the higher levels of microorganic contaminants, which characterized these two sites.

Contrary to SFG, the Survival in air test has frequently been used in association with other biomarkers to detect general physiological impairment at organism level due to its ease and quickness of execution as well as its affordability (see Table for reference details) Nasci et al. (2002) did not find differences between reference mussels in comparison with either transplanted or indigenous ones, suggesting that pollution did not affect this parameter. On the contrary, Nesto et al. (2004) found significantly higher survival times in the reference site compared to polluted ones (8 days vs 7 and 6 days, respectively), but only in samples performed in the warm season. These results were also confirmed by Moschino et al. (2011) in transplanted mussels, which exhibited significant differences in survival curves comparing references with polluted sites mainly in late spring, survival times ranging from 13 to 6 days in June and from 5 to 3 days in October), confirming the strong seasonal variability caused by the influence of natural factors such as seawater temperature and food availability as well as by physiological factors such as the reproductive cycle and the content of body reserves on this response (Eertman et al., 1993).

Ruditapes philippinarum

The Manila clam *Ruditapes* (*Tapes*) *philippinarum* has been suggested as a biomonitor for several transitional areas, mainly because of its wide spreading habits and its ecological and economic relevance (Ji et al., 2006; Li et al., 2006; Martin-Diaz et al., 2007). In particular, because of its feeding behavior confined at the sediment-water interface, this species has been recommended as a possible indicator of seawater as well as sediments, being at the same time exposed to both compartments (Li et al., 2006; Wang et al., 2010, 2012). This infaunal suspension feeder, inhabiting sandy-mud bottoms, extrudes its short inhalant siphon at or below the sediment surface to filter-feed in the near bottom waters (Kanaya et al., 2005). The major food source is represented by suspended materials, including phytoplankton and inorganic seston, which, in environments characterized by frequent bottom perturbing events, may largely originate from re-suspended sediments (Chong and Wang, 2000).

Native to the Indo-Pacific region, *R. philippinarum* was intentionally introduced into the North Adriatic lagoons and specifically in the Lagoon of Venice for aquaculture purposes in 1983. Its great capacity to adapt to the new environment soon became evident not only in relation to the individual high growth rates recorded, but also to its natural reproduction (Pellizzato and Da Ros, 2005; Pranovi et al., 2006).

R. philippinarum was first investigated for monitoring a polluted, i.e. Marghera, and a clean site, i.e. Palude della Rosa, in the late 1996 by Da Ros et al. (1998), who investigated LMS and SFG as well as a prevalence of parasites (*Perkinsus* sp.) in indigenous clams, obtaining promising results. Active biomonitoring was applied using this species only from 1998 (Nasci et al., 2000; Marin et al., 2001); afterward, this approach was abandoned as these studies had shown not only the great operational difficulties for transplanting clams in this environment, but also the low responsiveness of the analyzed biomarkers, in particular in relation to clearance and respiration rates and SFG (Marin et al. 2001). Some biochemical biomarkers (Bap-hydroxidase BPH, NADPH cytochrome c reductase, SOD, CAT, GPX) and LMS were investigated together with PCBs, PAHs and DDTs in sediments by Nasci et al. (2000). The chemical results showed a bioaccumulation activity from sediment in clams translocated to the most polluted area, and, as for biological effects, the only marked reduction of LMS in organisms from the same site. CAT and SOD were assessed in clams by Matozzo et al. (2012) in one reference site near the Lido inlet and one close to the northern inner border influenced by riverine inputs. Higher CAT and SOD activities were generally found in gills and digestive glands of clams from the reference site, highlighting higher efficiency of clams from this site to cope with oxidative stress than bivalves from the inner area, at the same time suggesting possible adaptation to the chronic exposure to high/moderate levels of contamination.

Inhibition of AChE activity in gills and induction of vitellogenin-like protein in hemolymph and digestive glands were evaluated in three monitoring surveys performed in 2003-2004 and 2007, when various lagoon sites were monitored including areas nearby with effluents from a sewage treatment plant, and highly polluted industrial and reference sites (Matozzo et al., 2005; Matozzo et al. 2010; Munari et al., 2010; Matozzo et al., 2012). Although inhibition of AChE activity was observed in clams from the most polluted sites of the central basin (such as Marghera, San Servolo and Fusina), the authors hypothesized the influence of other pollutants than organophosphates and carbamates in inhibiting the enzyme activities, the negative influence of pesticides from agricultural sources being excluded for these sites. The highest Vg-like protein levels were recorded in females from the reference site located in the southern basin (Cà Roman) and in the industrial area (Marghera), and in male clams collected locallynear sites affected by urban and industrial impacts. The authors supposed the influence of xenoestrogens that probably originated from the industrial area of Porto Marghera and the nearby sewage treatment plant, by sewage effluents from the city of Chioggia at Cà Roman, and by municipal untreated sewage from Lido Island and in the Venice city center.

MTs were evaluated in the study of Moschino et al. (2012) with results generally quite similar throughout the study areas (eight sites distributed in the four basins) and not reflecting the spatial variability of trace metals detected in both sediments and clam tissues. On the contrary, the results of this study showed a clear influence of gonadal development in seasonal metallothionein trends.

LMS in clams were evaluated in organisms collected in 1996 and in 1998 in the study of Da Ros and Nesto (2005) where a polluted site was close to the points of effluent discharge from the industrial zone. Two reference sites from the northern basin (Palude della Rosa and Val Dogà) were monitored, showing that LMS values were always significantly lower in samples from the polluted site than in the reference ones. Other lysosomal alterations (NL, LF and L/V) were assessed in clams seasonally sampled in various sites of the lagoon, equally distributed in the four basins, in the period 2003-2005 (Moschino et al., 2011; Moschino et al., 2012). The results highlighted a good general discrimination capability of NL and L/V only between the reference and polluted sites (survey performed in June), whereas low responsiveness was detected in lipofuscin accumulations in all the sampling campaigns. This response showed a high spatial variability, which however did not correlate to contaminants in sediments and tissues. In particular, in June 2004 and 2005 even though a higher accumulation was observed at sites opposite the industrial zone, the highest lipofuscin value was detected at the relatively unpolluted site located nearthe Lido inlet (S. Erasmo). These results were contradictory to those reported in the previous study of Moschino et al. (2010), which observed a clear-cut response in lipofuscin content in clams collected at Porto Marghera in comparison to organisms from a farming area.

Autometallography was evaluated in clams only in one study, showing that this biomarker has a good discrimination potential among sites differently affected by heavy metals in the sediments (Marin et al., 2006).

Similar to mussels, the most applied biomarker at organism level has been SOS (7 articles), whereas SFG was detected by Da Ros et al. (1998) and Marin et al. (2001). SOS, however, showed ambiguous results when comparing the various studies, a good responsiveness in relation to pollutant levels being detected only occasionally. The results obtained by Marin et al. (2001, 2006), Munari et al. (2010), Moschino et al. (2011, 2012) Matozzo et al. (2012) suggested that this physiological parameter is strongly influenced in clams by peculiar environmental conditions (i.e. salinity, total chlorophyll concentrations in seawater, grain size and organic matter in sediments) and by endogenous factors (i.e. reproductive conditions).

Recently, the reburrowing behavior of *R. philippinarum* was investigated as a possible biomarker in highlighting different susceptibility to pollutants (Moschino et al. 2010; Moschino et al. 2011; Moschino et al. 2012), giving promising results. In particular, strong differences were observed comparing the reburrowing time of clams collected at the polluted site of Marghera in comparison to that of clams from the farm area (139 min vs. 35 min, respectively; Moschino et al., 2010). Moreover, in Moschino et al. (2012) this response exhibited a strong correlation with most trace metals analyzed in clam tissues.

Zosterisessor ophiocephalus

The grass goby *Z. ophiocephalus* is a Mediterranean endemic species with a distribution mainly confined to the upper Adriatic Sea, the Black Sea, and the Sea of Azov. This cavity nester species, living on soft-bottoms in shallow brackish water and colonizing seagrass meadows in brackish waters, is particularly abundant in the Venice Lagoon. It can live up to 5 years (Malavasi et al., 2004) and exhibits epibenthic and territorial habits. The grass goby is commonly used as an indicator of sediment quality in biomonitoring studies of brackish environments, because it is able to accumulate a wide range of pollutants (Nasci et al., 1991; Livingstone et al., 1995).

The first ecotoxicological studies in using *Z. ophiocephalus* from the Venice Lagoon as bioindicator of the environmental quality were carried out by Nasci et al (1991). A series of laboratory exposure experiments were performed using low concentrations of PCBs (Aroclor 1254) (1-10 μg/l), which revealed Aroclor 1254 was quite effective in inducing MFO enzyme activity, expressed as benzo(a)pyrene hydroxilase activity and Cytocrome P450 content, in the liver of this species, which revealed high levels of PCBs (Nasci et al., 1991). Afterwards, several field studies were carried out in the framework of various national projects:UNESCO project in 1992-1993 (Lasserre and Marzollo, 2000); MURST project in 1997-1999; (MURST, 1999); Corila Research Programmes (Co.Ri.La 2003, 2006).

Antioxidant enzymes were first determined during a field monitoring study just from the beginning of the '90s. CAT activity showed a seasonal variability, ranging from 0.6 to 2.5 mmol min^{-1} g^{-1} ww, and significant differences between samples (only males) were collected in the most impacted area (Porto Marghera, central basin) with respect to a reference site in the northern basin, reflecting the different tissue contaminant levels, whereas no differences were observed in SOD activities, whose values ranged from 189 to 353 units g^{-1} ww (Livingstone et al., 1995). MDA showed generally significantly higher values in fish from the polluted sites of S. Giuliano-Campalto, (7.36-28.58 nmol g^{-1} ww) than the less impacted Sacca Sessola (4.39-29.31 nmol g^{-1} ww), both in the central basin, decreasing values being observed during summer in agreement with the seasonal trend of PCB concentration, although this trend may also be explained by physiological variation in the oxidative status (Nesto et al., 2007). No statistical difference between sites was observed in MDA measured in fish plasma (Pascoli et al., 2011), however, immunohistochemical and western blot analyses allowed the detection of other byproducts of lipid peroxidation, whose immunositivity was mainly localized in melanomacrophage centers of the spleen and kidney.

EROD activity was the biomarker usually used for detecting the response of the MFO system to organic pollutants in *Z. ophiocephalus*. In the framework of the UNESCO project, EROD activity appeared to vary with season and with sex of organism (range 3.0-39.6 pmol min^{-1} mg^{-1} protein). In general, the activity was always higher in fish collected at the polluted site of Porto Marghera than at the less impacted sites. Resulting differences correlated well with the levels of organic pollutants measured in fish (Livingstone et al., 1995), which were related to both the lipid content of tissues (Fossato et al., 2000) and to the organic pollutant level in the sediments (Secco et al, 2005). The high responsiveness of EROD activity in *Z. ophiocephalus* collected from sites representative of the three lagoon basins (northern, central and southern) was further demonstrated in a study carried out in 1999 (Franco et al., 2002). Fish from the central basin showed particularly high levels of EROD activity (up to 2000 pmol min^{-1} mg^{-1} proteins) reflecting the highest concentration of PCBs and dioxin in the surface sediments of this area (Critto and Marcomini, 2001). Other seasonal studies carried out during 2005-2006 confirmed EROD activities as very sensitive biomarkers for PCB contamination in *Z. ophiocephalus* (Nesto et al., 2007). In this study the activity levels results were generally lower than those reported in previous studies (range 6.1-48.8 pmol min^{-1} mg^{-1} proteins), confirming the improvement in the sediment quality asevidenced by chemical analyses of PCBs (Mugnai et al., 2011). The activity was always significantly higher in fish collected from the most polluted sites showing a decreasing trend during the reproductive period (winter months) when its induction is inhibited by estradiol, a hormone produced by mature females (Förlin and Anderson, 1984).

MT content was evaluated in the grass gobies only recently (Nesto et al., 2007), showing values from 21.9 to 66.5 nmol (GSH) g^{-1} ww, generally higher in fish from the polluted area of S. Giuliano-Campalto; however, no relationships with metal tissue concentrations were found contrasting with results reported elsewhere (Roch and McCarter, 1984; Olsson and Haux, 1986; Hylland et al., 1992; Olsvik et al., 2000). This was probably due to antagonistic effects occurring in the lagoon areas, as it is known that organic and metallic pollution may interact, altering the response of MT (Roméo et al., 1997).

In the already cited paper of Nesto et al. (2007), PAH bile metabolites were also measured in Lagoon fish for the first time using the methods described by Aas et al. (2000). Seasonal monitoring campaigns showed that the most abundant bile metabolite was 1-OH naphthalene with similar levels (from 153 to 443 fluorescence units) in fish from all monitored areas in the central basin of the Lagoon, indicating they were diffusely exposed to PAHs. Differently, 3- OH benzo(a)pyrene and 1-OH pyrene showed lower concentration levels, i.e. 2.256-21.57 for 3-OH benzo(a)pyrene and 14.46-86.54 for 1-OH pyrene, being significantly higher in fish collected from the most polluted areas (Nesto et al., 2007). To explain this behavior it has been hypothesized that, due to higher molecular weights and lipophilic characteristics, parental compounds benzo(a)pyrene and pyrene were more easily absorbed by sediment organic matter than naphthalene, thus affecting gobies predominantly (Aas et al., 2000). This is also suggested by the peaks shown by the two metabolites during the breeding season (February), when the fish live deeper in the sediments (Privileggi et al., 1997; Malavasi et al., 2002).

The immune system of *Z. ophiocephalus* was investigated in fish seasonally collected from various areas of the Lagoon (Pulsford et al., 1995). Results showed the increase of phagocytes in the blood of fish sampled from polluted areas, and simultaneously the reduction of lymphocytes and thrombocytes. Moreover, a significant reduction in the phagocytic capacity of isolated spleen adherent cells was recorded in the most polluted fish. The response of phagocytic cells was correlated with exposure to high levels of pesticides, PAHs, PCBs and metals at the most polluted site of Porto Marghera.

Among the various genotoxic biomarkers, DNA adducts were only determined in *Z. ophiocephalus*. Both laboratory and field studies proved that this biomarker is a reliable and sensitive index of biological effective exposure to genotoxic contaminants (Dolcetti et al., 2002; Venier and Zampieron, 2005). In particular, benzo(a)pyrene treated fish and environmental samples collected in differently impacted areas of the Lagoon revealed the presence of a number of DNA adducts in gills, digestive tubes and livers (environmental samples). In general, they were higher in samples from the most polluted areas (3.848-4.049 DNA adducts/10^8 total nucleotides) and lower adducts level were detected in liver than gills (Venier and Zampieron, 2005). Also the paper of Pascoli et al. (2011) showed the highest level of adducts in fish from high-pollution areas in the central basin.

Cellular studies on *Z. ophiocephalus* sampled from differently impacted areas of the Lagoon were performed by Lowe and Da Ros (2000) and Pascoli et al. (2011). The results showed a strong accumulation of neutral lipids in hepatocytes and of melanomacrophages (MMCs) in the spleen and kidney of polluted fish, reflecting the possible impairment of their metabolic function due to the presence of the contaminants.

In the paper of Franco et al. (2002), the use of anFA index was explored as a biomarker in fish collected from the three lagoon basins. Fish from the northern basin showed the highest level of asymmetry, followed by the central and southern ones, and no correlations were found with EROD results. The authors explained the contrasting responses, highlighting that these two biomarkers act at different biological levels and temporal scales and have different sensitivities to environmental stress. It is likely that the contamination level characterizing the central basin may have induced detoxification processes in fish, as revealed by high levels of EROD activity, but that pollution levels were not high enough to affect the morphology of fish. Moreover, the authors hypothesized that the higher values of FA in fish from the northern basin might be a sign of a different type of stress.

The liver somatic index (LSI) and gonad somatic index (GSI) were seasonally determined in *Z. ophiocephalus* from the Venice Lagoon (Nesto et al., 2007). In general, higher LSI values (ranging from 1.6-4.2) were observed in fish from polluted areas of the Lagoon confirming that the presence of pollutants is able to induce an increase of hepatic metabolic activities and to increase the liver weight. That study revealed that both male and female individuals showed a LSI during the reproductive period, suggesting a positive energy flow from the liver to the gonads during the breeding season and a reverse flow in the next period (Malavasi et al., 2004). Also a laboratory exposure experiment revealed increased values of LSI in fish exposed to high PCB concentrations when compared to the controls (Nasci et al., 1991). The gonad somatic index (GSI) were generally higher in winter and spring and lower in summer, showing lower values in males than in females, thus reflecting a different sexual development in the two sexes. Although GSI results were similar between samples collected in the differently impacted areas of the Lagoon, its evaluation is recommended during the monitoring campaign to identify the specific stage of sexual development, because it is known that some MFO enzymes may be depressed by hormones produced by mature female gonads (Förlin and Anderson, 1984).

Carcinus aestuarii

The Mediterranean shore crab (*Carcinus aestuarii*, Decapoda: Portunidae) is a euryhaline and eurytherm species occupying estuarine and lagoon waters of the Mediterranean Sea. This species has been proposed recently as bioindicator for the Lagoon of Venice because it can accumulate contaminants both from the water column and food reaching levels of bioaccumulation higher than those found in bivalves (Micheletti et al., 2007). The biomarkers detected in the Mediterranean shore crab have been measurements at the biochemical levels, i.e. MFO, AChE, EROD and Vg-like proteins. The MFO was investigated only by Fossi et al. (1998),for BPH activity in hepatopancreas and gills from 5 lagoon sites including areas of Chioggia, Porto Marghera and Pellestrina-S.Pietro in Volta, the latter as a reference site. Although the authors found higher values of BPH activity in crabs collected at Chioggia and Porto Marghera, this enzymatic induction probably responded to micro-organic contaminants which affect both areas; possible statistical differences between these sites and the reference site were concealed by the high inter-individual variability of this response in both hepatopancreas and gills. More recently, *C. aestuarii* has been suggested as a possible sentinel organism in a multi-species approach in association with a more classic species such as mussels (Locatello et al., 2009; Munari and Marin, 2010; Ricciardi et al., 2010). In particular, Locatello et al. (2009) proposed *C. estuarii* for biomonitoring large geographical areas, due to its movement capacity, instead of or in combination with sessile species. In these three studies, EROD, AChE and Vg-like proteins were evaluated in various lagoon sites, always including Ca' Roman, Valle di Brenta (located near Chioggia), Porto Marghera and Palude del Monte (a licensed area for clam culture in the northern basin). EROD activity measured in hepatopancreas did not reveal significant differences among sites during any sampling campaigns, questioning the efficacy of this response in this species. Inhibition of AChE activitiy was evaluated in both gills and hemolymph, but results were not directly related to organophosphorus and carbonates (which were not analyzed during the studies). In particular, Locatello et a. (2009) reported significant differences when comparing AChE activity in hemolymph in Palude del Monte and Porto Marghera with Valle di Brenta (Chioggia), whereas no differences were found in gills (from the lowest mean value of 3.29 to the highest of 4.48 nmol/min/mg prot). The authors hypothesized that the presence of other pollutants, such as heavy metals, explain the inhibition of activities in crabs from the industrial area and the near location of a riverine mouth for those from Palude del Monte, which is not directly influenced by pesticide sources. Ricciardi et al. (2010), although reporting significant differences among sites when comparing AChE activity detected in both hemolymph and gills, stressed that AChE activity measured in hemolymph was not significant as a descriptor of chemical pollution as demonstrated by the Distance-based multivariate multiple regression.

The induction of vitellogenin-like proteins was evaluated in hemolymph, digestive glands and gonads of crabs (Locatello et al., 2009; Munari and Marin, 2010; Ricciardi et al., 2010). The significantly higher values found in both digestive glands and hemolymph at Palude del Monte were possibly related to the higher impact by xenoestrogenic compounds, particularly pesticides (Locatello et al., 2009). Ricciardi et al. (2010), comparing the values detected in gonads and hepatopancreas, observed an inverse trend suggesting different timing in the responses between these two tissues. In particular, vitellogenin-like protein levels were lower in crab digestive glands and higher in gonads from sites with clear-cut PAH and PCB pollution, i.e. San Servolo-Sacca Sessola (26.5 mg PO4/mg prot in digestive glands and 239.8 mg PO4/mg prot), Fusina-Marghera (14 mg PO4/mg prot in digestive glands and 39 mg PO4/mg prot) and Dese (8.3 mg PO4/mg prot in digestive glands and 362 mg PO4/mg prot), all areas included in central basin. The same authors, however, stressed that there is no definitive evidence concerning tissues involved in crustacean vitellogenin synthesis, hypothesizing that in *C. aestuarii* males the digestive gland could be the production site, the hemolymph the carrier tissue, and the gonad the target tissue.

CONCLUSION

In the previous sections we have reported the main findings obtained by the biomarker approach deployed over the last years on various invertebrate and vertebrate species living in the Lagoon of Venice. The various investigations carried out in this complex environment were aimed basically at the same ultimate goal, i.e. to validate the use of exposure/effect biomarkers in field surveys, thus providing basic supplement information to the most popular chemical monitoring approach. Over the years, however, the need to adopt complementary biological tools has become more and more evident, in order to reach a better comprehension of the effects caused at sub-organism/organisms levels by the mixture of pollutants, often likely to remain unknown, which nowadays affect the Lagoon as well as other similarly urbanized coastal/estuarine environments. We are aware that biological-effect techniques, such as those adopted for biomarker analyses, may be questionable because of the lack of their clear ecological meaning. However it is quite reasonable to affirm that monitoring biological impairments at low levels of biological organization may offer an invaluable tool to detect early signs of possible negative effects, thus contributing to direct management actions towards the most challenging anthropogenic changes. In the case of the Lagoon, long-term and systematic monitoring surveys and biomarkers have not yet been incorporated into any institutional monitoring program. This makes it particularly complex to evaluate the obtained results as a whole, although some general conclusions may be outlined. First of all, most of the biological impairments evidenced by the deployed biomarkers confirm most of the chemical analyses of the sediments performed in the Lagoon of Venice, which at present highlight a low–medium and quite uniform general redistribution of pollutants, with a few hotspots persisting in the central inner areas, towards the industrial district still near its north-east edge.

Secondly, the seasonal fluctuations of endogenous and external parameters (linked to reproductivecycle, water temperature, food supply) have been recognized to have a strong influence on baseline physiological processes, particularly in mussels and clams, and their effects on several biomarker responses must be pinpointed. In particular, episodes of high water temperatures, like the one observed during the summer of 2003, might be carefully evaluated, considering that their likely increasing frequency over the next years will induce the shift of physiological cycle patterns. Also the high variability of the physical factors typically affecting estuarine environments can be heightened by unexpected and unusual atmospheric conditions causing further endogenous stress and abnormal organism reactions.

The organisms most often used for monitoring the Lagoon of Venice environment belong to the class of bivalvia, and specifically are the mussel *M. galloprovincialis* (14 papers) and the clam *R. philippinarum* (13 papers), whereas the use of grass goby *Z. ophiocephalus* (6 papers) and crab *Carcinus estuarii* (4 papers) is used less. The suitability of *Mytilus sp.* in evaluating environmental quality, particularly referring to the water column, is confirmed also for the Lagoon of Venice. In particular, the most responsive and valuable biomarkers are those highlighting genetic alterations (DNA adducts and micronucleous frequency), the responses at the cellular and tissue level, such as neutral lipid accumulation and lysosomal enlargement (only in the warm season), as well as the survival in air test at an organismic level. *R. philippinarum* has been employed as an indicator of the water-sediment interface but its use is still controversial. Biomarker results on this species are generally not linked to contamination levels, as for example for CAT, SOD and LF, and also when considering exposure biomarkers, such as AChE and Vg-like proteins, the correlation with the specific class of contaminants which theoretically affected these responses, is weak or not performed. The grass goby *Z. ophiocephalus* has been occasionally used as an indicator of water and sediments and its results are particularly responsive when considering DNA adducts at the genetic level, and AchE inibition and EROD activity at biochemical levels. Only recently, some studies have focused attention on the possibility ofusing other classes of invertebrates than bivalves, such as the crab *C. estuarii*, as a sentinel species. In this context, only biochemical biomarkers have been evaluated, however obtaining confused results, particularly when the responsiveness of AChE activity and Vg-like protein induction evaluated on different tissues was compared among the various studies.

REFERENCES

Aas E., Beyer J., Goksoyr A., 2000. Fixed wavelength fluorescence (FF) of bile as monitoring tool for polyaromatic hydrocarbon exposure in fish: an evaluation of compound specificity, inner filter effect and signal interpretation. *Biomarker* 5, 9–23.

Allenbach D. M., 2011. Fluctuating asymmetry and exogenous stress in fishes: a review. *Reviews in Fish Biology and Fisheries* 21, 355-376.

Amiard J.C., Amiard-Triquet C., Barka S., Pellerin J., Rainbow P.S., 2006. Metallothioneins in aquatic invertebrates: Their role in metal detoxification and their use as biomarkers. *Aquatic Toxicology* 76, 160-202.

Bancroft J.D., 1967. *An introduction to histochemical technique*. London: Butterworths: p. 268.

Bayne B.L., Newell R.C., 1983. Physiological energetics of marine molluscs. In: The Mollusca, vol. 4, Wilburg K.M., Saleuddinn A.S.M. (Eds) London: *Academic Press*, pp. 407-515.

Bayne B.L., Widdows J., 1978. The physiological ecology of two populations of *Mytilus edulis* L. *Oecologia,* 37, 137-162.

Bellucci L.G., Frignani M., Raccanelli S., Carraro C., 2000. Polychlorinated dibenzo-p-dioxins and dibenzofurans in surficial sediments of the Venice Lagoon (Italy). *Marine Pollution Bulletin* 40, 65-76.

Blaise C., Gagné F., Pellerin J., Hansen P.D., 1999. Determination of vitellogenin-like properties in Mya arenaria hemolymph (Saguenay Fjord, Canada): a potential biomarker for endocrine disruption. *Environmental Toxicology* 14, 455–65.

Bolognesi C, Hayashi M., 2011. *Micronucleus assay in aquatic animals. Mutagenesis*, 26 205-13.

Bolognesi C., Degan P., 2001. Genotoxicity biomarkers in aquatic organisms as indicators of carcinogenic marine pollutants. In: Talence, Ph. Garrigues, H. Barth, C.H. Walker and J.F. Narbonne (Eds) *Elsevier Science,* B.V. pp. 550.

Brehmer P., Chi T. D., Laugier T., Galgani F., Laloe F., Darnaude A. M., Fiandrino A., Mouillot D., 2011. Field investigations and multi-indicators for shallow water lagoon management: perspective for societal benefit. *Aquatic Conservation-Marine and Freshwater Ecosystems* 21, 728-742.

Bucheli T.D., Fent K., 1995. Induction of cytocrome P450 as a biomarker for environmental contamination in aquatic ecosystems. *Critical Review of Environmental Sciences and Technology* 25, 201-268.

Burke M.D., Mayer R.T., 1974. Ethoxyresorufine: direct fluorimetric assay of a microsomal O-dealkylation, which is preferentially inducible by 3-methylcholanthrene. *Drug Metabolism and Disposition* 2, 583–588.

Cajaraville M.P., Abascal I., 1995. Etxeberria M., Marigomez I., 1995. Lysosomes as Cellular Markers of Environmental-Pollution - Time-Dependent and Dose-Dependent Responses of the Digestive Lysosomal System of Mussels after Petroleum Hydrocarbon Exposure. *Environmental Toxicology and Water Quality 10*, 1-8.

Cajaraville M.P., Bebianno M.J., Blasco J., Porte C., Sarasquete C., Viarengo A., 2000. The use of biomarkers to assess the impact of pollution in coastal environments of the Iberian Peninsula: a practical approach. *Science of the Total Environment* 247, 295-311.

Cajaraville, M. P., Marigomez J.A., Angulo E., 1991. Automated Measurement of Lysosomal Structure Alterations in Oocytes of Mussels Exposed to Petroleum-Hydrocarbons. *Archives of Environmental Contamination and Toxicology* 21, 395-400.

Chapman P. M., Wang F.Y., 2001. Assessing sediment contamination in estuaries. *Environmental Toxicology and Chemistry* 20, 3-22.

Chong, K., Wang W.X., 2000. Bioavailability of sediment-bound Cd, Cr and Zn to the green mussel Perna viridis and the Manila clam *Ruditapes philippinarum. Journal of Experimental Marine Biology and Ecology* 255, 75-92.

Co.Ri.La 2003.Progetto: "Efficienza del metabolismo lagunare" *Coordinatore: Prof. A. Marcomini. Relazione finale.* Venice, Italy.

Co.Ri.La 2006. Progetto: "Indicatori e indici di qualità ambientale per la Laguna di Venezia: sviluppo e applicazione di una metodologia integrata per la determinazione di Indici di Qualità per la gestione ambientale della Laguna di Venezia" Coordinatore: *Prof. A. Marcomini. Relazione finale.* Venice, Italy.

Collavini F., Bettiol C., Zaggia L., Zonta R., 2005. Pollutant loads from the drainage basin to the Venice Lagoon (Italy). *Environment International 31*, 939-947.

Corsi I., Mariottini M., Sensini C., Lancini L., Focardi S., 2003. Cytochrome P450, acetylcholinesterase and gonadal histology for evaluating contaminant exposure levels in fishes from a highly eutrophic brackish ecosystem: the Orbetello Lagoon, Italy. *Marine Pollution Bulletin,* 46 (2), 203-212.

Cossu C., Doyotte A., Jacquin M.C., Babut M., Exinger A., Vasseur P., 1997. Glutathione reductase, selenium-dependent glutathione peroxidase, glutathione levels, and lipid peroxidation in freshwater bivalves, *Unio tumidus*, as biomarkers of aquatic contamination in field studies. *Ecotoxicology and Environmental Safety* 38, 122–131.

Critto A., Marcomini A., 2001. Ecological Risk and Chemical Pollution in the Lagoon of Venice (in Italian).Cafoscarina Editrice, Venice, Italy, 147 pp.

Da Ros L., Marin M.G., Nesto N., Ford S.E., 1998. Preliminary results of a field study on some stress-related parameters in *Tapes philippinarum naturally* infected by the protozoan Perkinsus sp. *Marine Environmental Research* 46, 249-252.

Da Ros L., Nesto N., 2005. Cellular alterations in *Mytilus galloprovincialis* (LMK) and Tapes philippinarum (Adams and Reeve, 1850) as biomarkers of environmental stress: field studies in the lagoon of Venice (Italy). *Environment International* 31,1078–88.

Da Ros, L., Meneghetti, F., Nasci C., 2002. Field application of lysosomal destabilisation indices in the mussel *Mytilus galloprovincialis*: biomonitoring and transplantation in the Lagoon of Venice (north-east Italy). *Marine Environmental Research* 54(3-5): 817-822.

Danscher G., 1981. Histochemical demonstration of heavy metals.A revised version of the sulphide silver method, suitable for both light and electron microscopy. *Histochemistry* 71, 1-16.

Danscher G., 1984. Autometallography.A new technique for light and electron microscopic visualization of metals in biological tissues (gold, silver, metal sulphides and metal selenides). *Histochemistry* 81, 331–5.

Danscher G., 1991. Histochemical tracing of zinc, mercury, silver and gold. In: GraumannW, Drukker J, editors. *Histo- and cytochemistry as a tool in environmental toxicology.* Progress in histo- and cytochemistry. Stuttgard: Fischer Verlag, pp. 273–85.

Danscher G., Montagnese C., 1994. Autometallographical localization of synaptic vesicular zinc, and lysosomal gold, silver and mercury. *Journal of Histotechnology* 17, 15–22.

Danscher G., Rytter Norgaard J.O., Baatrup E., 1987. Autometallography: tissue metals demonstrated by a silver enhancement kit. *Histochemistry* 86, 465–9.

de Fur P.L., Crane M., Ingersoll C., Tattersfield L., 1999. Endocrine Disruption in Invertebrates: Endocrinology, Testing and Assessment. Society of Environmental Toxicology and Chemistry Press, Pensacola, Florida, USA.

de Kock W.C., Kramer K.J.M., 1994. Active biomonitoring (ABM) by translocation of bivalve molluscs. In: Kramer, K.J.M. (ed.), Biomonitoring of Coastal Waters and Estuaries. CRC, Boca Raton, pp. 51–84.

de Zwaan A., Eertman R.H.M., 1996. Anoxic or aerial survival of bivalves and other euryoxic invertebrates as a useful response to environmental stress – a comprehensive review. *Comparative Biochemistry and Physiology* 113 C, 299-312.

Dolcetti L., Dalla Zuanna L., Venier P. 2002. DNA adducts in mussels and fish exposed to bulky genotoxic compounds. *Marine Environmental Research* 54, 481-486.

Domouhtsidou G.P., Dimitriadis V.K., 2001. Lysosomal and lipid alterations in the digestive gland of mussels, Mytilus galloprovincialis (L.) as biomarkers of environmental stress. *Environmental Pollution* 115, 123-137.

Eertman R.H.M., Wagenvoort A.J., Hummel H., Smaal A.C., 1993. "Survival in air" of the blue mussel *Mytilus edulis* L. as a sensitive response to pollution-induced environmental stress. *Journal of Experimental Marine Biology and Ecology* 70, 179-195.

Fitzpatrick P.J., Sheehan D., Livingstone D.R., 1995. Proc. 7th PRIMO. *Marine Environmental Research 39*, 241-244.

Förlin L., Andersson T., 1984. Influence of biological and environmental factors on hepatic steroid and xenobiotic metabolism in fish: interaction with PCB and b-naphthoflavone. *Marine Environmental Research* 14, 47–58.

Förlin L., Lemaire P., Livingstone D. R. (1995). Comparative studies of hepatic xenobiotic metabolizing and antioxidant enzymes in different fish species. *Marine Environmental* Research, 39, 201–204.

Fossato V.U. Campesan G., Craboledda L., Dolci F., Stocco G., 2000. Persistent chimica pollutants in mussels and gobies from the Venice Lagoon. In: The Venice Lagoon Ecosystem. Inputs and Interactions Between Land and Sea. Lasserre P., Marzollo A. (eds), UNESCO, *Paris and The Parthenon Publishing Group*, 421-437.

Fossi M.C., Savelli C., Casini S., 1998. Mixed function oxidase induction in *Carcinus aestuarii.* Field and experimental studies for the evaluation of toxicological risk due to Mediterranean contaminants. *Comparative Biochemistry and Physiology C-Toxicology* and Pharmacology 121, 321-331.

Franco A., Malavasi S., Pranovi F., Nasci C., Torricelli P., 2002. Ethoxyresorufin o-deethylase (EROD) activity and fluctuating asymmetry (FA) in *Zosterisessor ophiocephalus* (Teleostei, Gobiidae) as indicators of environmental stress in the Venice Lagoon. *Journal of Aquatic Ecosystem Stress and Recovery*, 9, 239-247.

Frignani M., Bellucci L., Langone L., Muntau H., 1997. Metal fluxes to the sediments of the northern Venice Lagoon. *Marine Chemistry* 58, 275-292.

Frignani M., Bellucci L.G., Favotto M., Albertazzi S., 2005. Pollution historical trends as recorded by sediments at selected sites of the Venice Lagoon. *Environment International* 31, 1011-1022.

Fulton M.H., Key P.B., 2001. Acetylcholinesterase inhibition in estuarine fish and invertebrates as an indicator of organophosphorus insecticide exposure and effects. *Environmental Toxicology and Chemistry*, 20, 37-45.

Gagnè F., Marcogliese D.J., Blaise C., Gendron A.D., 2001. Occurrence of compounds estrogenic to freshwater mussels in surface waters in an urban area. *Environmental Toxicology* 16, 260–268.

Galloway T.S., Millward N., Browne M.A., Depledge M.H., 2002. Rapid assessment of organophosphorous/carbamate exposure in the bivalve mollusk *Mytilus edulis* using combined esterase activities as biomarkers. *Aquatic Toxicology* 61, 169–180.

Gérard-Monnier D., Erdelmeier I., Régnard K., Moze-Henry N., Yadan J.C., Chaudiére J., 1998. Reactions of 1-methyl-2-phenylindole with malondialdehyde and 4-hydroxyalkenals. Analytical applications to a colorimetric assay of lipid peroxidation. *Chemical Research in Toxicology,* 11, 1176–1183.

Giusti L., Zhang H., 2002. Heavy metals and arsenic in sediments, mussels and marine water from Murano (Venice, Italy). *Environmental Geochemistry and Health* 24, 47-65.

Goksøyr A., Förlin L., 1992. The cytocrome P450 system in fish, aquatic toxicology and environmental monitoring. *Aquatic Toxicology*, 22, 287-312.

Guerlet E., Ledy K., Meyer A., Giamberini L., 2007. Towards a validation of a cellular biomarker suite in native and transplanted zebra mussels: A two-year integrative field study of seasonal and pollution-induced variations. *Aquatic Toxicology* 81, 377-388.

Guerzoni S., Raccanelli S., 2004. The sick lagoon. Dioxin and other Persistent Organic Pollutants (POPs) in the Lagoon of Venice. Lib. Ed. Cafoscarina, Venice. *ISBN* 88-7543-043-8.

Guerzoni S., Rossini P., Sarretta A., Raccanelli S., Ferrari G., Molinaroli E., 2007. POPs in the lagoon of Venice: budgets and pathways. *Chemosphere* 67, 1776-1785.

Habig W.H., Pabst M.J., Jakoby W.B., 1974. Glutathione S-transferases. *The Journal of Biological Chemistry* 249, 7130–7139.

Herwig H.J., Brands F., Kruitwagen E., Zandee D.I., 1989. Bioaccumulation and histochemical localization of cadmium in Dreissena polymorpha exposed to cadmium chloride. *Aquatic Toxicology* 15, 269–86.

Hugget R.J., Kimerle R.A., Mehrle P.M., Bergman H.L. (Eds.), 1992. Biochemical, Physiological and Histological Markers of Anthropogenic Stress. Lewis Publishers, Boca Raton, FL,pp. 356.

Hylland K., Haux C., Hogstrand C., 1992. Hepatic metallothionein and heavy metals in dab *Limanda limanda* from the German bight. *Marine Ecology Progress Series* 91, 89–96.

Irato P., Santovito G., Cassini A., Piccinni E., Albergoni V., 2003. Metal accumulation and binding protein induction in *Mytilus galloprovincialis*, *Scapharca inaequivalvis*, and *Tapes philippinarum* from the lagoon of Venice. *Archives of Environmental Contamination and Toxicology* 44, 476-484.

Ischiropoulos H., Zhu L., Chen J., Tsai M., Martin J.C., Smith C.D., Beckman J.S., 2002. Peroxynitrite-mediated tyrosine nitration catalyzed by superoxide dismutase. *Archives of Biochemistry and Biophysiology 298*, 431–437.

Ji J., Choi H.J., Ahn I., 2006. Evaluation of Manila clam *Ruditapes philippinarum* as a sentinel species for metal pollution monitoring in estuarine tidal flats of Korea: Effects of size, sex, and spawning on baseline accumulation. *Marine Pollution Bulletin* 52, 447–468.

Kanaya G., Nobata E., Toya T., Kikuchi E., 2005. Effects of different feeding habits of three bivalve species on sediment characteristics and benthic diatom abundance. *Marine Ecology Progress-Series* 299, 67-78.

Kennish M.J., Paerl H.W, 2010. Coastal Lagoons Critical Habitats of Environmental Change. In: Coastal Lagoons Critical Habitats of Environmental Change, Keddish M.J. and Pearl H.W (Eds), CRC Press, London, pp. 1-15.

Kohler A., Wahl E., Soffker K., 2002. Functional and morphological changes of lysosomes as prognostic biomarkers of toxic liver injury in a marine flatfish (Platichthys flesus (L.)). *Environmental Toxicology and Chemistry* 21, 2434-2444.

Lasserre P., Marzollo A., 2000. *The Venice Lagoon ecosystem. Inputs and interactions between land and sea.* UNESCO, Paris and The Parthenon Publishing Group, Lancs. and N.Y. Press.

Lasserre P., Marzollo A., 2000. The Venice Lagoon Ecosystem. *Inputs and Interactions Between Land and Sea.* UNESCO, Paris and The Parthenon Publishing Group, Lancs and NY Press, 508 pp

Li Q., Osada M., Suzuki R., Mori K., 1998. Changes in vitellin during oogenesis and effect of estradiol 17-β on vitellogenesis in the Pacific oyster *Crassostrea gigas. Invertebrate Reproductive Development* 33, 87–93.

Li Y., Yu Z. M., Song X.X., Mu Q.L., 2006. Trace metal concentrations in suspended particles, sediments and clams (*Ruditapes philippinarum*) from Jiaozhou Bay of China. *Environmental Monitoring and Assessment* 121, 491-501.

Livingstone D.R., Archibald S., Chipman J.K., Marsh J.W., 1992a. Antioxidant enzymes in liver of dab *Limanda limanda* from the North Sea. *Marine Ecology progress Series*, 91, 97-104.

Livingstone D.R., Lamaire P., Matthews A., peters l.D., Porte C., Fitzpatrick P.J., Förlin L., Nasci C., Fossato V.U., Wotton N., Goldfarb P., 1995. Assessment of the impact of organic pollutants on goby (*Zosterisessor ophiocephalus*) and mussel (*Mytilus galloprovincialis*) from the Venice lagoon, Italy: biochemical studies. *Marine Environmental Research,* 39, 235-240.

Livingstone D.R., Lips F., Garcia Martinez P., Pipe R.K., 1992b. Antioxidant enzymes in the digestive gland of the common mussel *Mytilus edulis. Marine Biology* 112, 265-276.

Livingstone D.R., Moore M.N., Lowe D.M., Nasci C., Farrar S.V., 1985. Responses of the cytocrome P-450 monooxygenase system to diesel oil in the common mussel, *Mytilus edulis* L. and the periwinkle, *Littorina littorea* L., *Aquatic Toxicology*, 7 (1-2), 79-91.

Livingstone D.R., Nasci C., 2000. Biotransformation and antioxidant enzymes as potential biomarkers of contaminant exposure in goby (*Zosterisessor ophiocephalus*) and mussel (*Mytilus galloprovincialis*) from the Venice Lagoon. In: The Venice Lagoon ecosystem. Inputs and interactions between land and sea. UNESCO, Paris and The Parthenon Publishing Group, Lancs. and N.Y. Press, 357-373.

Locatello L., Matozzo V., Marin M.G., 2009. Biomarker responses in the crab *Carcinus aestuarii* to assess environmental pollution in the Lagoon of Venice (Italy). *Ecotoxicology* 18, 869-877.

Lowe D.M., Fossato V.U., 2000. The influence of environmental contaminants on lysosomal activity in the digestive cells of mussels (*Mytilus galloprovincialis*) from the Venice Lagoon." *Aquatic Toxicology* 48, 75-85.

Lowe D.M., Fossato V.U., Depledge M.H., 1995. Contaminant-induced lysosomal membrane damage in blood cells of mussels *Mytilus galloprovincialis* from the Venice Lagoon: An in vitro study." *Marine Ecology-Progress Series* 129, 189-196.

Malavasi S., Coppola J., Pranovi, F., Granzotto A., Franco A., Torricelli P., 2002. Habitat riproduttivo di *Zosterisessor ophiocephalus*Pall. (Pisces, Gobiidae) in Laguna di Venezia e osservazioni sulle caratteristiche dei riproduttori. Lavori della Societa` Veneta di Scienze Naturali 27, 47–56.

Malavasi S., Fiorin R., Franco A., Torricelli p., 2004. Somatic Energy storage and reproductive investment in the grass goby *Zosterisessor ophiocephalus. Journal of Marine Biological Association of the U.K.*, 84, 455-459.

Marigomez I., Baybay-Villacorta L., 2003. Pollutant-specific and general lysosomal responses in digestive cells of mussels exposed to model organic chemicals. *Aquatic Toxicology* 64, 235-257.

Marigomez I., Orbea A., Olabarrieta I., Etxeberria M., Cajaraville M.P., 1996. Structural changes in the digestive lysosomal system of sentinel mussels as biomarkers of environmental stress in mussel-watch programmes. *Comparative Biochemistry and Physiology C-Pharmacology Toxicology and Endocrinology* 113, 291-297.

Marin M.G., Boscolo R., Cella A., DegettoS., Da Ros l:, 2006. Field validation of autometallographical black silver deposit (BSD) extent in three bivalve species of the Lagoon of Venice, Italy (Mytilus galloprovincialis, Tapes philippinarum, Scapharca inaequivalvis) for metal bioavailability assessment. *Science of the Total Environment* 371, 156-167.

Marin, M.G., Nesto N., Da Ros L., 2001. Evaluation of biological stress indices in *Tapes philippinarum* from the Lagoon of Venice, through monitoring of natural populations and transplantation experiments. In: Structures and Processes in the Mediterranean Ecosystem, Faranda, F.M., L. Guglielmo and G. Spezie (Eds), Springer-Verlag, Italia, Milan, pp. 91-94.

Marnett L.J., 1999. Lipid peroxidation – DNA damage by malondialdehyde. *Mutation Research*, 424, 83–95.

Martin-Diaz M.L., Blasco J., Sales D., DelValls T.A., 2007. Biomarkers study for sediment quality assessment in Spanish ports using the crab *Carcinus maenas* and the clam *Ruditapes philippinarum. Archives of Environmental Contamination and Toxicology* 53, 66-76.

Matozzo V., Binelli A., Parolini M., Locatello L., Marin M.G., 2010. Biomarker responses and contamination levels in the clam *Ruditapes philippinarum* for biomonitoring the Lagoon of Venice (Italy). *Journal of Environmental Monitoring* 12, 776-786.

Matozzo V., Binelli A., Parolini M., Previato M., Masiero L., Finos L., Bressan M., Marin M.G. 2012. Biomarker responses in the clam *Ruditapes philippinarum* and contamination levels in sediments from seaward and landward sites in the Lagoon of Venice. *Ecological Indicators* 19, 191-205.

Matozzo V., Gagne F., Marin, M.,G., Ricciardi F., Blaise C., 2008. Vitellogenin as a biomarker of exposure to estrogenic compounds in aquatic invertebrates: *A review. Environment International* 34, 531-545.

Matozzo V., Tomei A., Marin M.G., 2005. Acetylcholinesterase as a biomarker of exposure to neurotoxic compounds in the clam *Tapes philippinarum* from the Lagoon of Venice. *Marine Pollution Bulletin* 50, 1686-1693.

Micheletti C., Critto A., Carlon C., Marcomini A.2004. Ecological risk assessment of persistent toxic substances for the clam *Tapes philipinarum* in the Lagoon of Venice, Italy." *Environmental Toxicology and Chemistry* 23, 1575-1582.

Micheletti C., Critto A., Marcomini A., 2007. Assessment of ecological risk from bioaccumulation of PCDD/Fs and dioxin-like PCBs in a coastal lagoon. *Environment International* 33, 45-55.

Moore M.N., 1976. Cytochemical demonstration of latency of lysosomal hydrolases in digestive cells of common mussel, *Mytilus edulis*, and changes induced by thermal-stress. *Cell and Tissue Research* 175, 279-287.

Moore M.N., 1988. Cytochemical Responses of the Lysosomal System and Nadph-Ferrihemoprotein Reductase in Molluscan Digestive Cells to Environmental and Experimental Exposure to Xenobiotics. *Marine Ecology-Progress Series* 46, 81-89.

Moore M.N., Allen J.I., McVeigh A., 2006. Environmental prognostics: An integrated model supporting lysosomal stress responses as predictive biomarkers of animal health status. *Marine Environmental Research* 61, 278-304.

Moore M.N., Willows R.I., 1998.A model for cellular uptake and intracellular behaviour of particulate-bound micropollutants. *Marine Environmental Research* 46, 509-514.

Moschino V., Delaney E., Da Ros L. 2012. Assessing the significance of *Ruditapes philippinarum* as a sentinel for sediment pollution: Bioaccumulation and biomarker responses. *Environmental Pollution* 171, 52-60.

Moschino V., Delaney E., Meneghetti F., Da Ros L., 2011. Biomonitoring approach with mussel *Mytilus galloprovincialis* (Lmk) and clam *Ruditapes philippinarum* (Adams and Reeve, 1850) in the Lagoon of Venice. *Environmental Monitoring and Assessment* 177, 649-663.

Moschino V., Meneghetti F., Da Ros, 2010. Use of biomarkers to assess the welfare of the edible clam, *Ruditapes philippinarum*: may it be a tool for proving areas of origin? *Aquaculture International* 18, 327-337.

Mugnai C., Giuliani S., Bellucci L.G., Carraro C., Favotto M., Frignani M., 2011. Polychlorinated biphenyls in two salt marsh sediments of the Venice Lagoon. *Environmental Monitoring and Assessment* 181, 243–254.

Munari M., Matozzo V., Marin M.G., 2010. Vitellogenin induction in the clam, *Ruditapes philippinarum*, and the crab, *Carcinus aestuarii*, from the Lagoon of Venice: a comparative study. *Fresenius Environmental Bulletin* 19(10A): 2312-2317.

Murphy S.D., 1986. Pesticides. In: Doul, J., Klassen, C.D., Anders, M.O. (Eds.), The Basic Science of Poisons. Macmillan, New York, 519-581.

MURST (Ministero dell'Università e della ricerca scientifica e tecnologica) 1999."Conservazione della biodiversità e gestione sostenibile dei biotopi salmastri delle coste italiane". Linea di ricerca: studio di comunità ittiche e di invertebrati bentonici della laguna di Venezia ai fini gestionali e di salvaguardia ambientale.

Nasci C., campesan G., Fossato V.U., Tallandini L., Turchetto M., 1991. Induction of cytocrome P-450 and mixed function oxygenase activity by low concentrations of polychlorinated biphenyls in marine fish *Zosterisessor ophiocephalus* (Pall). *Aquatic Toxicology*, 19, 281-290.

Nasci C., Da Ros L., Nesto N., Sperni L., Passarini F., Pavoni B., 2000. Biochemical and histochemical responses to environmental contaminants in clam, *Tapes philippinarum*, transplanted to different polluted areas of Venice Lagoon, Italy. *Marine Environmental Research* 50, 425-430.

Nasci C., Nesto N., Monteduro R.A., Da Ros L., 2002. Field application of biochemical markers and a physiological index in the mussel, *Mytilus galloprovincialis*: transplantation and biomonitoring studies in the lagoon of Venice (NE Italy). *Marine Environmental Research* 54, 811-816.

Nesto N., Romano S., Moschino V., Mauri M., Da Ros L., 2007. Bioaccumulation and biomarker responses of trace metals and micro-organic pollutants in mussels and fish from the Lagoon of Venice, Italy. *Marine Pollution Bulletin* 55, 469-484.

Nesto N., Romano S., Moschino V., Mauri M., Da Ros L., 2007. Bioaccumulation and biomarker responses of trace metals and microorganic pollutants in mussels and fish from the lagoon of Venice, Italy. *Marine Pollution Bulletin,* 55: 469-484.

Nesto, N., Bertoldo M., Nasci C., Da Ros L., 2004. Spatial and temporal variation of biomarkers in mussels (*Mytilus galloprovincialis*) from the Lagoon of Venice, Italy. *Marine Environmental Research* 58, 287-291.

Olsson P.E., Haux C., 1986. Increased hepatic metallothioneins content correlates to cadmium accumulation in environmental exposed perch (*Perca fluviatilus*). *Aquatic Toxicology*, 9, 231–242.

Olsvik P.A., Gundersen A R.A., Zachariassen K.E., 2000. Metal accumulation and metallothionein in two populations of brown trout, *Salmo trutta*, exposed to different natural water environments during run-off episode. *Aquatic Toxicology* 50, 231–242.

Pampanin D.M., Camus L., Gomiero A., Marangon I., Volpato E., Nasci C., 2005c. Susceptibility to oxidative stress of mussels (*Mytilus galloprovincialis*) in the Venice Lagoon (Italy). *Marine Pollution Bulletin* 50, 1548-1557.

Pampanin D.M., Marangon I., Volpato E., Campesan G., Nasci C., 2005a. Stress biomarkers and alkali-labile phosphate level in mussels (*Mytilus galloprovincialis*) collected in the urban area of Venice (Venice Lagoon, Italy). *Environmental Pollution* 136, 103-107.

Pampanin D.M., Volpato E., Marangon I., Nasci C., 2005b. Physiological measurements from native and transplanted mussels (*Mytilus galloprovincialis*) in the canals of Venice. Survival in air and condition index." *Comparative Biochemistry and Physiology Molecular and Integrative Physiology* 140, 41-52.

Pascoli F., Negrato E., Di Giancamillo A., Bertotto D., Domeneghini C., Simontacchia C., Mutinelli F., Radaelli G., 2011. Evaluation of oxidative stress biomarkers in *Zosterisessor ophiocephalus* from the Venice Lagoon, Italy. *Aquatic Toxicology* 101,512–520.

Pavoni B., Donazzolo R., Marcomini A., Degobbis D., Orio A.A., 1987. Historical development of the Venice Lagoon contamination as recorded in radiodated sediment cores. *Marine Pollution Bulletin* 18, 18-24.

Pearson W.H., Woodruff D.L., Sugarman P.C., Olla B.L., 1981. Effects of oiled sediment on predation on the littleneck clam, Protothaca staminea, by the Dungeness crab, *Cancer magister*. Estuarine, *Coastal and Shelf Science* 13, 445-454.

Pellizzato M., Da Ros L., 2005. Clam farming quality as a management tool: A proposal based on recent studies in Northern Adriatic lagoons. *Aquaculture International* 13, 57-66.

Petrovic S., Ozretic B., Krajnovic-Ozretic M., Bobinac D., 2001. Lysosomal membrane stability and metallothioneins in digestive gland of mussels (*Mytilus galloprovincialis* Lam.) as biomarkers in a field study. *Marine Pollution Bulletin* 42, 1373–1378.

Phelps H.L., 1989. Clam burrowing bioassay for estuarine sediment. *Bulletin Of Environmental Contamination and Toxicology* 43, 838-845.

Phelps H.L., Pearson W.H., Hardy J.T., 1985. Clam Burrowing Behavior and Mortality Related to Sediment Copper. *Marine Pollution Bulletin* 16, 309-313.

Phillips D.H., Castegnaro M., 1999. Standardization and validation of DNA adduct post-labelling methods: report of interlaboratory trials and production of recommended protocols. *Mutagenesis,* 14, 301– 15.

Ponti M., Fava F., Abbiati M., 2011. Spatial-temporal variability of epibenthic assemblages on subtidal biogenic reefs in the northern Adriatic Sea. *Marine Biology* 158, 1447-1459.

Pranovi F., Franceschini G., Casale M., Zucchetta M., Torricelli P., Giovanardi O., 2006. An ecological imbalance induced by a non-native species: The Manila clam in the Venice Lagoon. *Biological Invasions* 8, 595-609.

Privileggi N., Ota D., Ferrero E.A., 1997. Embryonic and larval development of the grass goby *Zosterisessor ophiocephalus* (Teleostei, Gobiidae). *Italian Journal of Zoology* 64, 201–207.

Puinean A.M., Labadie P., Hill E.M., Osada M., Kishida M., Nakao R., et al.2006. Laboratory exposure to 17β-estradiol fails to induce vitellogenin and estrogen receptor gene expression in the marine invertebrate *Mytilus edulis*. *Aquatic Toxicology* 79, 376–83.

Pulsford A.L., Thomas M.E., Lemaire-Gony S., Coles J.A, Fossato V.U., Pipe R.K.., 1995. Studies on the immune system of the goby *Zosterisessor ophiocephalus*, from the Venice Lagoon. *Marine Pollution Bulletin* 30 (9), 586-591.

Raccanelli S., Libralato S., Tundo P., 2009. Fate of persistent organic pollutants in the Venice Lagoon: from the environment to human beings through biological exploitation? In: Bahadir, A.M., Duca, G., Eds.,*Role of ecological chemistry in pollution research and sustainable development*. Book Series: NATO Science for Peace and Security Series C - Environmental Security, 15-25.

Regoli F., 2000. Total oxyradical scavenging capacity (TOSC) in polluted and translocated mussels: a predictive biomarker of oxidative stress. *Aquatic Toxicology* 50 (4), 351-361.

Regoli F., Winston G.W., 1999. Quantification of total oxidant scavenging capacity (TOSC) of antioxidants for peroxynitrite, peroxyl radicals and hydroxyl radicals. *Toxicology and Applied Pharmacology* 156, 96–105.

Ricciardi F., Matozzo V., Binelli A., Marin M. G., 2010. Biomarker responses and contamination levels in crabs (*Carcinus aestuarii*) from the Lagoon of Venice: An integrated approach in biomonitoring estuarine environments. *Water Research* 44, 1725-1736.

Rickwood C. J., Galloway T. S. 2004. Acetylcholinesterase inhibition as a biomarker of adverse effect - A study of *Mytilus edulis* exposed to the priority pollutant chlorfenvinphos. *Aquatic Toxicology* 67, 45-56.

Ringwood A.H., Conners D.E., Hoguet J., 1998. Effects of natural and anthropogenic stressors on lysosomal destabilization in oysters *Crassostrea virginica*. *Marine Ecology-Progress Series* 166, 163-171.

Roch M., McCarter J.A., 1984. Hepatic metallothionein production and resistance to heavy metals by rainbow trout (*Salmo gairdneri*) held in a series of contaminated lakes. *Comparative Biochemistry and Physiology* 77C, 77–82.

Roex E.W. M, Keijzers R.; van Gestel C.A.M., 2003. Acetylcholinesterase inhibition and increased food consumption rate in the zebrafish, *Danio rerio*, after chronic exposure to parathion. *Aquatic Toxicology* 64 (4), 451-460.

Roméo M., Cosson R.P., Gnassia-Barelli M., Risso C., Stien X., Lafaurie M., 1997. Metallothionein determination in the liver of the sea bass *Dicentrarchus labrax* treated with copper and B(a)P. *Marine Environmental Research* 44 (3), 275–284.

Romero-Ruiz A, Amezcua O, Rodríguez-Ortega MJ, Muñoz JL, Alhama J, Rodríguez-Ariza A, Gómez-Ariza JL, López-Barea J. 2003 Oxidative stress biomarkers in bivalves transplanted to the Guadalquivir estuary after Aznalcóllar spill. *Environ Toxicol Chem.* 22(1), 92-100.

Ryvolova M., Krizkova S., Adam V., Beklova M., Trnkova L., Hubalek J., Kizek R., 2011. Analytical Methods for Metallothionein Detection. *Current Analytical Chemistry* 7, 243-261.

Secco T., Pellizzato F., Sfriso A., Pavoni B., 2005. The changing state of contamination in the Lagoon of Venice. Part 1: organic pollutants. *Chemosphere* 58, 279–290.

Slooff W., van Kreijl C.F., Baars A.J., 1983. Relative liver weights and xenobiotic-metabolizing enzymes of fish from polluted surface waters in Netherlands. *Aquatic Toxicology* 4, 1-14.

Smaal A.C., Widdows J., 1994. The Scope for Growth in bivalves as an integrated response parameter in biological monitoring. In: Biomonitoring of coastal water and estuaries, Kramer, K.J.M. (Ed), Boca Raton, FL: CRC Press, pp. 247-267.

Solidoro C., Bandelj V., Bernardi Aubry F., Camatti E., Ciavatta S., Cossarini G., Facca C., Franzoi P., Libralato S., Melaku Canu D., Pastres R., Pranovi F., Raicevich S., Socal G., Sfriso A., Sigovini M., Tagliapietra D., Torricelli P., 2010. Response of the Venice lagoon ecosystem to natural and anthropogenic pressures over the last 50 years. In: Kennish, M.J., Paerl, H.W. (Eds.), Coastal Lagoons: *Critical Habitats of EnvironmentalChanges.* CRC Press, London, pp. 483-511.

Soto M., Zaldibar B., Cancio I., Taylor M.G., Turner M., Morgan A.J., et al. 2002. Subcellular distribution of cadmiumand its cellular ligands inmussel digestive gland cells as revealed by combined autometallography and X-ray microprobe analysis. *Histochemistry Journal* 34, 273–80.

Tagliapietra D., Ghirardini A. V., 2006. Notes on coastal lagoon typology in the light of the EU Water Framework Directive: Italy as a case study. *Aquatic Conservation. Marine and Freshwater Ecosystems* 16, 457-467.

Talesa V., Romani R., Antognelli C., Giovannini E., Rosi G., 2001. Soluble and membrane-bound acetylcholinesterase in *Mytilus galloprovincialis* (Pelecypoda: Filibranchia) from the northern Adriatic Sea. *Chem. Biol. Interact.* 134, 151–166.

van der Oost R., Beyer J., Vermeulen N.P.E., 2003. Fish bioaccumulation and biomarkers in environmental risk assessment: a review. *Environmental Toxicology and Pharmacology* 13, 57-149.

Veldhuizen-Tsoerkan M.B., Holwerda D.A., de Bont A.M.T., Smaal A.C., Zandee D.I., 1991. A field study on stress indices in the sea mussel, *Mytilus edulis*: application of the "Stress Approach" in biomonitoring. *Archives of Environmental Contamination and Toxicology* 21, 497-504.

Venier P., Zampieron C., 2005. Evidence of genetic damage in grass gobies and mussels from the Venice lagoon. *Environment International* 31, 1053-1064.

Viarengo A., Burlando B., Ceratto N., Panfoli I., 2000. Antioxidant role of metallothioneins: A comparative overview. *Cellular and Molecular Biology* 46, 407-417.

Viarengo A., Canesi L., Pertica, M., Mancinelli G., Accomando R., Smaal A.C., Orunesu M., 1995. Stress on Stress-Response - a Simple Monitoring Tool in the Assessment of a General Stress Syndrome in Mussels. *Marine Environmental Research* 39, 245-248.

Viarengo A., Dondero F., Pampanin D.M., Fabbri R., Poggi E., Malizia M., Bolognesi C., Perrone E., Gollo E., Cossa G.P., 2007. A biomonitoring study assessing the residual biological effects of pollution caused by the HAVEN wreck on marine organisms in the

Ligurian sea (Italy). *Archives of Environmental Contamination and Toxicology* 53, 607-616.

Viarengo A., Nott J.A., 1993. Mechanisms of heavy metal cation homeostasis in marine invertebrates. *Comparative Biochemistry and Physiology* 104C, 355-372.

Viarengo A., Ponzano E., Dondero F., Fabbri, R., 1997. A simple spectrophotometric method for metallothionein evaluation in marine organisms: An application to Mediterranean and Antarctic molluscs. *Marine Environmental Research* 44, 69-84.

Wallace R.A., 1985. Vitellogenesis and oocyte growth in non-mammalian vertebrates. In: Developmental Biology: A Comprehensive Synthesis, vol. 1. Browder L.W. (Ed), New York, Plenum Press, p. 127–77.

Wang, X., Zhou, Y., Yang, H., Wang, Q., Liu S., 2010. Investigation of heavy metals in sediments and Manila clams Ruditapes philippinarum from Jiaozhou Bay, China. *Environmental Monitoring and Assessment 170*, 631-643.

Wang, Z., Yan, C., Vulpe, C.D., Yan, Y., Chi, Q., 2012. Incorporation of in situ exposure and biomarkers response in clams Ruditapes philippinarum for assessment of metal pollution in coastal areas from the Maluan Bay of China. *Marine Pollution Bulletin* 64, 90-98.

Wester P.W., Vethaak D., van Muiswinkel W.B., 1994. Fish as biomarkers in immunotoxicology. *Toxicology 86*, 213-232.

Widdows J., 1985. Physiological measurements and physiological procedures. In: The effects of stress and pollution on marine animals, Bayne B.L., Brown D.A., Burns K., Dixon D.R., Ivanovici A., Livingstone D.R., Lowe D.M., Moore M.N., Stebbing A.R.D., Widdows J. (Eds), New York: Preager, pp. 3-45.

Widdows J., Salked P.N., 1992. Practical procedures for the measurement of scope for growth. In: Techniques for monitoring biological effects of pollutants on marine organisms. Plymouth Marine Laboratory, U.K., FAO/UNEP Training Workshop, Nice, France, pp. 1-23.

Widdows, J., C. Nasci, Fossato, V. U. (1997). "Effects of pollution on the scope for growth of mussels (*Mytilus galloprovincialis*) from the Venice lagoon, Italy." *Marine Environmental Research* 43(1-2): 69-79.

Winston G.W., Di Giulio R.T., 1991. Prooxidant and antioxidant mechanisms in aquatic organisms. *Aquatic Toxicology* 19,137–161.

Yoshida, Y., Itoh, N., Hayakawa, M., Piga, R., Cynshi, O., Jishage, K., Niki, E., 2005. Lipid peroxidation induced by carbon tetrachloride and its inhibition by antioxidant as evaluated by an oxidative stress marker, HODE. *Toxicology and Applied Pharmacology* 208, 87–97.

In: Lagoons
Editor: Mwinyikione Mwinyihija
ISBN: 978-1-62808-092-6

Chapter 6

LAGOONS FISH COMMUNITIES IN FLOOD-PULSED FLOODPLAINS: HETEROGENEITY IN A HIGHLY DYNAMIC SYSTEM? THE CASE OF THE OKAVANGO DELTA

***K. Mosepele*[1,*]*, M. Murray-Hudson*[1]*, I. Mosie*[1] *and K. Sethebe*[2]**

[1]University of Botswana, Okavango Research Institute, Maun, Botswana
[2]Botswana International University of Science and Technology, Bontleng, Gaborone, Botswana

ABSTRACT

Lagoons maintain floodplain fish biodiversity and serve as fish refuges at low flood levels. They also maintain the largest fish biomass in tropical floodplains making them key fisheries for riparian communities. Therefore, a comprehensive understanding of the fish communities in these systems is critical towards the development of a holistic management paradigm of floodplain fisheries. This chapter synthesizes previous studies from the delta (based on seven lagoons) and describes the nature and form of lagoon fish communities in the Delta. Data from previous studies shows that there are significant differences in the environment of some lagoons in the delta. This chapter highlights the dynamicity of lagoon habitats in the Okavango Delta and their impact o floodplain fisheries. Analysis of depth measurements from 17 lagoons shows that upper delta are generally deeper than lower delta lagoons ($\approx R^2 = 77\%$). Significant differences in mean temperature (ANOVA: $F=53.41$, $p= 0.02$, $DF = 53$), mean turbidity (ANOVA: $F=13.59$, $p= 0.00$, $DF = 37$) and mean DO (ANOVA: $F=15.14$, $p= 0.00$, $DF = 48$) were found among some lagoons in the Delta. Moreover, because of the flood-pulsed nature of the system, analysis of hydrological data revealed that there was a time lag in discharge among the delta's lagoons. It was within this background that analysis of fish community structure among seven lagoons revealed spatial variations. SIMPER analysis revealed differences in fish species assemblages among several lagoons where *Marcusenius*

[*]Corresponding author: mosepelek@gmail.com.

macrolepidotus and *Brycinus lateralis* are some of the key fish species that contributed to the dissimilarity among the lagoons. Single factor ANOVA showed that *Tilapia rendalli*(ANOVA: F = 7.61, p = 0.02, DF = 6) and *T. sparrmanii*(ANOVA: F = 11.79, *p* = 0.002, DF = 10) juveniles' exhibited significant differences in growth among three lagoons. Results also show that there were significant differences in dry season biomass (ANOVA: F = 5.9, p = 0.0, DF = 44) among four lagoons in the Delta. Multiple regression analysis revealed that fish dynamics within the lagoons were driven by environmental variability where discharge, depth, DO and temperature were some of the key variables that had significant relationships with various fish indices (such as diversity, catch rates, mean length, etc). Fish communities in some lagoons (e.g. Guma) undergo annual fish kills which targets larger sized individuals. These larger sized fish are possibly trapped in the lagoon and ultimately die as a result of anoxic water conditions. This study also showed that lagoons fish communities are a key fishery for small-scale commercial fishers in the delta. This study highlights that despite the dynamicity of the lagoon fish communities, the fish stocks are managed based on a classical approach that is not feasible and practical. The paper highlights the need for a holistic management approach that integrates this dynamicity into a coherent fisheries management of the delta's fish community.

INTRODUCTION

The Okavango Delta in Botswana is one of the largest inland deltas in the world (Allanson *et al.* 1990). The delta is the most resource-rich ecosystem in Botswana (Mmopelwa, 2006) with fish being one of the top priorities. These fish species are distributed heterogeneously throughout the different habitats in the system (Mosepele, 2008; Mosepele et.al 2009) and their biology and ecology is driven largely by the Delta's seasonal flood pulse (Mosepele, 2008). Approximately 71 different fish species have been identified in this system (Ramberg et al. 2006) and these support a well-established small scale commercial fishery (Mosepele and Ngwenya, 2010), a vibrant subsistence fishery (Ngwenya and Mosepele, 2008) and a world class recreational fishery (Merron and Bruton, 1988). Fishing occurs in different habitats using different fishing gears (Mmopelwa, et.al. 2007) where lagoons have been identified as key fishing habitats (Merron and Bruton, 1988; Mosepele et al., 2011). This observation is based on the fact that lagoons have been identified as important habitats for cichlids, which are important food species (Mosepele, et al 2011). Moreover, Mosepele et al (2009) highlighted that lagoons are key habitats for fish production in the delta.

Lagoons are important for maintaining floodplain biodiversity (Rodrigues *et al,* 2002). They act as fish refuges at low flood levels when shallow areas of the floodplain have dried up (Henderson and Crampton 1997; Welcomme 2005). They also sustain the largest fish biomass in tropical floodplain systems (Silvano *et al.* 2000; Welcomme 1979), and therefore constitute an important fishing resource for peripheral communities (Pauly 1975; Soares and Junk 2000). The seasonal connection between floodplains and lagoons is important for maintenance of biodiversity (Junk, 2000). The study by Henderson & Crampton (1997) on floodplain lakes (lagoons) in the Amazon addressed the question of whether different lagoons harbored different fish communities. Despite the recognized importance of lagoons/ lakes in floodplains, their role in sustaining fish populations in the Okavango Delta has never been properly investigated.

Apart from fishing, there are environmental factors that drive the biology and ecology of fish communities in the delta's lagoons. Previous studies indicate that the flood regime is the most important factor influencing fish distribution in the Delta (Lindholm 2006). Merron (1991) also highlighted the critical role that the flood regime plays in floodplain fish ecology of the Okavango Delta. Skelton *et al.* (1985) recommended that in view of different management challenges and threats facing the Okavango, there is a clear and present need to establish the role of the flood regime in maintaining fish biotic diversity. This agrees with Merron and Bruton (1995) who also advocated for a model that can simulate the effect of hydrological alterations on fish production. Notwithstanding, while some references have been made on the relationship between water flow and fish population dynamics (Merron and Bruton 1988: Merron 1991), no direct relationship has been established in the Okavango system (Mosepele 2000). This study will assess dynamics in the fish community of several lagoons from the delta, distributed spatially from the panhandle to the lower delta (figure 1). This study concludes with a discussion of the potential effect of fishing on the fish community of these systems, and then highlights the prevailing management paradigm. A more comprehensive and management approach is then proposed, which accounts for the heterogenous nature of the fish communities and the dynamic attributes of the floodplain ecosystem.

LAGOONS (*MADIBA*) OF THE OKAVANGO DELTA

Lagoons are characterised as stretches and areas of open water where depth (possibly in combination other factors) prevents the growth of aquatic vegetation are common in the Okavango Delta. They vary in size between a few tens of metres across to hundreds of metres, and have highly varied and irregular shapes. They appear to have their origins either as relict sections of now dysfunctional large channels (classic oxbow shapes) or as areas in which peat burning or oxidation have reduced the elevation of former floodplain soils to create large, deep depressions (McCarthy et al. 1987, Porter et al. 1989). Locally known as *madiba* or lagoons, they form important foci for fauna and also for human activities.

Porter and Muzila (1989) investigated two *madiba* on the defunct Thaoge system in the west of the Delta. They found a mean depth of around 3.2m in June 1987(clearly this would vary according to general flood stage). Water balance estimations for Guma lagoon found that the total volume was in the order of 4.3×10^6 m^3, and that this volume was about 1% of the estimated annual through-flow for this lagoon. That is equivalent to a turn-over time of 3-4 days; lagoons in the system are dynamic, through-flow entities in which flow velocities are not high enough to move bed-load sediment, but can move particulate organic matter. Guma lagoon is also the site of numerous observations of fish die-offs which are apparently related to catastrophic rapid declines in dissolved oxygen (DO) related to the flushing of large quantities of organic matter from the underneath of papyrus mats which overly the primary inflow channel. This phenomenon appears to be a very transitory event, and in general terms, lagoons are habitat for many of the larger predatory fish species, and consequently their predators, too (fish eagle, fishing owl, crocodile, etc). Furthermore, figure 2 illustrates that generally, upper delta lagoons (panhandle) are deeper than lower delta (seasonal floodplains) lagoons. This trend is illustrated by a relatively good fitting linear trend line

($\approx R^2$ = 77%) to mean depths of 17 lagoons in the delta. These spatial differences in mean depth are important because they suggest that the shallower lower delta lagoons are more productive than the deeper, upper delta lagoons. This observation is based on Gai et al (2012) conclusion that shallower lagoons have "better light penetration, faster nutrient recycling and higher primary productivity" making them more productive.

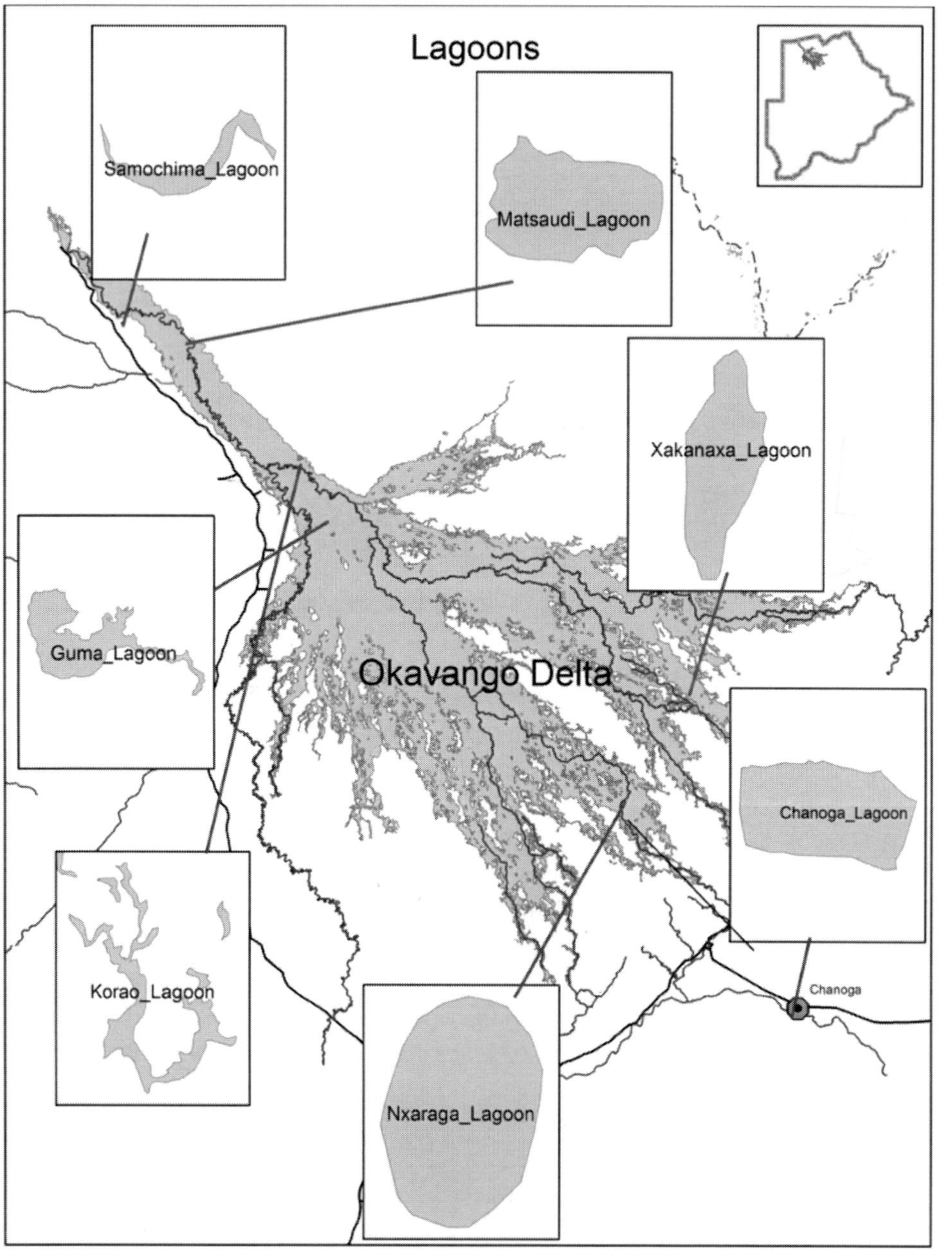

Figure 1. Some of the lagoons that were studied from the Okavango Delta.

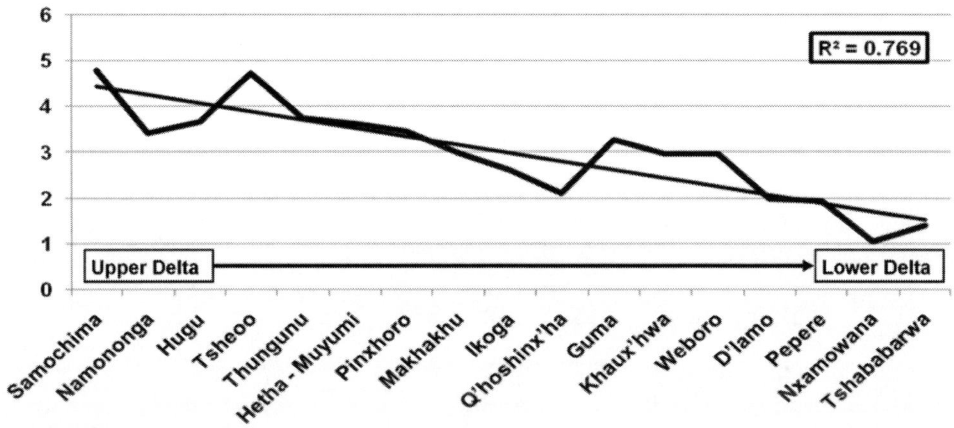

Source: Mosepele, 2001a.

Figure 2. Spatial variations in mean depth from upper delta to lower delta of 17 lagoons from the Okavango Delta.

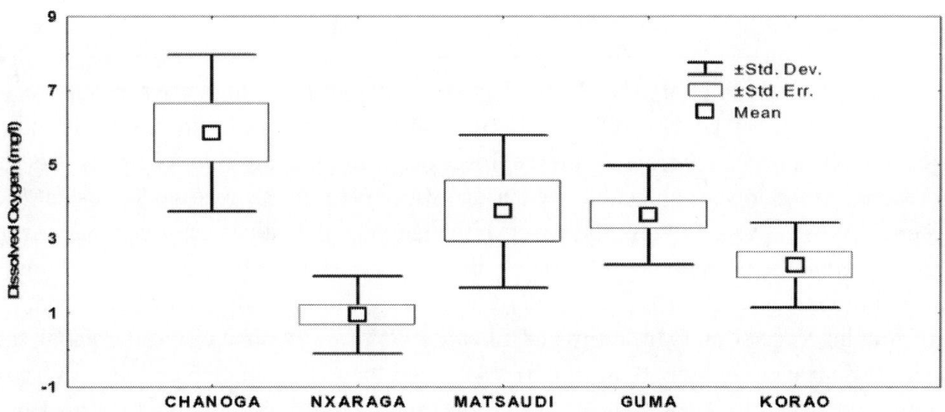

a

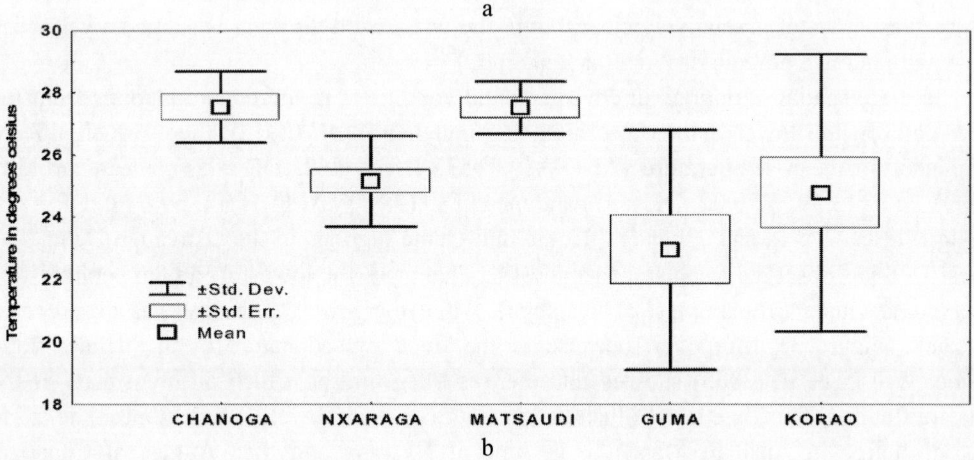

b

Figure 3. (Continued)

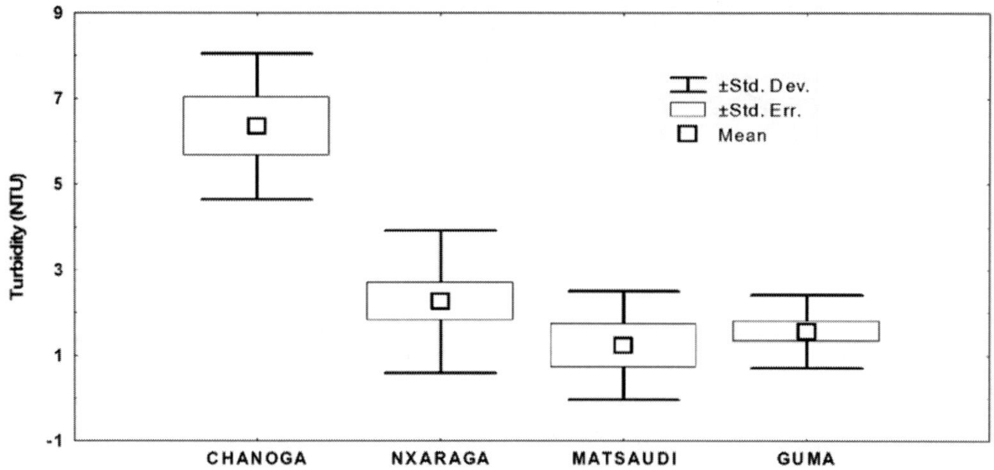

c

Figure 3 a-c. Spatial variations in mean temperature, dissolved oxygen and turbidity among three lagoons (i.e. Chanoga, Nxaraga and Matsaudi) in the Okavango Delta. Adapted from Amutenya et al (2012), Mosepele et al (2012) and Sethebe (2011).

According to Ashton et al., (2003), the panhandle region has a higher water quality, low concentrations of dissolved salts, 65 – 95 % of dissolved oxygen, with low concentration of nitrogen and phosphorus compared with the lower delta. In the main river channels, from the upper reaches to the lower panhandle, the surface water pH remains relatively constant (6.76 – 7.05 and 7.0 – 7. 2 respectively), however in the lagoons, pH values can reach 5.5 (Ashton et al. 2003). In the lagoons at the lower delta, dissolved oxygen rarely exceeds 3 mg/L, and at night often drops below 1 mg/L due to the water passing through the swamps or beneath the mats of floating vegetation (McCarthy and Ellery, 1995) and this is close to the lethal value for many fish species. Ashton et al. (2003) also found the electrical conductivity (EC) to be generally low, 33 – 75 μS/cm, which is aan indication of low total dissolved solids. During the dry season, water is relatively clear, but in the wet season the river is stained a yellowish brown with humic compounds (Ashton et al. 2003).

There are spatial variations in environmental conditions in the delta's lagoons from the upper delta to the lower delta. According to Amutenya et al. (2012), there are significant differences in mean temperature (ANOVA: $F=53.41$, $p= 0.02$, $DF = 53$), mean turbidity (ANOVA: $F=13.59$, $p= 0.00$, $DF = 37$) and mean DO (ANOVA: $F=15.14$, $p= 0.00$, $DF = 48$) among Matsaudi, Nxaraga, Chanoga, Korao and Guma lagoons in the Okavango Delta. The lowest DO was observed at Nxaraga (highest in Chanoga), the lowest mean temperature was observed at Guma (The highest at Chanoga), while the lowest turbidity was observed at Matsaudi (Figure 3). Moreover, because of the flood pulsed nature of the delta's flood regime, Wolski et al. (2005) showed that there is a lag time at which different parts of the delta are flooded. This time lag is illustrated in figure 4 where there is a two month time lag in peak discharge in April at Matsaudi, in June at Nxaraga and then August at Chanoga. Qualitatively, these differences suggest that the different lagoons in the delta are bound to maintain different fish communities. Similar fish species in these lagoons are also bound to have different life history strategies, due to different environmental factors in each lagoon.

Fish Species Assemblage

Previous research from the Okavango Delta has shown that there are spatial variations in fish species assemblage among the Delta's lagoons (Mosepele and Mosepele, 2005; Mosepele et al., 2011). A similar observation is made in this study as illustrated in Table 1. While ANOVA revealed lack of a significant difference in species diversity among four lagoons in the delta, SIMPER analysis showed differences in assemblage structure among the lagoons (Mosepele et al., 2011). However, Table 2 shows that some fish families like Hepsetidae, Schilbeidae, Cichlidae, Clariidae (expect for *C. theodorae*) and Mockokidae are relatively ubiquitous among (five) lagoons in the delta. One characin (*H. vittatus*) some cyprinids (e.g. *B. barnardi*, *B. barotseensis*, *B. bifrenatus*, etc), some myrids (e.g. *H. ansorgii*) and a bagrid (*P. ngamensis*) have relatively restricted distribution among the lagoons.

Mosepele et al. (2011) study revealed that species assemblage structure in distant lagoons was more dissimilar than lagoons in close proximity. This agrees with Rodriguez and Lewis (1990) who made similar observations in the Orinoco floodplains in Venezuela. They (Rodriguez and Lewis, 1994) attributed some of the spatial variability in species assemblage among different lagoons to differences in (e.g.) water conductivity among lagoons. A similar argument may be made for the Okavango Delta that environmental variability among lagoons may account for some of the observed variability in fish species assemblages in the system. However, Nxaraga and Guma lagoons are also similar at around 80% similarity, even though they are fairly distant from each other as shown in figure 5. This suggests therefore, that proximity is not the only factor that accounts for similarities in fish community structure between lagoons.

Table 1. SIMPER analysis results summarizing % dissimilarity in fish species assemblage of five lagoons from the Delta. Data used is from Amutenya et al. (2012), Mosepele et al. (2010) and Sethebe (2011)

Test	Species	% contribution	% Dissimilarity
Chanoga vs. Nxaraga	*M. macrolepidotus*	23.64	77.64
	B. lateralis	21.17	
Chanoga vs. Guma	*B. poechii*	25.23	66.31
	M. macrolepidotus	22.76	
Chanoga vs. Korao	*B. poechii*	28.5	56.34
	S. intermedius	21.37	
Nxaraga vs. Matsaudi	*B. lateralis*	39.65	44.96
	M. macrolepidotus	35.88	
Chanoga vs. Matsaudi	*B. poechii*	37.32	43.4
	S. intermedius	24.52	
Nxaraga vs. Korao	*M. macrolepidotus*	40.89	36.44
	B. lateralis	30.31	
Guma vs. Matsaudi	*M. macrolepidotus*	37.14	34.64
	B. lateralis	32.01	
Guma vs. Korao	*M. macrolepidotus*	47.45	24.51
	B. lateralis	17.57	
Guma vs. Nxaraga	*B. lateralis*	33.32	20.22
	M. macrolepidotus	16.17	
Matsaudi vs. Korao	*B. lateralis*	35.94	18.87
	P. catastoma	19.58	

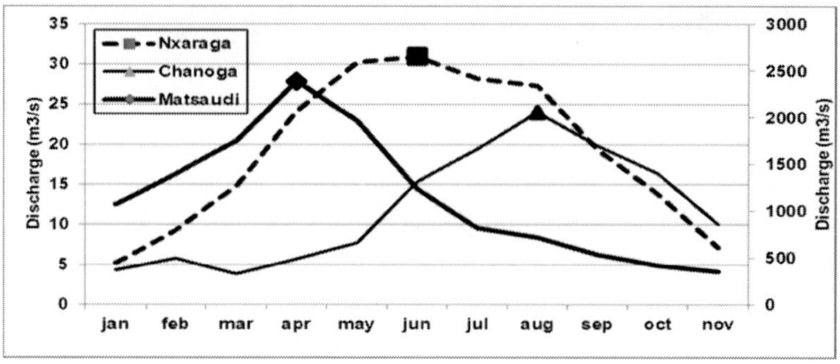

Figure 4. Mean monthly discharge (m3/s) of feeder streams into Matsaudi, Nxaraga and Chanoga lagoons showing a two month time lag in peak discharge among the three lagoons (Data from the Okavango Research Institute long term monitoring database at http://168.167.30.198/ori/monitoring/water/index.php).

Table 2. Fish species checklist from 5 lagoons in the Okavango Delta

Fish Order	Fish Family	Fish Species	Chanoga	Guma	Nxaraga	Matsaudi	Korao
Characiformes	Characidae	*Brycinus lateralis*	√	√	√	√	√
		Hydrocynus vittatus	-	√	-	√	√
	Citharinidae	*Nannocharax macropterus*	-	-	√	-	-
	Hepsetidae	*Hepsetus odoe*	√	√	√	√	√
Cypriniformes	Cyprinidae	*Barbus barnardi*	√	-	-	-	-
		Barbus barotseensis	-	-	√	-	-
		Barbus bifrenatus	-	√	-	-	-
		Barbus codringtonii	-	√	-	-	-
		Barbus fasciolatus	-	√	-	-	-
		Barbus paludinosus	√	√	√	-	√
		Barbus poechii	√	√	√	√	√
		Barbus radiates	-	√	√	√	-
		Barbus thamalakanensis	√	-	-	-	-
		Barbus unitaeniatus	√	√	√	√	-
		Labeo lunatus	-	√	√	-	-
Osteoglossi-formes	Momyridae	*Hippopotamyrus ansorgii*	-	√	-	-	√
		Hippopotamyrus discorhynchus	-	√	√	√	√
		Marcusenius macrolepidotus	√	√	√	√	√
		Momyrus lacerda	-	√	√	√	√
		Petrocephalus catastoma	√	√	√	√	√
		Pollimyrus castelnaui	-	√	√	-	√
Perciformes	Anabantidae	*Ctenopoma multispine*	-	√	√	-	-
	Cichlidae	*Oreochromis andersonii*	√	√	√	√	√
		Oreochromis macrochir	√	√	√	√	√
		Pharyngochromis acuticeps	√	√	-	√	√
		Pseudocrenilabrus philander	-	√	-	√	√

Fish Order	Fish Family	Fish Species	Chanoga	Guma	Nxaraga	Matsaudi	Korao
		Sargochromis carlottae	√	√	√	√	√
		Sargochromis codringtonii	√	√	√	√	√
		Sargochromis giardia	-	√	√	√	√
		Sargochromis greenwoodii	√	√	√	-	√
		Serranochromis altus	√	√	√	√	√
		Serranochromis angusticeps	√	√	√	√	√
		Serranochromis longimanus	-	-	√	√	-
		Serranochromis macrocephalus	√	√	√	√	√
		Serranochromis robustus	√	√	-	√	√
		Serranochromis thumbergi	√	√	√	√	√
		Tilapia rendalli	√	√	√	√	√
		Tilapia ruweti	-	√	√	√	-
		Tilapia sparrmanii	√	√	√	√	√
Siluriformes	Bagridae	*Parauchenoglanis ngamensis*	-	√	-	-	√
	Clariidae	*Clarias gariepinus*	√	√	√	√	√
		Clarias ngamensis	√	√	√	√	√
		Clarias theodorae	√	-	√	√	-
	Mochokidae	*Synodontis leopardinus*	√	√	√	√	√
		Synodontis macrostigma	√	√	√	√	√
		Synodontis macrostoma	√	√	√	√	√
		Synodontis nigromaculatus	√	√	√	√	√
		Synodontis thamalakanensis	-	√	√	√	√
		Synodontis vanderwaali	√	√	√	√	√
		Synodontis woosnami	-	√	√	√	√
	Schilbeidae	*Schilbe intermedius*	√	√	√	√	√
Species Richness	11	51	31	37	40	37	36

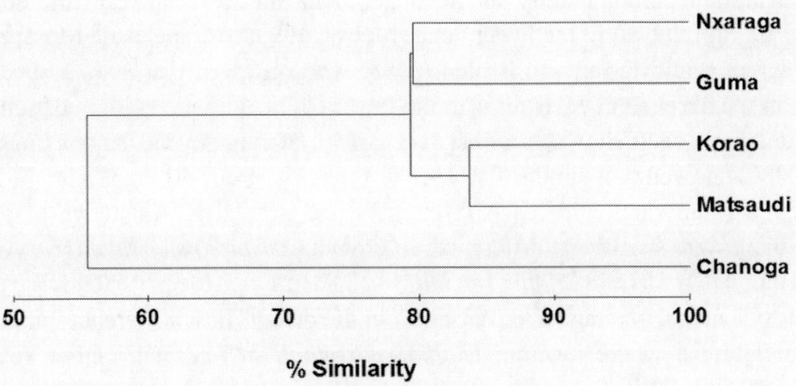

Figure 5. Cluster analysis showing spatial variation in species assemblage among five lagoons in the Delta. Chanoga data is based on Mosepele et al 2010 while data for the other four lagoons is based on Mosepele et al 2011.

As highlighted by Mosepele et al. (2011), environmental variability is another factor that can account for similarities in fish communities among different lagoons. They (i.e. Mosepele et al. 2011) highlight that both these two lagoons (i.e. Nxaraga and Guma) are subject to intense environmental variability. Subsequently, this suggests that the lagoons' fish communities are subject to dramatic/ dynamic perturbations. It can therefore be surmised that the fish species found in these two lagoons have a high tolerance for sharp changes in environmental variability.

In their study, Mosepele et al. (2011) compared the fish community of four lagoons (Korao, Matsaudi, Guma and Nxaraga; Figure 1) in the Okavango Delta. Their study revealed that species diversity is highest in the upper Delta lagoons (environmentally/ hydrologically stable habitats), which agrees with observations by Lowe-McConnell (1987) that fish species diversity is highest in environmentally stable habitats. Moreover, Albaret and Lae (2003) discuss that fish communities with high species diversity are more stable than those with lower diversity. Mosepele et al. (2011) and Tweddle et al. (2003) observed that Guma lagoon has very low species diversity, which suggests that its fish community is possibly more stressed and hence more unstable than other lagoons. This indeed is true because high annual fish mortalities have been observed in Guma lagoon before (Bills 1996; Tweddle et al. 2003). These annual kills have been attributed to seasonal changes in water quality, with a subsequent release of hydrogen sulphide gas (H_2S) in the lagoon (Bills, 1996). These notwithstanding, Sethebe (2011) revealed that a combination of H_2S, pH, DO and turbidity contribute to the annual fish kills in Guma lagoon. In essence then, Guma lagoon fish are exposed to not only this form of mortality, but also fishing mortality from both commercial and recreational fishing. Henderson and Crampton (1997) made a similar observation when they concluded that white water lagoons in the Amazon have unstable fish communities because they suffer annual fish kills.

Growth Rates

Different environmental and biological factors drive the biology and ecology of fish populations among the different lagoons in the delta. One of the delta's keystone fish species, *H. vittatus,* is found predominantly in the upper delta and at relatively low abundances (during years of high floods) in the lower delta (Merron 1993; Mosepele and Mosepele 2005). Indeed this agrees with Merron and Bruton (1988) who observed that *H. odoe* becomes the top predator in the absence of *H. vittatus* in the lower Delta. Variations in predation pressure among lagoons have been shown to affect fish growth among several lagoons in the delta. Mosepele (2000) showed that similar fish species (cichlids) from different parts of the Delta have different growth rates and reach different sizes. Similarly, Amutenya et al. (2012) showed that there were significant differences in growth of *Tilapia sparrmanii* (ANOVA: $F = 11.79$, $p = 0.002$, $DF = 10$) and *Tilapia rendalli* (ANOVA: $F = 7.61$, $p = 0.02$, $DF = 6$) among three lagoons (Chanoga, Nxaraga and Matsaudi) in the delta. These differences in growth are illustrated in Figure 6 where juvenile *Tilapia sparrmanii* in Matsaudi lagoon (panhandle) grew significantly ($p<0.05$) faster than juvenile *T. sparrmanii* in Nxaraga (lower delta) and Chanoga (outlet river) lagoons. Furthermore, Figure 7 also illustrates that juvenile *T. rendalli* in Matsaudi (Ngarange) grew significantly faster than *T. rendalli* in the other two lagoons.

Essentially, this observation suggests that fish from the upper delta grow significantly faster than the same species from the lower delta.

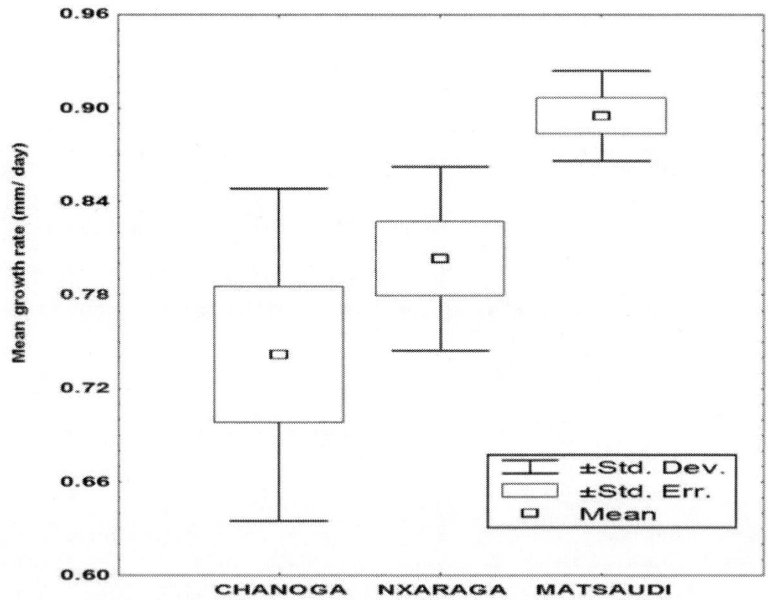

Figure 6. *Tilapia sparrmanii* growth rates from three different lagoons in the Okavango delta (Amutenya et al 2012).

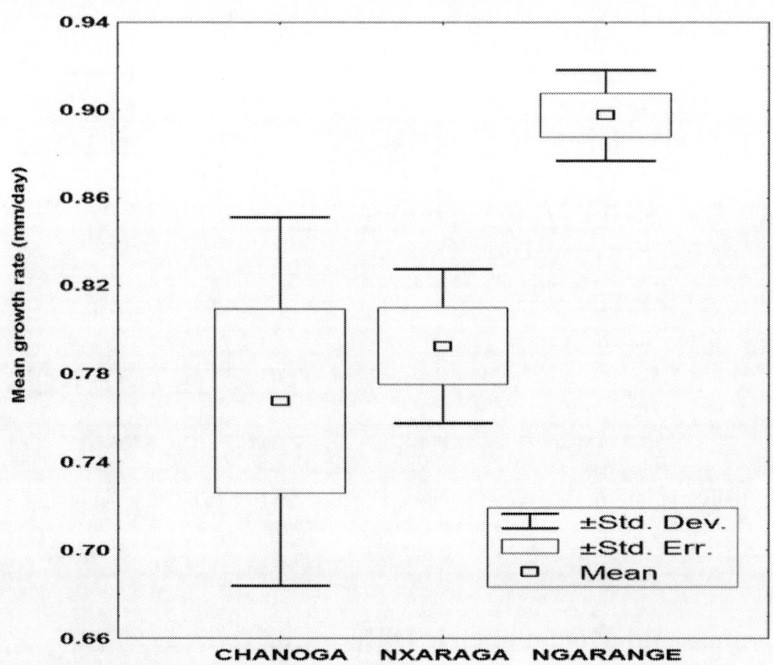

Source: Amutenya et al 2012.

Note: Matsaudi lagoon is at Ngarange.

Figure 7. *Tilapia rendalli* growth rates from three different lagoons in the Okavango delta.

These differences in growth are attributed to different biophysical factors among the delta's lagoons which were found to have a regulatory effect on growth of *Tilapia sparrmanii* and *T. rendalli* (Amutenya et al. 2012). They (i.e. Amutenya et al. 2012) showed that while a multiple regression model showed that several physical variables (i.e. DO, Temperature, Turbidity, Conductivity, and Discharge) contributed significantly to growth rates of *T. sparrmanii* among three lagoons in the delta (Matsaudi, Nxaraga and Chanoga), the presence of a predator like *Schilbe intermedius* (Ngarange and Chanoga) and *Hepsetus odoe* (Nxaraga) produced a better fitting regression model (R^2> 90%).

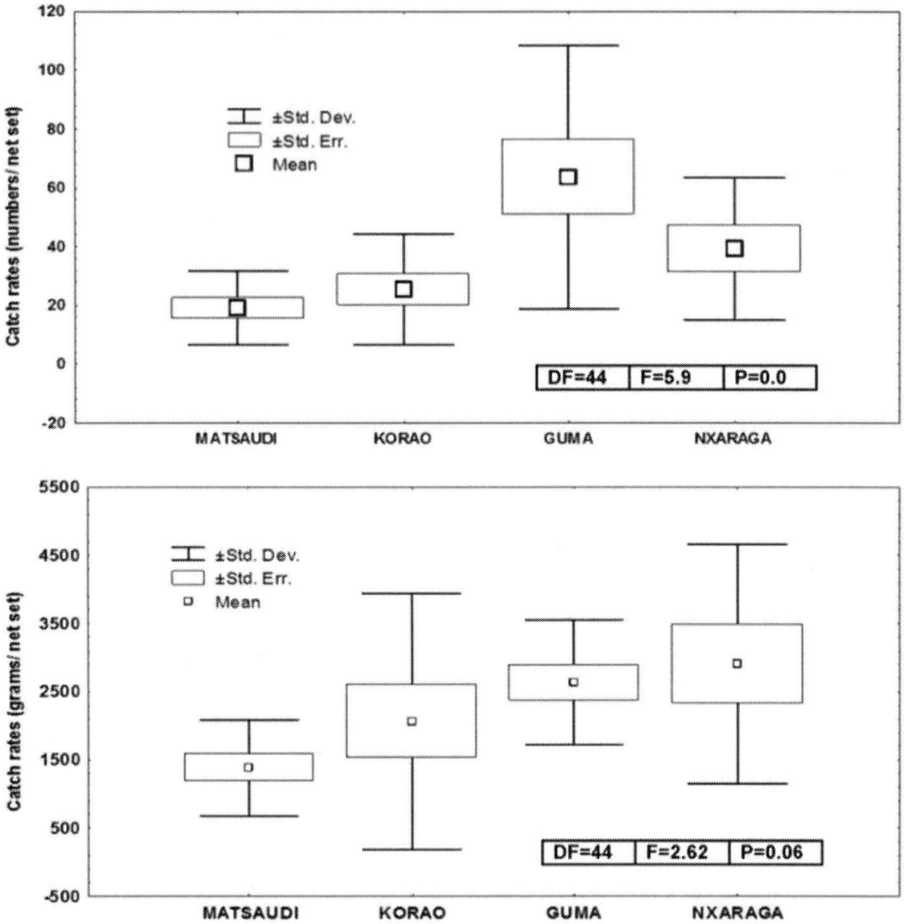

Figure 8a-b. Spatial variations in catch rates (numbers/ net set and grams/ net set) among four lagoons in the Delta.

Relative Biomass and Morphometric Differences

While Mosepele et al. (2011) revealed that there were no significant differences in mean fish catch rates among four lagoons in the Delta, a re-analysis of the data using mean catch rates expressed as numbers of fish caught/ net set revealed significant differences (p=0.0) in abundance (figure 8). However, this study has also revealed that differences in mean catch rates expressed as grams/ net set were significant at a 90% confidence level. Moreover, Mosepele et al. (2011) also showed that there were significant differences in dry season lagoon fish biomass where Nxaraga had the highest biomass while those of Korao and Matsaudi were the lowest. While these observations may agree with Fox (1976) that lower delta lagoons are more productive (in fish biomass production) than upper delta lagoons, it may also be argued that lower delta lagoons are a major source of fish refuge during the dry season. This argument is based on the observation that lagoons in the lower delta are the only major sources of water that can sustain fish populations while riverine channels in the upper delta still maintain large fish populations during the dry season. This may suggest, therefore, that lagoons are crucial fish habitats in the lower delta than the upper delta, especially during the dry season.

Moreover, Mosepele (et al. 2011) showed that there are morphometric differences among some of the major lagoon species between the upper and lower Delta lagoons. These are illustrated by significant differences in body size (length) for *S. intermedius* and for several other fish species. This observation agrees with Mosepele et al. (2005) who showed that several fish species (*Oreochromis andersonii, Tilapia rendalli* and *O. macrochir*) from different parts of the Delta follow different life history strategies. Gopelomang (2010) also revealed that fish from Nxaraga lagoon were on average significantly bigger than those from Matsaudi lagoon. Mosepele et al. (2011) observed that flood duration in the floodplains is perhaps not the only determining factor in fish life history strategies. This is based on the observation that while Matsaudi and Korao lagoons are both situated in the permanent portion of the Delta, and are exposed to relatively similar levels of fishing, the mean lengths of *C. gariepinus* and *M. macrolepidotus* in the two lagoons are significantly different.

Temporal Variations in Species Assemblage in Lagoons

Fish population communities (such as diversity, catch rates, composition, etc) in the delta's lagoons undergo temporal variations. This agrees with several other studies elsewhere that found that fish populations in floodplains pulse in tandem with the flood regime (Henderson and Crampton 1997; Petry et al. 2003). Moreover, Merron and Bruton (1988) highlighted that different fish species undergo lateral and longitudinal migrations in the Delta as a consequence of the flood regime.

Nxaraga Lagoon (Species Diversity and Community Structure)

According to Hoberg et al. (2002), fish species diversity in Nxaraga Lagoon peaks in June which corresponds to high water levels. This observation agrees with Mosie (2007) who concluded that while variations in species diversity among years is stochastically different, mean diversity over a three year period peaked in June. Mosie (2007) highlighted that these inter-annual differences in mean diversity suggest that no particular year can be representative of fish species variability in Nxaraga lagoon. Regression analysis from this study revealed that the combined effects of temperature, pH, TDS and discharge contributed to a well-fitting model ($R^2 = 0.90$) that explains the variations in fish species diversity in this lagoon. The analysis revealed that discharge had the most significant effect ($p = 0.01$) on species diversity (Figure 9).

This suggests that there is some species immigration into the lagoon from upstream habitats (or from other water bodies that may become disconnected from the river channel, and hence the lagoon, at low flood levels), which increases species diversity in the lagoon during this time. These dynamic changes in species diversity are also illustrated in figure 10 below where the fish community changes dramatically over time. As described in Mosie (2007) and illustrated in figure 10, *Hepsetus odoe* was the most dominant fish species in the lagoon during January and November only (i.e. 2 months). Thereafter, the community was dominated by *Marcusenius macrolepidotus* from March – June. This was then replaced by *Schilbe intermedius* in July, but replaced by *Clarias gariepinus* in August. Then *Schilbe intermedius* dominated the community again between September and October.

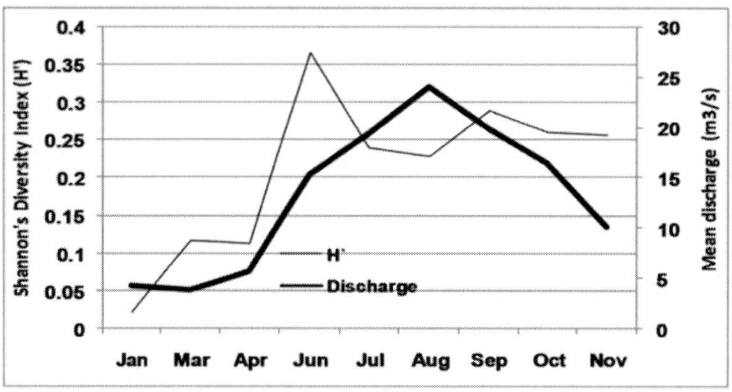

	p
Temperature	0.2
pH	0.44
TDS	0.23
Discharge	0.01
R^2	0.9
F	9.35

Figure 9. Seasonal variations in species diversity at Nxaraga lagoon in relation to mean monthly discharge (From Mosie, 2007) and results of a multiple regression analysis to assess the relationship between diversity and selected environmental factors.

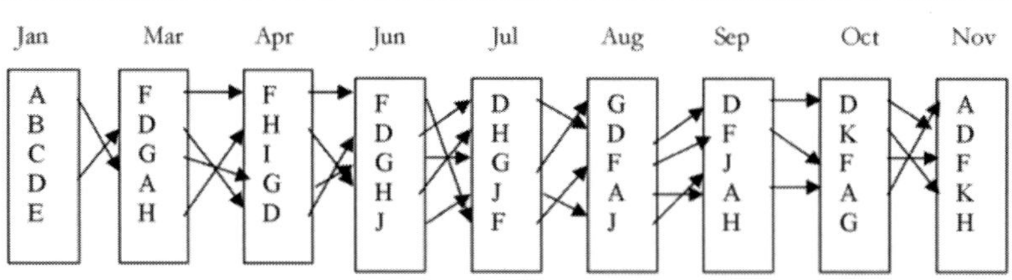

Legend:
A = *Hepsetus odoe*; B = *Serranochromis altus*; C = *Serranochromis angusticeps*; D = *Schilbe intermedius*; E = *Tilapia ruweti*; F = *Marcusenius macrolepidotus*; G = *Clarias gariepinus*; H = *Clarias ngamensis*; I = *Petrocephalus catastoma*; J = *Brycinus lateralis*; K = *Barbus poechii*.

Source: Mosie, 2007.

Figure 10. Intra annual variations of the top five most important species (based on IRI) at Nxaraga Lagoon.

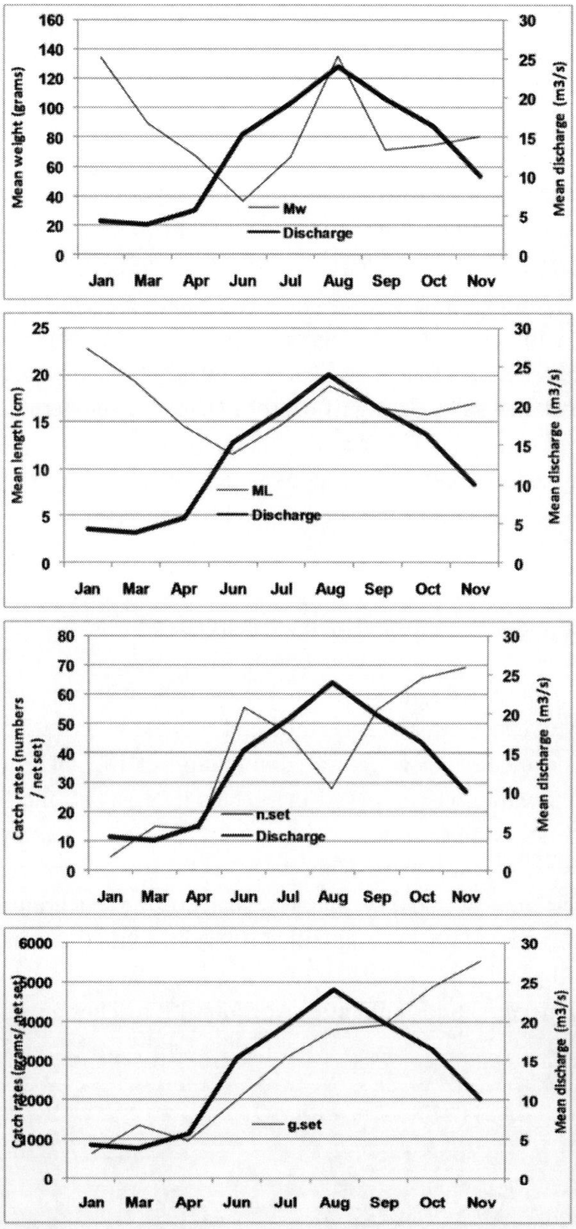

	p
DO	0.16
Temperature	0.053
Discharge	0.14
R^2	0.61
F	2.6

	p
DO	0.15
Temperature	0.1
pH	0.87
Discharge	0.03
R^2	0.79
F	3.71

	p
DO	0.22
Temperature	0.32
TDS	0.15
Depth	0.02
R^2	0.86
F	6.32

	p
DO	0.06
Temperature	0.03
R^2	0.58
F	4.21

Source; Mosie, 2007.

Figure 11. Seasonal variations in several fish indices from Nxaraga lagoon and results of a multiple regression analysis to assess the nature of the relationship between these and several environmental factors.

Nxaraga Lagoon (Relative Biomass (Catch Rates) and Other Metrics)

Mosie (2007) observed that seasonal flooding has a significant effect on fish dynamics in Nxaraga lagoon, which is also illustrated in figure 11. According to figure 11, while no environmental variable (in this study) had any significant effect ($p>0.05$) on changes in the mean weight of fish in the lagoon, a combination of DO, temperature and discharge explained 61% of the variations in mean fish weight in the lagoon. Conversely, while a combination of DO, temperature, pH and discharge explained 79% of the variations in mean fish length in the lagoon, discharge had a significant effect (p=0.03) with mean fish length in the lagoon. Variations observed in both mean fish weight and length in the lagoon suggest that there are some fairly large sized fish species which are resident in the lagoon and do not migrate frequently. This observation is validated by high mean fish weight (≈ 120 – 140 grams) and length (≈ = 20 – 25 cm) in January at low discharge and (≈ 15 – 20 cm) in August at high discharge. Mean depth had a significant relationship with fish relative abundance (expressed as numbers/ net set) where a combination of DO, temperature, TDS and depth explained 86% of the variations observed in relative abundance. On the other hand, mean fish relative abundance expressed as grams/ net set was significantly related to temperature (p=0.03) where a combination of temperature and DO could explain about 58% of the variations observed in this index. The intrinsic importance of lagoons generally in fish production, and as key fish habitats during the dry season is described in detail by Mosepele et al. (2009).

Guma Lagoon (Fish Diversity and Community Structure)

Figure 12 illustrates that fish species diversity in Guma lagoons undergoes temporal variations where it is lowest between February and April and highest in June. Results of the (best fitting) multiple regression model shows that a combination of DO, pH, turbidity and conductivity explain 61% of the variations observed in species diversity in the lagoon. The results illustrated in figure 12 show that dissolved oxygen (DO) had the strongest effect (p=0.02) on changes in diversity. According to Sethebe (2011), the fish species diversity crashes in February due to the annual fish kill that is usually observed around this time according to Bills (1996). Both Sethebe (2011) and Bills (1996) attribute this annual fish kill to anoxic water that is flushed from underneath papyrus by incoming new water into the lagoon. The fish community then starts to re-build after April until it reaches a peak/ stabilises in June.

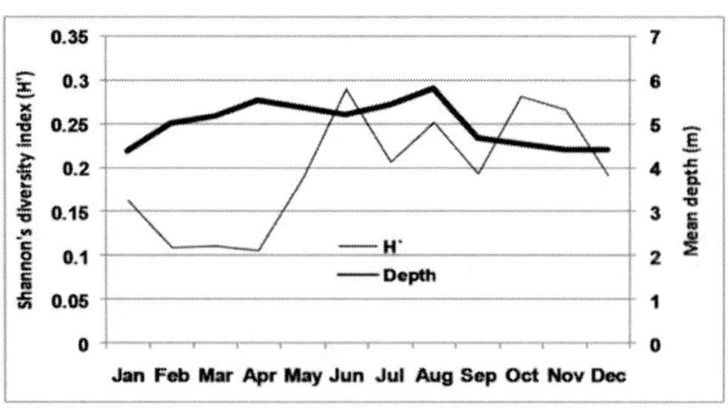

	p
DO	0.02
Turbidity	0.48
pH	0.21
Conductivity	0.5
R^2	0.61
F	2.71

Source: Sethebe, 2011.

Figure 12. Temporal variations in species diversity at Guma lagoon with the results of a multiple regression model summarising the relationship between fish species diversity and several environmental factors.

Guma Lagoon (Relative Biomass (Catch Rates) and Other Metrics)

According to Sethebe (2011),Table 3 below shows the relationship between environmental variables and fish biological variables in Guma lagoon during the study period. While depth, DOC, DO and pH explained approximately 81% of the variations in mean length in the fish community, pH had the strongest significant effect on this variable. The best regression model was observed for changes in mean weight where 96% of the variations observed could be explained by a combination changes in pH, DO, Temperature, DOC and depth. These results also showed that conductivity (EC), temperature, and turbidity were not key variables in the fish community because they were not significantly correlated to any of the biological variables studied.

Table 3 below summarises the 10 most important fish species collected during an annual fish kill in Guma lagoon in (February) 2008 (Sethebe, 2011). According to the table, *Oreochromis andersonii* underwent the highest mortality and Sethebe (2011) highlighted that cichlids generally were the most vulnerable species during the annual fish kill. Another interesting observation made from the fish kill was that most of the fish that died were relatively large as shown in the table. An interesting dynamic that Sethebe (2011) was that the fish kill occurred in February when water levels in the lagoon were increasing (see figure 12). It was therefore surmised (by Bills, 1996; Sethebe, 2011) that the onset of new floods in the lagoon flushes out anoxic water from below the fringing papyrus into the lagoon. According to Sethebe (2011), this period coincided with increased turbidity in the lagoon, which then impairs the fish's ability to optimally utilise the DO in the water column. Hence, a combination of H_2S gas (Bills, 1996; Sethebe, 2011) and anoxic water (Sethebe, 2011) creates a deadly environment to larger sized fish (e.g. *O. andersonii*≈ 33 cm, *M. lacerda*≈ 35 cm, S. *intermedius*≈ 29 cm, *H. vittatus*≈ 41 cm, etc) in the lagoon. It was on the basis of this scenario, that Sethebe (2011) concluded that large fish are essentially trapped in the lagoon at increasing water levels, and this exposes them to unfavourable conditions that result in increased mortality of adult fish.

New flood waters enter Guma lagoon at around February and slowly start filling it (Bills, 1996; Sethebe, 2011). By this time, the peripheral floodplains are either still dry or have very little water in them because the new floods fill the main river channels and lagoons first before over-spilling into the peripheral seasonal floodplains. At this time, only the smaller sized fish species are able to undergo both lateral and longitudinal migrations (Merron and Bruton, 1988) out of the lagoon. The water levels at this time are however not yet deep enough for the big fish to swim out of the lagoon and are hence trapped. Furthermore, the high mortality of disproportionately large fish in this lagoon suggests that lagoons are also major habitats for large sized fish. This observation agrees with Mosepele et al.'s (2009) conceptual model of the delta's fish communities and their relationships with lagoons. They (Mosepele et al. 2009) highlight that new floods periodically connect the lagoons and the rest of the delta into one complex that ultimately facilitates fish production through migrations to new habitats that are opened up.

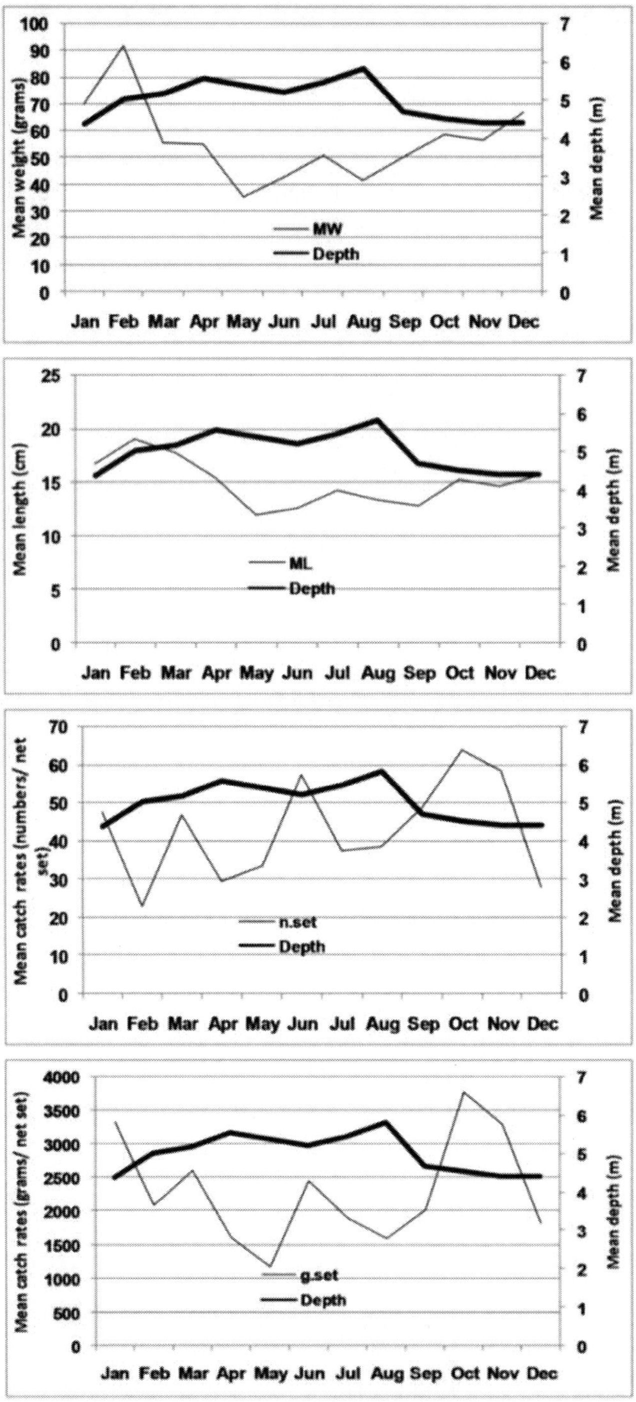

	p
DO	0.22
Temperature	0.17
Depth	0.08
DOC	0
pH	0.02
R^2	0.96
F	27.15

	p
DO	0.06
Depth	0.34
DOC	0.19
pH	0.047
R^2	0.81
F	7.36

	p
Depth	0.04
Turbidity	0.12
DOC	0.02
pH	0.21
R^2	0.68
F	3.79

	p
DO	0.52
Depth	0.046
Turbidity	0.26
DOC	0.23
R^2	0.57
F	2.28

Source: Sethebe, 2011.

Figure 13. Relationship between mean depth (m) and several fish metrics (mean weight, mean length, numbers/ set and grams/ set) with results of a multiple regression analysis at Guma lagoon. Korao Lagoon (Fish Species Diversity and Community Structure).

Table 3. Species Composition and Abundance of species collected during an annual fish kill at Guma Lagoon based on the 10 most important species collected

Species	NO	%NO	Weight (kg)	%Weight	FO	%FO	IRI	%IRI	ML
O. andersonii	362.000	34.908	242.609	0.052	0.349	34.908	1220.424	24.530	32.763
M. lacerda	283.000	27.290	113.275	0.024	0.273	27.290	745.426	14.983	34.917
Synodontisspp	177.000	17.068	15.265	0.003	0.171	17.068	291.389	5.857	20.047
S. intermedius	67.000	6.461	14.606	0.003	0.065	6.461	41.764	0.839	28.554
S. macrocephalus	51.000	4.918	24.553	0.005	0.049	4.918	24.213	0.487	29.308
H. vittatus	43.000	4.147	31.152	0.007	0.041	4.147	17.222	0.346	41.063
H. odoe	21.000	2.025	9.065	0.002	0.020	2.025	4.105	0.083	37.385
S. carlottae	9.000	0.868	1.976	0.000	0.009	0.868	0.754	0.015	23.333
S. angusticeps	7.000	0.675	2.420	0.001	0.007	0.675	0.456	0.009	28.400
S. altus	5.000	0.482	3.587	0.001	0.005	0.482	0.233	0.005	33.620

Source: Sethebe, 2011.

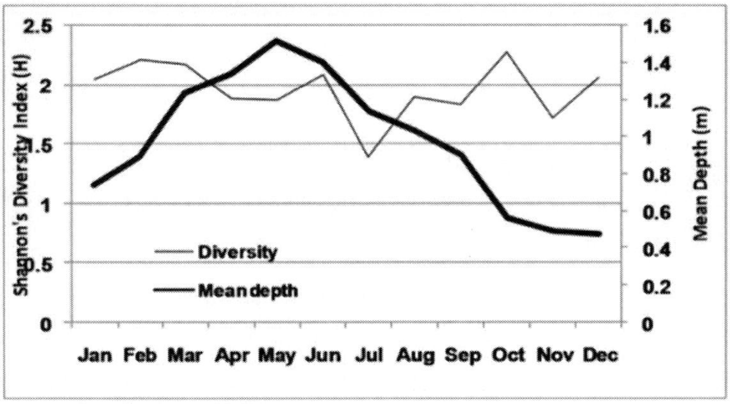

	p level
TDS	0.35
pH	0.39
DO	0.36
R^2	0.3
F	0.87

Source: Gopelomang, 2010.

Figure 14. Temporal variations in species diversity at Korao lagoon with the results of a multiple regression model summarising the relationship between fish species diversity and several environmental factors.

Figure 14 illustrates that a multiple regression model revealed that none of the environmental variables studied (in the lagoon) had any significant effect on fish species diversity in this lagoon. Even though Gopelomang (2010) highlighted that there were no significant seasonal variations in species in fish species diversity (p=0.43), the highest diversity was observed in August while the lowest was observed in July. The regression model also showed that variations in TDS, pH and DO could explain only 30% of the observed variation in species diversity. The regression model is relatively poorly fitting (F=0.87) which suggests that there are other factors (apart from mean depth, and temperature which did not have any significant contribution to the regression model) that regulate/ affect changes in species diversity.

Korao lagoon is a major fishing area for commercial fishers from the nearby village of Seronga (Mosepele, 2001b). Therefore, while the direct effects of fishing effort on this lagoon's fish community was not assessed, it is possible that fishing effort is a major contributor to changes in fish species diversity in the lagoon. A selective harvesting regime (from commercial fishing) through the use of gill net of a narrow range of mesh sizes invariably selects out the larger fish which are usually predators. This selective harvesting of top predators (e.g. the large predatory cichlids like *S. robustus*, *A. angusticeps*, etc and the targeted removal of other predators like *H. vittatus* and *C. gariepinus*) can upturn a fish community from a predominantly top-controlled system (predation controlled) to a bottom up controlled system. This invariably changes the dynamics in the fish community and may hence account for change in fish species diversity. This observation is based on Mosepele et al.'s (2012) discussion that biological factors (predation) had a significant effect on the fish community of the delta. Similarly, Amutenya et al. (2012) also revealed that several predatory species (e.g. *S. intermedius*, *H. odoe*, etc) had a significant effect on the growth rates of *T. sparrmanii* and *T. rendalli* in Matsaudi, Nxaraga and Chanoga lagoons.

Relative Biomass (Catch Rates) and Other Metrics

There is a significant relationship between the mean fish relative abundance (grams/ set and numbers/ set) and mean depth at Korao lagoon (figure 15). When relative abundance is expressed as numbers/ set, the best fitting regression model includes TDS, mean depth and DO as key variables which collectively account for approximately 81% of the variation observed in this metric. However, when catch rates expressed as grams/ set is used, mean depth, TDS and temperature are the key variables that contribute approximately 79% of the variations observed in this metric. What is particularly noteworthy is that catch rates are highest on the 11th month of the year (November) and lowest around the fourth (numbers/ set) or fifth (grams/ set) month of the year. The period of lowest catch rates corresponds to the deepest water levels in the lagoon which is a classic example of the "dilution" and "concentration" effects of floodplain fisheries as described by Welcomme (1999). Moreover, figure 15 illustrates that mean depth had the greatest (and significant) effect on variations in mean length and mean weight of fish species in the lagoon. Regression models for these two morphometric variables were relatively good where 93% of the observed variation in mean weight could be explained by a combination of changes in mean depth, TDS, pH, DO and temperature. On the other hand, 86% of the observed variations in mean length could be explained by a combination of changes in mean depth, pH, DO and temperature (figure 15).

According to Gopelomang (2010), fish catch rates in Korao lagoon undergo dynamic variations as a consequence of migrations driven by seasonal flooding. The high catch rates observed in the dry summer months (October - December) as illustrated in figure 15 are possibly caused by what Merron (1993) described as the catfish run. Mosepele et al (2009) also illustrates the positive effect of this catfish run on fish catch rates in the Okavango Delta. During this period, catfish (Clarias spp.), especially the sharp-tooth catfish (*C. gariepinus*) undertake frenzied feeding where they congregate in packs along channel (or lagoons in this case) inlets where they feed voraciously on smaller sized fish species (especially *Marcusenius macrolepidotus*) which are back-migrating from the now-drying up shallow inundated floodplains. Merron (1993) further discussed that packs of hunting *Hydrocynus vittatus* and other predatory cichlids (e.g. *Serranochromis angusticeps*, *S. altus*, *S. robustus*, etc) usually follow these pack hunting catfish and feed on any prey fish species that may have escaped the catfish.

FISH EXPLOITATION IN LAGOONS

Approximately 87 lagoons used by fishers were identified in the Delta by Mosepele (2001a). The mean depth of the lagoons was 2.89 m while the mean transparency (based on secci disk readings) was 1.425 m. These lagoons range from those which are connected to the main channel and are permanently flooded in the panhandle, to seasonal floodplain lagoons that remain dry for some parts of the year and are disconnected from the river channels. Mosepele (2001b) highlights that 73% of gill net fishers fish from lagoons, 91% of hook and line fishers also fish from lagoons, while 94% of basket fishers fish from the fringes of lagoons. These two reports highlight the intrinsic importance of lagoons towards the delta's small-scale commercial and subsistence fishery. Mmopelwa et al. (2009) showed that gill net fishers usually harvest larger sized fish while subsistence (basket) fishers generally harvest smaller-sized fish species (e.g. *Aplocheilichthys johnstoni* which is one of the smallest fish species that occurs in the delta.), which agrees with Mosepele et al.'s (2003) earlier observations.

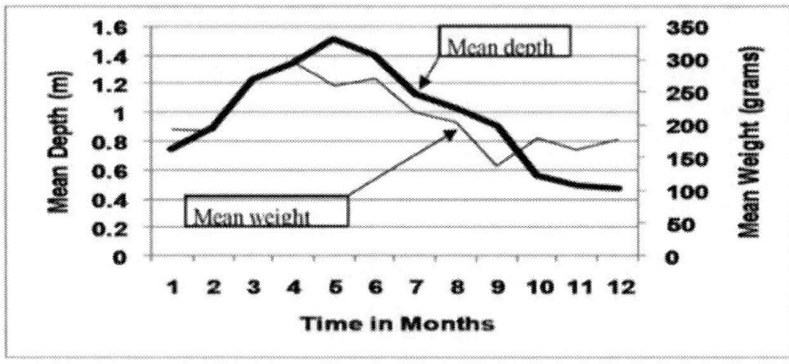

Mean Depth	0.006
TDS	0.206
pH	0.033
DO	0.072
Temperature	0.032
R^2	0.93
F	10.25

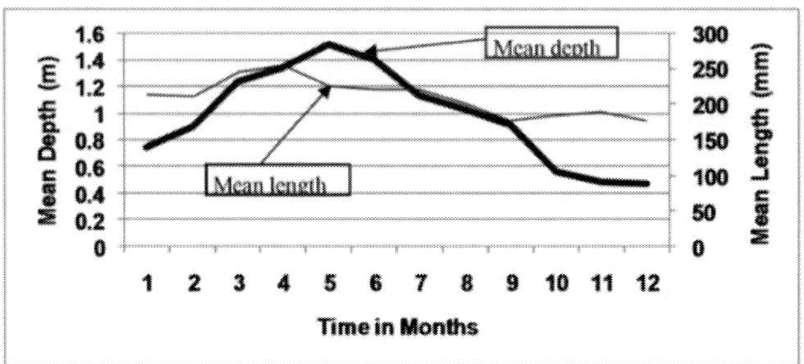

Mean Depth	0.005
pH	0.02
DO	0.1
Temperature	0.03
R^2	0.86
F	7.59

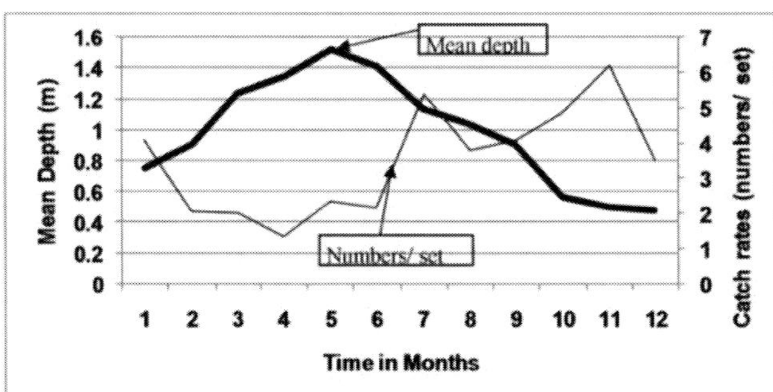

Mean Depth	0.003
TDS	0.637
DO	0.074
R^2	0.81
F	8.58

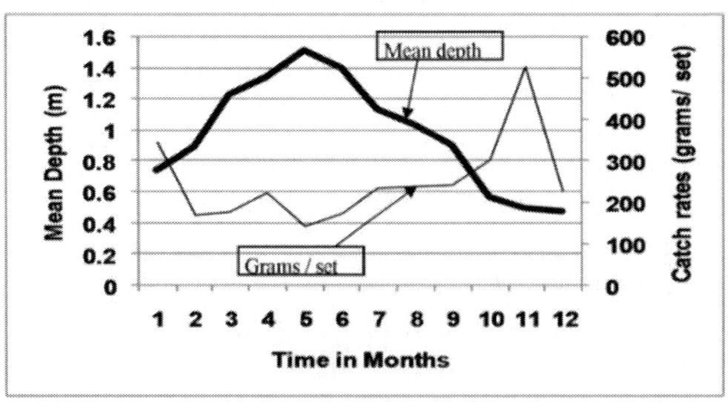

Mean Depth	0.003
TDS	0.331
Temperature	0.318
R^2	0.79
F	7.38

Source: Gopelomang, 2010; Mosepele et al 2012.

Figure 15. Relationship between mean depth (m) and several fish metrics (mean weight, mean length, numbers/ set and grams/ set) with results of a multiple regression analysis at Korao lagoon.

According to Mosepele et al. (2007), small-scale commercial fishers used to undertake "drive fishing" (before it was banned under the Fish Protection Regulations promulgated in 2008) in lagoons during the dry season (at low water levels between August and November in the upper delta lagoons). The key fish species targeted using this fishing method was normally *Tilapia rendalli*, which, according to Bills (1996), is otherwise difficult to harvest using normal gill net setting methods. Lagoons are key fishing habitats, especially those from the lower/ seasonal delta which are shallower and hence have higher fish productivity as observed by Fox (1976). This observation is more evident in a recent study (Mosepele, 2013) on Lake Ngami which revealed that after being dry for around 20 years, the lake system currently supports a vibrant and relatively profitable small-scale commercial fishery.

Perhaps what makes lagoons key commercial fishing habitats is the dominance of Cichlids in these systems as observed by Mosepele et al. (2011). This makes lagoons key repositories of livelihoods for riparian communities. Over and above this, the presence of species such as *Hydrocynus vittatus* (Tiger-fish) makes these systems potential recreational fishing hotspots. This is based on Skelton (2001) who observed that *H. vittatus* is a key recreational fish species. However, due to the variability in spatial distribution of fish species among the delta's lagoons, only those in the upper delta such as Guma, Korao and Matsaudi are potentially recreational fishing areas while those in the lower delta may not necessarily be appealing due to low incidences of key recreational fishing species.

Conclusion – An Overview

This study has shown that lagoon fish communities in the Okavango Delta are not only diverse but also dynamic. The delta's fish species are distributed heterogeneously at both spatial and temporal scales. This paper has shown that environmental variability contributes to this spatio-temporal variability in fish species in the lagoons. Therefore, some management regulations that have been implemented in the delta based on classical fisheries management approaches may not be feasible at all, or even add value to the sustainable utilization of the delta's fish stocks.

This observation suggests that implementing closed/ restricted fishing areas (Welcomme 2001) as a management tool (e.g. fish refuges, etc.) might not only be impractical but also irrelevant in the Delta. Since Nxaraga Lagoon is located within a protected area where fishing is restricted, it follows that this management approach does not add any significant value towards fisheries management in the Delta. This study has shown that the Nxaraga lagoon fish community is different from those of the other lagoons in the Delta; therefore conserving the fish community in these lagoons does not add value towards fish conservation in the Delta. This argument agrees with Mosepele and Mosepele (2005) observations. The differences in fish community structure and in basic fish population parameters among lagoons present management challenges in this highly diverse and dynamic environment. This suggests that classical management approaches, such as mesh size regulations (Welcome 2001), might again be impractical in this highly dynamic and complex environment (where there are significant differences in morphometric characteristics of similar species from different lagoons in the delta!). This is based on the observation that similar species in different habitats follow different life history strategies, an observation that has already been made by Mosepele et al. (2005) and observed in this study. According to Mosepele (2008), there should be a paradigm shift in floodplain fisheries management, where dynamic variations such as those observed in this study are taken into account.

The biggest management challenge in floodplain fisheries is how to manage fish stocks in fluctuating systems. Moreover, this study has revealed that the Delta is not a homogenous system with species distributed evenly around it. Rather, the Delta's fish stocks in the lagoons are distributed heterogeneously throughout the entire system which allows for a more dynamic management approach. Management of these systems cannot be based on static management approaches, but should be formulated to address the fluidity of the system. Mosepele (2008) argued that a classical management approach in these fluctuating and highly dynamic fisheries is not the best management approach. He argues that there is a need for a paradigm shift in the management of floodplain fisheries, where the fluxes and dynamism of the system are incorporated into the management regime.

REFERENCES

Allanson, B. R., Hart, R. C., O'Keeffe, J. H.& Robarts, R. D. (1990). *Inland Waters of Southern Africa; An Ecological Perspective*. Kluiwer Academic Publishers, The Netherlands.

Amutenya, K., Mosepele, K.& Masamba, W. R. L. (2012). The effects of water quality (Temperature, Dissolved Oxygen and pH) on fish growth rate by using otoliths. *Technical Report*. Okavango Research Institute, Botswana.

Bills, R. (1996). *Fish Stock Assessment of the Okavango River.* J.L.B. Smith Institute of Ichthyology, Grahamstown, South Africa. Investigational Report No. 56. 110.

Fox, P. J. (1976). Preliminary observations on fish communities of the Okavango Delta., 125-130. In: *Proceedings of the symposium on the Okavango Delta and its future utilization.* National Museum, Gaborone, Botswana.

Gopelomang, A. (2010). *Spatial variations on the fish community of the Okavango delta;* a comparative study of Matsaudi and Chanoga lagoons' fish community. Report. Botswana Wildlife raining Institute, Botswana.

Henderson, P. A. & Crampton, W. G. R. (1997).A comparison of fish diversity and abundance between nutrient-rich and nutrient-poor lakes in the Upper Amazon. *Journal of Tropical. Ecology, 13* (2), 175–198.

Hoberg, P., Lindholm, M., Ramberg, L.& Hessen, D. O. (2002). Aquatic food web dynamics on a floodplain in the Okavango Delta, Botswana. *Hydrobiologia,470*, 23–30.

Junk, W. J. (2000). The Central Amazon river floodplain: Concepts for the sustainable use of its resources. 75 – 94. In: Junk W.J., Ohly J.J., Piedade M.T.F., Soares M.G.M., (Eds.) *The central Amazon floodplain: Actual use and options for a sustainable management.* Backhuys Publishers, Leiden.

Lindholm, M. (2006). Zooplankton diversity and successions in the Okavango delta, Botswana - diversity and ecosystem interactions in land/water ecotones. *PhD Thesis,* University of Oslo.

McCarthy, T. S., Ellery, W. N. & Rogers, K. H. (1987). Observations on the abandoned Nqoga Channel of the Okavango Delta. *Botswana Notes and Records,19*, 83–89.

Merron, G. S.& Bruton, M. N. (1988). *The ecology and management of the fishes of the Okavango delta,* Botswana, with special reference to the role of the seasonal floods. J.L.B. Smith Institute of Ichthyology, Grahamstown, South Africa. Investigational Report, No. 29.

Merron, G. S.& Bruton, M. N. (1995). Community ecology and conservation of the fishes of the Okavango Delta, Botswana. *Environmental Biology of Fishes*, *43*, 109-119.

Merron, G. S. (1991). *The ecology and management of the fishes of the Okavango Delta, Botswana, with particular reference to the role of the seasonal floods.* Rhodes University, PhD thesis. Grahamstown, S. Africa.

Merron, G. S. (1993). The diversity, distribution and abundance of the fishes in the Moremi Wildlife Reserve, Okavango Delta, Botswana. *South African Wildlife Research,23*(4), 115 – 122.

Mmopelwa, G., Mosepele, K., Mosepele, B., Moleele, N.& Ngwenya, B. (2009). Environmental variability and the fishery dynamics of the Okavango Delta, Botswana: The case of subsistence fishing. *African Journal of Ecology*, *47* (Suppl. 1), 1–9.

Mmopelwa, G. (2006). The value of the Okavango Delta: A natural resource accounting approach. University of Pretoria, South Africa. PhD Thesis.

Mosepele, K. (2001b). *Description of the Okavango delta fishery.* Technical Report to the Fisheries Section, Ministry of Agriculture, Gaborone, Botswana.

Mosepele, K.& Ngwenya, B. (2010). *Socio-economic survey of commercial fishing in the Okavango Delta,* Botswana. Bay Publishing: Gaborone, Botswana.

Mosepele, K., Kolding, J., Bokhutlo, T.& Wolski, P. (2012). *Fisheries dynamics in a fluctuating environment: The seasonal flood pulse – a major driver of change in the Okavango delta's fishes and fishery?* Technical Report. Okavango Research Institute, Botswana.

Mosepele, K., Mosepele, B., Wolski, P. & Kolding, J. (2012). Dynamics of the feeding ecology of selected fish species from the Okavango River delta, Botswana. *Acta Ichthyologica et Piscatoria*, *42* (4), 271 – 289. DOI: 10.3750/AIP2012.42.4.01.

Mosepele, K., Moyle, P. B., Merron, G. S., Purkey, D.& Mosepele, B. (2009). Fish, floods and ecosystem engineers; interactions and conservation in the Okavango Delta, Botswana. *Bioscience*, *59* (1), 53–64.

Mosepele, K. (2000). Length based fish stock assessment of the main exploited fish stocks of the Okavango delta, Botswana. MPhil Thesis with the University of Bergen, Bergen, Norway.

Mosepele, K. (2001a). *An inventory of the fishing lagoons of the Okavango Delta.* Technical Report to the Fisheries Section, Ministry of Agriculture, Gaborone, Botswana.

Mosepele, K. (2008). Flood pulse in a subtropical floodplain fishery and the consequences for steady state management. 56 – 62. In Totolo, O (Editor). Proceedings of the Second IASTED Africa Conference Water Resource Management (Africa WRM 2008) September 8-10, 2008 Gaborone, Botswana.

Mosepele, K. (2010). *Fish and fisheries of Boteti River.* Technical Report. Okavango Research Institute, Botswana.

Mosepele, K. (2013). *The fishes and fishery of Lake Ngami.* Technical Report. Okavango Research Institute, Botswana.

Mosepele, K., Williams, L. & Mosepele, B. (2005). Assessment of the feeding ecology of the Silver Catfish (Schilbe intermdius, Ruppel, 1932) in a seasonal floodplain of the Okavango Delta, Botswana. *Botswana Notes and Records*, *37*, 208-217.

Mosie, I. (2007). Environmental variability and fishery dynamics in Nxaraga Lagoon, Botswana. MSc Thesis. University of the Orange Free State, South Africa.

Ngwenya, B.& Mosepele, K. (2008). Socio-economic survey of subsistence fishing in the Okavango Delta, Botswana. Bay Publishing: Gaborone, Botswana.

Pauly, D. (1975). On the ecology of a small West African lagoon. *Berichte der Deutschen Wissenschaftlichen Kommission für Meeresforschung*,*24*(1), 46-62.

Petry, A. C., Agostinho, A. A. & Gomes, L. C. (2003). Fish assemblages of tropical floodplain lagoons: Exploring the role of connectivity in a dry year. *Neotropical Ichthyology*,*1* (2), 111–119.

Porter, J. W. & Muzila, I. L. (1989). Aspects of swamp hydrology in the Okavango. *Botswana Notes and Records*,*21*, 73-91.

Ramberg, L., Hancock, P., Lindholm, M., Meyer, T., Ringrose, S., Silva, J., Van As, J.& Vanderpost, C. (2006). Species diversity of the Okavango Delta, Botswana. *Aquatic Sciences*,*68*, 310-337.

Rodrigues, L. C., Train, S., Roberto, M.,do, C. & Pagioro, T.A. (2002). Seasonal fluctuation of some limnological variables on a floodplain lake (Patos Lagoon) of the Upper Parana River, Mato Grosso do Sul State, Brazil. *Brazilian Archives of Biology and Technology*,*45* (4), 499–513.

Sethebe, K. (2011). *Fish Dynamics and Water Quality in the Okavango Delta, Botswana: the Case of Guma Lagoon. MSc Thesis.* University of the Free State, South Africa.

Silvano, R. A. M., Do Amaral, B. D. & Oyakawa, O. T. (2000). Spatial and temporal patterns of diversity and distribution of the Upper Jurua River fish community (Brazilian Amazon). *Environmental Biology of Fish*, *57*, 25–35.

Soares, M. G. M. & Junk, J. W. (2000). Commercial fishery and fish culture of the state of Mazonas: Status and perspectives. 433-461. In: Junk W.J., Ohly J.J., Piedade M.T.F., Soares M.G.M., (Eds.) *The central Amazon floodplain: Actual use and options for a sustainable management.* Backhuys Publishers, Leiden.

Tweddle, D., Bills, R., van Der Waal, B.,Skelton, P., Kolding, J. & Nengu, S. (2003). Fish diversity and fisheries in the Okavango Delta, Botswana. Pp. 97 – 110. In: Alonso L.E., Nordin L-A. (Editors), A rapid assessment of the aquatic ecosystem of the Okavango Delta, Botswana: high water survey. *RAP Bulletin of Biological Assessment*. 27.

Welcomme, R. L. (1999). A review of a model for qualitative evaluation of exploitation levels in multi-species fisheries. *Fisheries Management and Ecology*, (*6*), 1-19.

Welcomme, R. L. (1979). *Fisheries ecology of floodplain rivers*. Longman, London.

Welcomme, R. L. (2005). Impacts of fishing on inland fish populations in Africa. 81 – 85. In: Thieme M.L., Abell R., Stiassny M.L.J., Skelton P., Lehner B., Teugels G.G., Toham A.K., Burgess N., Olson D., 2005, *Freshwater ecoregions of Africa and Madagascar, a conservation assessment*. Island Press, Washington, DC.

In: Lagoons
Editor: Mwinyikione Mwinyihija
ISBN: 978-1-62808-092-6

Chapter 7

VEGETATION AND HABITAT ECOLOGY OF LAGOONS IN NORTH EGYPT: ROLE OF EDAPHIC AND ANTHROPOGENIC FACTORS

Mohamed A. El-Sheikh*
Damanhour University, Faculty of Science, Botany Department, Damanhour, Egypt

ABSTRACT

The main habitats of lagoons in the Nile Delta of Egypt include the Mediterranean Deltaic coast, the sand bar between the northern shore and Mediterranean Sea, the water courses that drain into the lakes and the wetland around it. These lagoon habitats represent highly dynamic ecosystems in Egypt that have been undergoing continuous changes until the present time due to increasing human impact. Increasing threats to plant diversity of lagoons in Egypt include loss of natural habitats through increased inhabitants which require more land for housing, roads, cultivation, industry, and waste disposal. In order to identify the plant communities, species diversity, vulnerable habitats, threatened species and environmental factors that affect their distribution, a multivariate (TWINSPAN, DCA and CCA) and statistical analysis of the vegetation data were applied. The vegetation was classified into different groups (i.e. plant communities). Many groups were dominated by halo-tolerant plants (e.g. *Arthrocnemum macrostachyum*, *Sarcocornia fruticosa*, *Halocnemum strobilaceum*, *Salsola kali*, *Senecio glaucus*, *Cakile maritima*, *Zygophyllum album*, *Bassia indica*, *Mesembryanthemum nodiflorum*, *Limbarda crithmoides*, etc) and common reed (*Phragmites australis*). The other few groups dominated by the emergent (e.g. *Typha domingensis*, *Juncus acutus*, *Cyperus alopecroides*, etc.) and hydrophytes plants (e.g. *Potamogeton pectinatus*, *Ceratophyllum demersum*, *Echhornia crassipes*, etc). Moisture, salinity, sedimentation and anthropogenic factors were the main factors that governed the plant succession in these lagoons.

* Corresponding author: el_sheikh_eg@yahoo.co.uk, King Saud University, College of Science, Department of Botany and Microbiology, P.O. Box 2455; Riyadh 11451, Kingdom of Saudi Arabia.

Ecology Characteristics of Lagoons Habitat in Nile Delta

Coastal Lagoons

Phleger (1969), defined coastal lagoons as shallow inland marine waters, usually oriented parallel to the coast, separated from the ocean or sea by a barrier, and connected to the ocean by one or more restricted inlets. To this should be added that the ocean or sea entrance (s) can at times be closed off by sediment deposition as a result of wave action and littoral drift. A workable definition of a coastal lagoon is thus ***a shallow coastal water body separated from the ocean or sea by a barrier, connected at least intermittently to the ocean or sea by one or more restricted inlets, and usually oriented shore-parallel.*** Therefore, "Lagoon" a shallow, coastal body of water separated from the ocean or sea by a sand bar, which may periodically breach, opening the lagoon to the sea for a time. Lagoons can form where a river meets the sea or ocean (an estuarine lagoon), or without the influence of a river. Coastal lagoons do not usually have a free connection with the open sea and may only be inundated with sea water at irregular intervals (Mc Lusky 1990). Lagoon also is a small world composed of environmental features and living organisms, bound and organized by interdependence and interrelationships (Forbes 1987). Coastal lagoons occupy 13% of coastal areas worldwide, and are often impacted by both natural and anthropogenic influences (Mee, 1978; Sikora & Kjefie, 1985). Depending on local climatic conditions, lagoons exhibit salinities which range from completely fresh to hypersaline (Moore & Slinn, 1984; Kjefie, 1986; Kjerfve & Magill, 1989; Knoppers et al., 1991). Coastal lagoons experience forcing from river input, wind stress, tides, precipitation to evaporation balance, and surface heat balance, and respond differently to these forcing functions. Water and salt balances, lagoon water quality, and eutrophication depend critically on lagoon circulation, salt and material dispersion, water exchange through the ocean or sea canal (s), and turnover, residence, or flushing times. The understanding of physical, chemical, geological, and ecological dynamics of lagoons is important for planning and implementation of coastal management strategies in coastal lagoons. Kjerfve (1986) sub-divided coastal lagoons into three geomorphic types according to water exchange with the coastal ocean: Choked, Restricted and Leaky Lagoons (Figure 1).

Ecological Characteristics of Lagoons

Coastal lagoons are highly productive ecosystems. They contribute to the overall productivity of coastal waters by supporting a variety of habitats including salt marshes, seagrasses, and mangroves (Anthony, et al., 2009). They also provide essential habitat for many fish and shellfish species (Bertness, 2007). Such beds play an important role in influencing the shape and stability of the shoreline, regulating dissolved oxygen (Nixon & Oviatt 1972) and filtering suspended matter (Bertness, 2007). They can enhance the biodiversity of a lagoon by providing a physical refuge from predation and also serve as nursery and feeding habitat to a variety of organisms (Heck & Thoman 1984, Harris et al., 2004). On the Egypt coast, salt marshes are one of the most prevalent habitats in lagoons and are one of the most productive natural vascular plant communities in the world (Whittaker 1975; Bertness 2007; El-Sheikh et al. 2012). Because of their relatively low flushing rates, coastal lagoons are favorable habitats for primary producers (phytoplankton and aquatic plants). Nutrients are transported to lagoons from surface and ground water flows, and through exchange with the sea. Because nutrient availability often limits primary productivity, coastal lagoons can foster high rates of primary production, thereby supporting high rates of secondary production as compared to other aquatic ecosystems (Nixon 1982, Nixon 1995). However, primary production that exceeds the demands of consumers can lead to eutrophication (Valiela et al. 1992). Eutrophication is characterized by excessive phytoplankton and macroalgae blooms and subsequent hypoxia, reduced light penetration (McGlathery 2001; Anderson et al. 2002), stress and die-offs of marine organisms, loss of seagrass beds, changes in food web interactions and community structure, and loss of biodiversity (National Research Council 2000). Lagoons and their associated ecosystems are highly valued by society. These values reflect the commercial, recreational, and tourism uses of lagoons as pleasant swimming beaches. Pragmatic values also include ecosystem services that indirectly support human uses. For example, salt marshes provide nursery habitat for juvenile fish that support commercial fisheries and protect developed shorelines by reducing the impact of severe storms (Anthony et al. 2009).

The lakes and lagoons are a major estuarine water body located in Egypt whose remaining natural resources need to be protected. The lagoonal ecosystem has been stressed through the past many years due to many anthropogenic influences. Altered hydrology of the system allows massive freshwater discharges into the lagoon. These discharges carry large influxes of nutrients, suspended and dissolved organic matter, contaminants, and toxins into the lagoon, affecting the flora and fauna. Additional pressures in this urbanized coastal area include boating and fishing pressures, as well as loss of natural habitat through physical alterations to the system (Galal, et al. 2012).

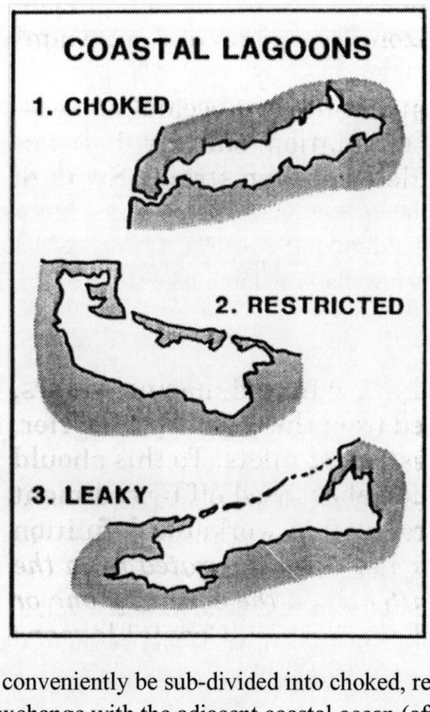

Figure 1. Coastal lagoons can conveniently be sub-divided into choked, restricted, and leaky systems based on the degree of water exchange with the adjacent coastal ocean (after Kjerfve, 1986).

LAGOON ECOLOGY OF CENTRAL COASTAL FISHES

The lagoons at the mouths of Nile Delta have been highly modified by adjacent agricultural and urban development. In addition, they receive the accumulated impacts of water diversion, sedimentation and pollution discharges within the watersheds. Despite historical impacts, these estuaries can provide potentially valuable habitat for aquatic invertebrates, fishes and the wildlife dependent upon them. The relative value of individual estuaries varies with size, tidal action, depth, salinity and water quality. These features not only vary between estuaries, but also vary within estuaries on a seasonal and year-to-year basis. Based upon studies of many estuaries in Nile Delta, this chapter outlines some of the important factors in the functioning of lagoons. It also describes the vegetation structure and habitat ecology of lagoons in Egypt.

HABITAT TYPES OF LAGOONS IN EGYPT

The Mediterranean coastal region of Egypt extends for about 970 km from Sallum in the west (31° 34` N, 25° 09`E) to Rafah in the east (34° 20`N, 31° 25`E) (Table 1) (Galal et al. 2012). Many smaller lagoons and lakes extend along this coast: Mariut (Western Coast), Edku, Burullus and Manzala (Deltaic Coast) and Bardawil (Sinai Coast). All lakes have a brackish water nature except Bardawil has saline water (Figure 2). These lagoon possibly represented as interdistributary bays, were closed-in by the long shore growth of spit from the various cuspate subdeltas. They are separated from the Mediterranean Sea by strips of sand bars. They also may have developed out of flood basins, especially during stages of large Nile floods. The lagoon expansion during the last 2000 years has been irregular, mainly due to subsidence behind a stable beach barrier. There is a record of rapid spread of swamps and lagoons since the 5-10th centuries, in coincidence with the major earthquakes (Abu Al-Izz 1971; Ben Menchem 1979). Shaltout & Al-Sodany (2008) described six main types of habitat in the Nile Delta coast and Burullus wetland: (1) coastal salt marshes covered with water, (2) sand formations that cover the surface of the marine bar (sheets, hillocks and dunes), (3) lake cuts that represent the recent lands resulting from the drying process of the outermost western and eastern fringes of the lake during 1984–1997, (4) drains that discharge at the southern shore, (5) the lake proper which is classified into lake shore and open water, and (6) the islets scattered within the water body.

SOILS

The soil of the Nile Delta coastal belt are genetically related to the major morphological units of alluvial plain deposition: sand dunes and coastal sandy plains, Nile levees and channels (silt to fine sand), Nile flood basins composed of silt and areas of marshes. Modifications have been introduced by continued agricultural activity (Balba 1981). Deposited in the brackish water of lagoon, lakes and swamps, the clay soils of the fluvio-marine marshlands contain some lime but much salt and gypsum (Galal, et al. 2012). The soils are saline sodic with a high concentration of salts on the surface.

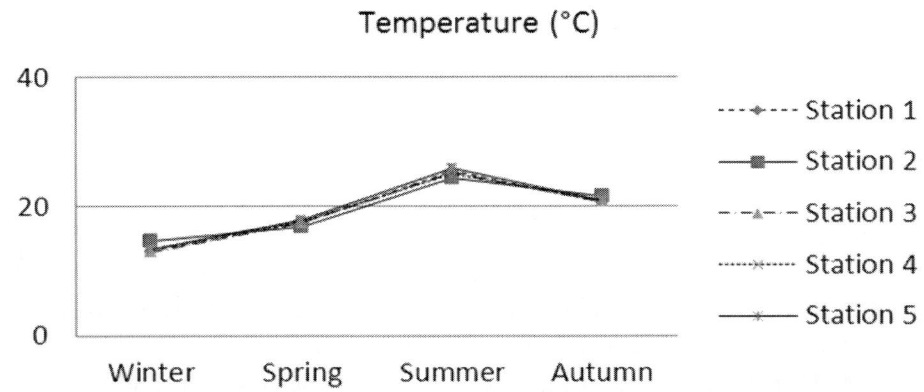

Figure 2. Seasonal variations of temperature water in five stations.

Table 1. Morphometry of the five northern lakes under study after (Galal, et al. 2012)

Lake	Coordinates		Dimensions			Area (km^2)	Fish Production (Ton / ha)
	Latitude (N)	Longitude (E)	Mean depth (m)	Maximum length (km)	Maximum width (km)		
Mariut	31°2′ - 31°12′	29°51′- 29°59′	*1.2*	*8.8*	7.7	*63*	0.50
Edku	31°13′ - 31°16′	30°07′ - 30°14′	*1.0*	21.0	*6.0*	126	1.10
Burullus	31°25′ - 31°35′	30°30′ - 31°10′	1.0	64.0	16.0	410	*1.20*
Manzala	31°00′ - 31°30′	31°16′ - 32°20′	1.1	64.5	*49.0*	*1200*	0.70
Bardawil	31°03′ - 31°14′	32°40′ - 33°30′	*1.0*	*75.0*	22.0	650	*0.03*

CLIMATE

According to the map of the world distribution of arid regions (UNESCO, 1977), the northern Mediterranean part of the Nile Delta belongs to the arid region. The climatic conditions are warm summers (20 to 30 °C) and mild winters (10 to 20 °C). The aridity index (P/PET: where P is the annual precipitation and PET is the potential evapo-transpiration) ranges between 0.03 and 0.2 in the north Delta (arid region), and less than 0.03 in the south hyper-arid region (Shaltout & Al-Sodany 2008).

STUDY DESIGN OF NORTH EGYPT LAGOONS

Many lagoons were surveyed in the last few years by many authors such as: Shaltout & Al-Sodany (2008) selected 227 stands which represented the different habitats at Burullus coastal lagoon; El-Sheikh et al. (2012) selected 22 stands to represent the different successional stages through 10 years of the land fill of lagoon at Burullus lake and Galal, et al. (2012) selected 148 stands in Mariut and 150 stands in Edku lake. The sampled stands about 20 m^2 was observed seasonally. The species present life forms and chorotypes were recorded. Soil and water samples were collected and their physical and chemical characteristics were analyzed. The total cover of each species (%) was estimate according (Canfield 1941). Two-way indicator species analysis (TWINSPAN), detrended correspondence analysis (DECORANA) and (CANOCO) were applied to stand, the plant cover and environmental variable matrices (species x stands x environment) (Hill, 1979 a, b; Ter Braak & Smilauer, 2002).

Species richness (α-diversity) of the vegetation cluster was calculated as the average number of species $stand^{-1}$. Equitability or evenness of the relative importance values of species was expressed according to the Shannon-Wiener index ($\hat{H} = -\Sigma$ pi log pi). A relative concentration of dominance was expressed by the Simpson index ($C = \Sigma pi^2$): where (pi) is the relative importance value (i.e., relative cover) of the ith species (Whittaker, 1972; Pielou, 1975; Magurran, 1988). The simple linear correlation (r) and analysis of variance ANOVA one were calculated for soil and vegetation variables according to SAS software (1989-196).

Table 2. Endemic species recorded in the five lagoons according to the IUCN (1994) Red list categories (after El-Hadidi and Hosny 2000 and Galal, et al., 2012)

Species	Life form	Nortern lakes				
		Mariut	Edku	Burullus	Manzala	Bardawil
Endemics						
Sinapis allionii Jacq.	Therophyte	+		+		
Bromus aegyptiacus Tausch	Therophyte	+			+	
Sonchus macrocarpus Boulos & C. Jeffrey	Chamaephyte			+		
Astragalus camelorum Barbey	Chamaephyte					+
Bellevalia salah-eidii Taekh. & Boulos	Geophyte-Helophyte					+
Near endemics						
Carduncellus mareoticus (Delile) Hanelt	Chamaephyte	+				
Biarum olivieri Blume	Geophyte-Helophyte					+
Iris mariae Barbey	Geophyte-Helophyte					+

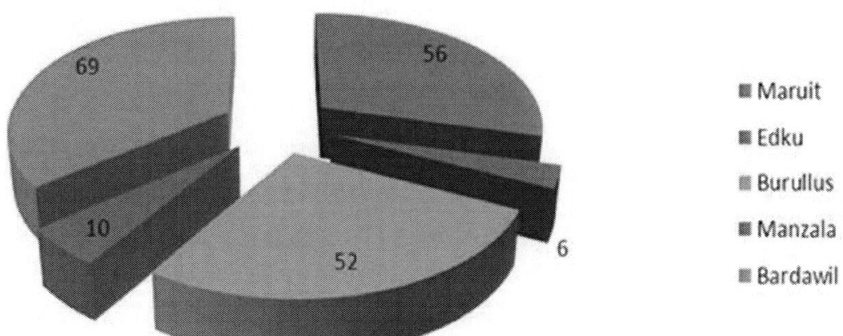

Figure 3. Number of species unique to each lake.

Floristic Diversity

A total of about 402 species belonging to 248 genera and 68 families were recorded in these lakes (Galal, et al., 2012). The grasses (66 spp.) have the highest contribution to the total flora, followed by composites (44 spp.) and chenopods (38 spp.). Floristically, the richest lake was Burullus. Burullus had the highest species number (224 spp. = 56%), followed by Mariut (148 spp.= 49%), Manzala (144 spp. = 36%), Edku (120 spp. = 30%) and Bardawil (136 spp. = 34%). A total of 189 species (48% of the total species) were recorded in only one lagoon "a unique species", 95 of them are annual and 94 perennials. Also, 56 species of them were recorded in Mariut only (as: *Anchusa hispida*, *Carrichtera annua, Mellilotus messanensis*, *Arisarum vulgar* and *Fagonia cretica*), 6 in Edku, 52 in Burullus, 10 in Manzala and 69 in Bardawil lagoon (Figure 3). The life forms of the recorded species indicated the predominance of therophytes (203 spp. = 51%) followed by geophytes–helophytes (61 spp. =15 %), chamaephytes (56 spp. =14 %), hydrophytes (48 spp. =12%) and hemicryptophytes (42 spp. =10%) (Figure 4). The global distribution of the recorded species indicated that seven species are endemic (Table 2) as: *Zygophyllum aegyptium*, *Sinapis allionii*, *Sonchus macrocarpus*, *Astragalus camelorum*, *Allium papillare*, *Bellevalia salah-eidii* and *Iris mariae* and most of the recorded species are Saharo-Arabian (57 spp.) of monoregionals. Fourteen species are threatened according to IUCN (1994) Red list categories (after El-Hadidi and Hosny 2000): *Astragalus camelorum*, *Bellevalia salah-eidii*, *Biarum olivieri*, *Chlamydophora tridenata*, *Clerodendrum acerbianum*, *Cynomorium coccineum*, *Iris mariae*, *Juncus bufonius*, *Lobularia arabica*, *Nymphaea caerulea*, *Nymphaea lotus*, *Salsola tetragona*, *Sinapis allionii* and *Sonchus macrocarpus* (Shaltout & Khalili 2005; Shaltout & Al-Sodany 2008;Galal, et al., 2012; El-Sheikh et al., 2012).

Vegetation and Habitat Composition Analysis

Burullus Lagoon: As a Case Study

Habitat analysis

The application of multivariate analysis techniques on the plant spp. 197 spp. x 10 habitats (after Shaltout and Al-Sodany 2008) (Figure 5) and the similarity DCA ordination (Figure 6) indicate a distinction of four clusters. These clusters were separated along the prevailing edaphic gradient from the shoreline to the open water. Cluster A: comprises the sand formation, cluster B: include the salt marshes, cluster C: include the shorelines and cluster D comprises lagoon open water habitats. The application of DCA and CCA confirmed the separation between these clusters and indicated relationships between environmental gradients and topographic aspects of lagoons (Figure 6 and 7a,b). CCA ordination was used to verify the correlation analysis between the dominant environmental factors and CCA axes (Figure 7a and b and Table 3). Correlation analysis indicated that the separation of the species along the first axis is strongly affected positively by pH, silt, organic matter, nitrogen, Na and K content and negatively by clay, P and Ca contents. On the other hand, Mg and silt content are correlated positively with the second axis and negatively with sand. Therefore, the cluster A that occupies the sand formation and B that occupies the salt marshes were separated on the left hand of axis 1 from the cluster C and D which inhabited the shoreline and open water zones on the right hand of axis 2 (Figure 7a). Halophytes and geophytes species occupied the sand formation and salt marshe habitats of communities (e.g. *Salsola kali, Senecio glaucus, Mesem.nodiflorum, Mesem. Crystallinum, Phragmites australis, Arthrocnemum macrostachyum, Sarcocornia fruticosa, Suaeda vera, Tamarix nilotica, Halocnemum strobilaceum,Cynodon dactylon,Zygophyllum album* var. *album, Inula crithmoides, Juncus acutus*) on the negative part of axis 1 are correlated with sand, clay, P, Ca and EC content. On the other hand, the hydrophytes species occupied open water habitat of communities (e.g. *Potamogeton pectinatus, Eichhornia crassipes, Ceratophyllum demersum, Lemna perpusilla,Ludwigia stolonifera, Lemna gibba, Potamogeton crispus, Azolla filiculoides, Wolffia hyaline, Ceratophyllum submersum, Najas marina v. armata* and *Najas minor*) on the positive part of axis 1 are correlated with pH, silt, organic matter, nitrogen, Na and K content (Figure 7b).

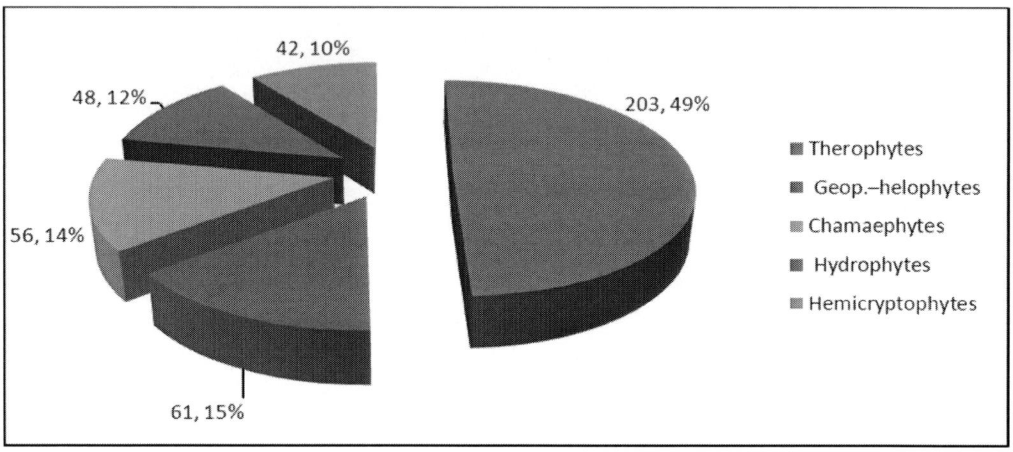

Figure 4. Life form spectra of the recorded species in the five lagoons North Egypt, after Galal, et al., (2012).

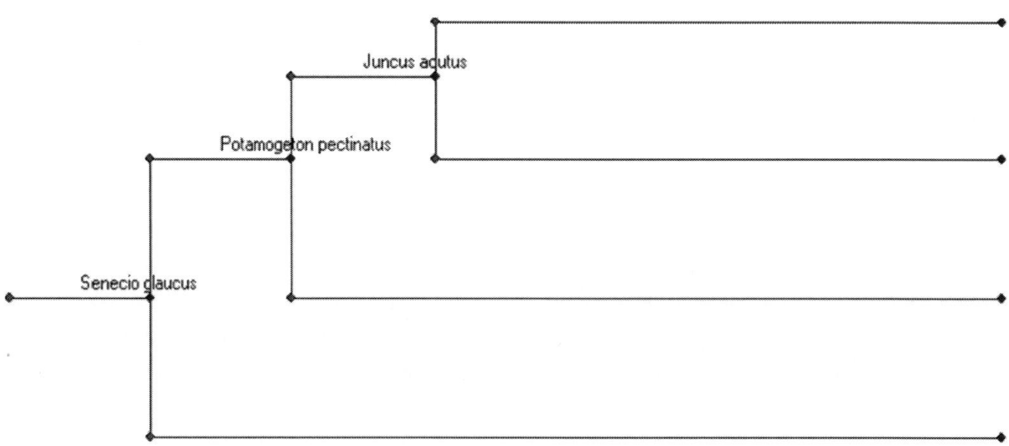

Figure 5. TWINSPAN classification of 10 habitats led to 4 habitat groups. 1=SM: salt marshes, 2=SS: sand formations, 3=LG: lake cuts, 4=TD: terraces, 5=SD: slopes, 6=ED: water edges and 7=OD: open water zones of the drains, 8=LS: lake shores. 9=LO: open water of the lake and 10= IS: lake islets.

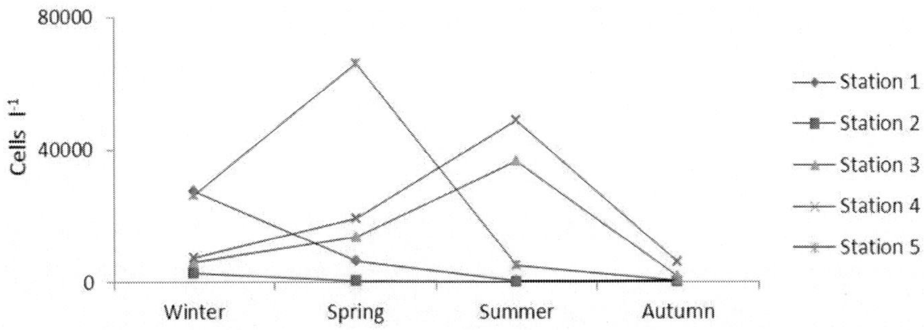

Figure 6. Seasonal changes of dinoflagellate community in five stations of Ghar El Melh lagoon.

LAKE BURULLUS: VEGETATION COMPOSITION

After the study of Shaltout and Al-Sodany (2008), the application of TWINSPAN on the cover estimates of 197 species recorded in 227 stands led to the recognition of 13 vegetation groups at the 6th level of classification. The application of DCA on the same set of data indicated a reasonable segregation between many of these groups along the ordination plane of axes 1 and 2 (Figure 3). Two groups mainly occupied the lake islets (*Arthrocnemum macrostachyum - Juncus acutus* and *Typha domingensis - Ceratophyllum demersum*), three inhabited the lake open water (*Potamogeton pectinatus*, *Phragmites australis* and *Phragmites australis-Potamogeton pectinatus*) and three inhabited the lake shores (*Juncus acutus*, *Sarcocornia fruticosa* and *Phragmites australis - Typha domingensis*). On the other hand, five groups occupied a wide habitat gradient (*Phragmites australis-Suaeda pruinosa*, *Phragmites australis - Arthrocnemum macrostachyum*, *Suaeda vera-Inula crithmoides*, *Halocnemum strobilaceum* and *Salsola kali*).

Table 3. Inter-set correlations of environmental variables with axes CCA of the 10 Burullus habitats. Significant values at * ≤0.05, ** ≤0.01 and* ≤0.001**

Variable	AX1	AX2
pH	0.71***	0.06
EC	-0.19	0.19
Clay	-0.56**	-0.04
Silt	0.34*	0.63**
Sand	-0.24	-0.64**
OM	0.38*	0.63**
$CaCO_3$	-0.05	0.13
N	0.68**	0.21
P	-0.82***	0.01
Na	0.44*	0.33*
K	0.68**	0.40*
Ca	-0.75***	0.05
Mg	-0.01	0.38*
Fe	-0.10	-0.02

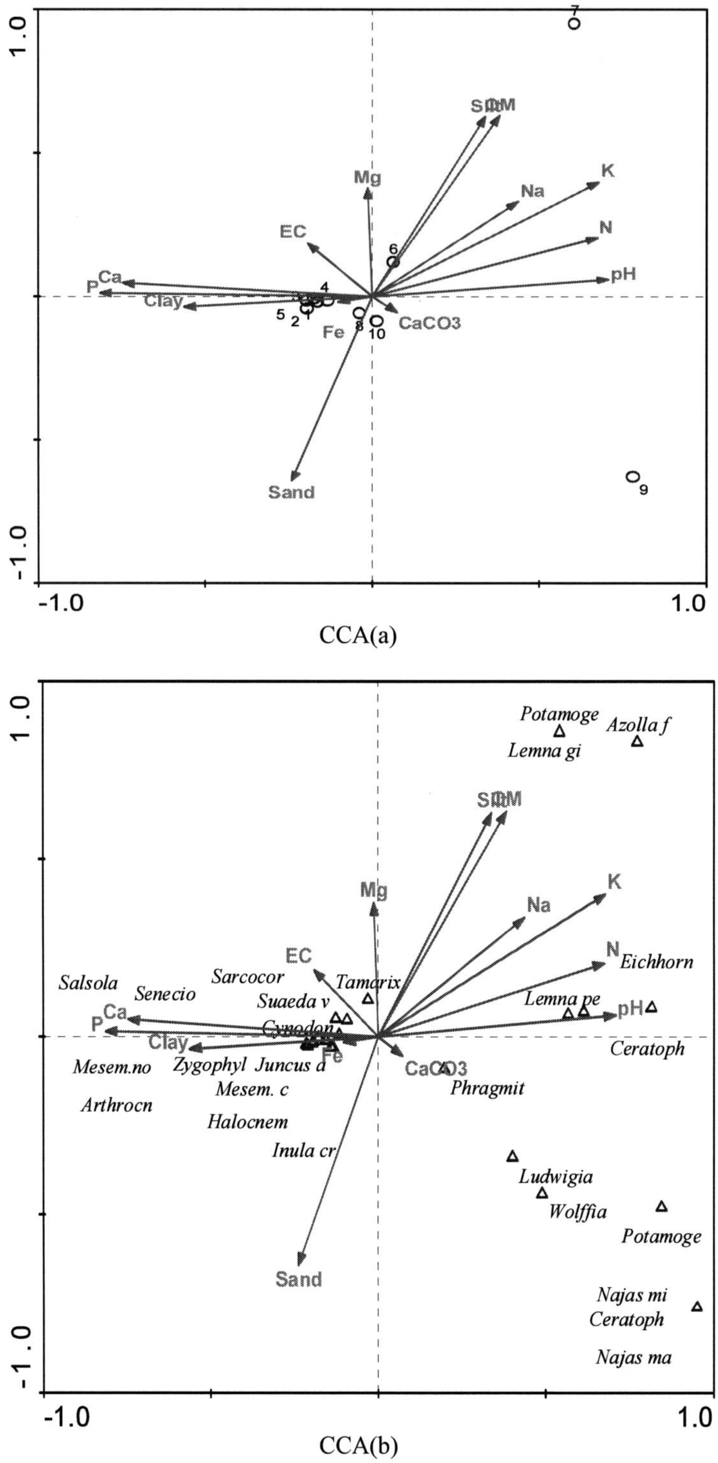

Figure 7a,b. CCA ordination of common dominant species in 10 habitats 1=SM: salt marshes, 2=SS: sand formations, 3=LG: lake cuts, 4=TD: terraces, 5=SD: slopes, 6=ED: water edges and 7=OD: open water zones of the drains, 8=LS: lake shores. 9=LO: open water of the lake and 10=IS: lake islets.

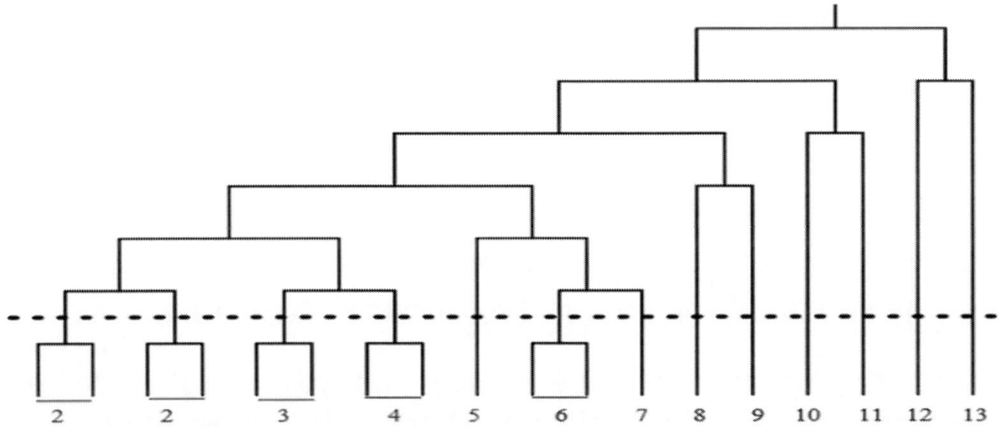

a: TWINSPAN

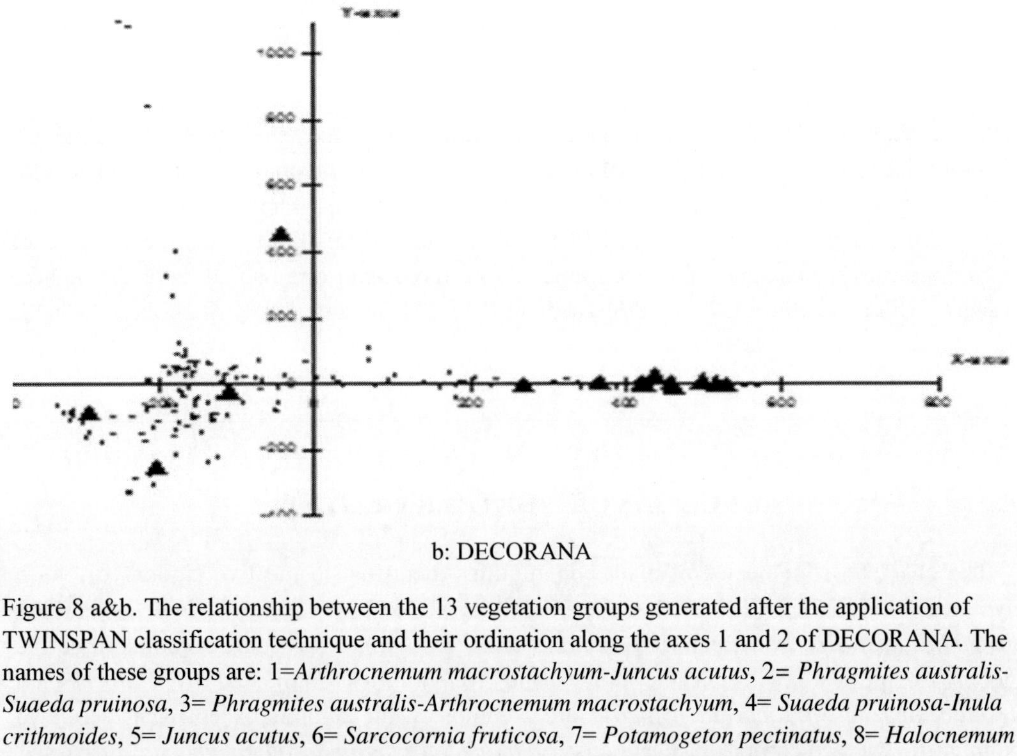

b: DECORANA

Figure 8 a&b. The relationship between the 13 vegetation groups generated after the application of TWINSPAN classification technique and their ordination along the axes 1 and 2 of DECORANA. The names of these groups are: 1=*Arthrocnemum macrostachyum-Juncus acutus*, 2= *Phragmites australis-Suaeda pruinosa*, 3= *Phragmites australis-Arthrocnemum macrostachyum*, 4= *Suaeda pruinosa-Inula crithmoides*, 5= *Juncus acutus*, 6= *Sarcocornia fruticosa*, 7= *Potamogeton pectinatus*, 8= *Halocnemum strobilaceum*, 9= *Salsola kali*, 10= *Phragmites australis*, 11= *Phragmites australis-Potamogeton pectinatus*, 12=*Typha domingensis-Ceratophyllum demersum*, and 13= *Phragmites australis-Potamogeton pectinatus* (After, Shaltout & Al-sodany 2008).

Phragmites australis is common in aquatic and moist places throughout north lagoons except Bardawil lagoon (Täckholm, 1974, Zahran & Willis 1992, Boulos 2005, Shaltout & Al-sodany 2008; El-sheikh 2012). Its growth in Burullus lagoon represents one of the most important reed bed in the Mediterranean basin, where this type of habitat is becoming rare and threatened, due to many human impacts in these north lagoon of Egypt as: discharge of polluted liquid, soil and organic refuses, grazing, cutting, burning and drying for cultivation. Wintering and migrant birds are strongly dependent on the reed beds for forage, refuge and breeding (Shaltout et al. 2004; Shaltout et al. 2005a; Shaltout & Khalil 2005; Shaltout & Al-Sodany 2008 and El-Sheikh et al. 2012). *Potamegaton pectinatus* is submerged macrophytes of nearly cosmopolitan distribution. It is the most dominant submerged plant in Burullus lagoon. It can also tolerant high salinity, pH, alkalinity and eutrophic waters, and may be grows in these polluted water and accumulate them in its leaves and stem (Demirezen & Aksoy 2004; Shaltout & Khalil, 2005).

VEGETATION COMPOSITION IN 5 LAGOONS OF NORTHERN EGYPT

After Galal, et al. (2012) multivariate analysis is used to analyze the vegetation and environmental variables of 298 sampled stands in 5 lagoons. 17 vegetation groups in Lake Mariut (188 spp. x148 samples), 15 in lagoon Edku (112 spp. x 150 samples), 13 in lagoon Burullus (197 spp. x 227 samples), 8 in Manzala (37 spp. x 100 samples) and 9 in Bardawil (45 spp. x 150 samples) were identified. These groups were separated along the prevailing moisture and salinity gradient from the shoreline to the open water.

The plant communities were listed with their species diversity and some common edaphic factors in (Table 4). The diversity of plant communities generated from multivariate analysis in Maruit lagoon characterize by different habitats (habitat diversity). Communities that inhabit open water habitat are less divers than other communities. The high specificity of the species of these communities to their habitats may be support this finding. On the other hand, the communities that extend their occurrence along more than one habitat have the highest diversity. These results are coinciding with that of (Al-Sodany 1998; Shaltout & Khalil, 2005).

DISTRIBUTION OF COMMON PLANT COMMUNITIES IN DIFFERENT LAGOONS OF NORTHERN EGYPT

The characteristic species of plant communities identified in the five lagoons of North Egypt are indicated in (Table 5) (After Galal, et al., 2012). *Ceratophyllum demersium-Phragmites australis*, *Potamegaton pectinatus* and *Typha domingensis* are common in all the lagoons except Bardawil. On the other hand communities dominated by *Halocnemum strobilaceum* and *Sarcocornia fruticosa* are common in all lagoons except Manzala. Two communities were recorded in 3 lagoons: *Echinochloa stagnina* and *Eichhornia crassipes.* Five communities were recorded in 2 lagoons: *Arthrocnemum macrostachyum*, *Juncus acutus*, *Najas marina* var. *armata*, *Limbarada crithmoides*, *Saueda vera* and *Zygophyllum album*. A total of 32 communities were recorded only in one lagoon, 8 of these communities were recorded in Mariut only and 7 in Edku. *Salsola kali* and *Suaeda vera* were restricted in Burullus, 4 were recorded in Manzala and 11 has a unique occurrence in Bardawil lagoon (Table 5).

Table 4. The vegetation groups and associated species recorded in the five lagoon in North of Egypt with their species diversity and some edaphic variables after (, Shaltout & Khalil 2005; Galal, et al. 2012; Shaltout & Al-Sodany 2008; El-Sheikh et al. 2012)

Community name	Associated species	Species diversity indices			Edaphic variables		
		Total cover of 1st dominant sp %	Sp. richness	Shannon H	Simpson C	pH	EC $mScm^{-1}$
Lake Maruit							
1. *Arthocnemum macrostachym-Juncus rigidus*	*Sarcocornia fruticosa-Sonchus oleraceus*	50	4	0.5	1.3	7.9	27
2. *Sarcocornia fruticosa-Halocnemum strobilaceum*	*Atriplex semibaccata-Phragmites australis*	50	10	2.5	5.5	7.7	15
3. *Atriplex halimus-Tamarix nilotica*	*Phragmites australis- Sarcocornia fruticosa*	29	11	3.3	12.9	7.9	6
4. *Bromus rubens-Launaea nudicaulis*	*Cynanchum acutum-Atriplex semibaccata*	13	23	3.2	21.0	7.8	5
5. *Phragmites australis*	*Cynanchum acutum-Plucheadiscorides*	73	12	2.5	3.8	7.9	4
6. *Atriplex semibaccata-Phragmites australis*	*Hordium murinum* subsp. *Leporinum-Suaeda vera*	6	24	3.9	27.3	7.8	4
7. *Cynanchum acutum-Phragmites australis*	*Cynodon dactylon-Bassia indica*	8	25	4.2	20.8	7.8	4
8. *Voltaria tubiflorae*	*Bassia indica-Malva parviflora*	3	19	3.1	10.0	7.9	7
9. *Najas marina* var *armata*	*Phragmites australis-Ceratophyllum demersum*	92	5	0.5	1.2	6.8	2
10. *Phragmites australis-Typha domingensis*	*Myriophyllum spicatum-Ceratophyllum demersum*	47	6	1.7	3.4	7.9	8
11. *Phragmites australis*	*Eichhornia crassipes-Echinochloa stagnina*	82	5	1.7	2.3	7.4	6
12. *Eichhornia crassipes*	*Phragmites australis-Ludwigia stolonifera*	86	6	1.0	1.6	7.3	5
13. *Echinochloa stagnina-Phragmites australis*	*Ceratophyllum demersum-Typha domingensis*	24	10	2.3	7.2	8.0	6
14. *Ceratophyllum demersum*	*Eichhornia crassipes-Azolla filiculoides*	24	7	1.9	5.3	7.2	5

Table 4. (Continued)

Community name	Associated species	Species diversity indices			Edaphic variables		
		Total cover of 1st dominant sp %	Sp. richness	Shannon H	Simpson C	pH	EC $mScm^{-1}$
15. *Potamogeton pectinatus-Eichhornia crassipes*	*Myriophyllum spicatum-Ceratophyllum demersum*	60	5	1.4	2.7	7.2	6
16. *Potamogeton pectinatus- Phragmites australis*	*Echinochloa stagnina*	75	3	0.7	1.7	7.1	11
17. *Potamogeton pectinatus*	*Phragmites australis*	99	1.2	0.1	1.0	7.1	15
Lake Edku							
1. *Eichhornia crassipes*	*Echinochloa stagnina-Rununclus scleratus*	63	7.3	1.4	2.5	7.9	1.4
2. *Echinochloa stagnina- Eichhornia crassipes*	*Ludwigia stolonifera-Rununclus scleratus*	31	6.6	1.3	2.7	7.8	1.5
3. *Ceratophyllum demersum-Eichhornia crassipes*	*Echinochloa stagnina-Potamogeton pectinatus*	21	8.0	2.0	7.0	7.7	2.8
4. *Potamogeton pectinatus*	*Phragmites australis-Eichhornia crassipes*	41	6.0	1.5	3.6	7.7	2.9
5. *Cyperus articulates*	*Panicum coloratum- Typha domingensis*	73	3.0	0.8	1.7	7.8	9.4
6. *Typha domingensis*	*Phragmites australis-Eichhornia crassipes*	65	4.0	0.7	1.9	7.7	11.0
7. *Phragmites australis*	*Typha domingensis-Echinochloa stagnina*	91	4.0	0.5	1.2	7.7	2.4
8. *Arundo donax*	*Cynanchum acutum-Plucheadiscorides*	96	4.0	0.2	1.1	7.8	1.5
9. *Juncus acutus-Typha domingensis*	*Atriplex protulacoides-Limbarda crithmoides*	18	9	1.7	4.2	7.8	10
10. *Phragmites australis-Typha domingensis*	*Limbarda crithmoides-Sarconcornia fruticosa*	38	13	1.6	3.9	7.8	3.1
11. *Halocnemum strobilaceum-Sarconcornia fruticosa*	*Phragmites australis-Scirpus litoralis*	80	5	0.6	1.5	7.5	14.2
12. *Sauaeda verra-Rumex dentatus*	*Sarconcornia fruticosa-Chenopodium murale*	46	14	2.1	4.0	7.8	3.7
13. *Bassia indica-Limbarda crithmoides*	*Atriplex halimus-Suaeda vera*	17	24	2.8	15	7.9	3.1

Community name	Associated species	Species diversity indices			Edaphic variables		
		Total cover of 1st dominant sp %	Sp. richness	Shannon H	Simpson C	pH	EC $mScm^{-1}$
14. *Centauria calcitrapa*	*Limbarda crithmoides-Cynodon dactylon*	42	15	1.7	3.8	7.9	12.4
15. *Cynodon dactylon-Medicago polymorpha*	*Pluchea discorides-Cyperus rotundus*	28	33	2.8	10	7.9	0.3
Lake Burullus							
1. *Arthrocnemum macrostachyum-Juncus acutus*	*Atriplex protulacoides-Limbarda crithmoides*	41	8.1	1.5	0.04	7.7	4.6
2. *Phragmites australis-Sauaeda pruinosa*	*Arthrocnemum macrostachyum-Malva parvifora*	59	134	1.5	0.05	7.5	2.8
3. *Phragmites australis-Arthrocnemum macrostachyum*	*Salsola kali-Limbarda crithmoides*	17	7.7	1.3	0.06	7.7	2.7
4. *Suaeda vera-Limbarda crithmoides*	*Atriplex protulacoides-Arthrocnemum macrostachyum*	28	7.8	1.7	0.03	7.9	2.9
5. *Juncus acutus*	*Phragmites australis-Scirpus litoralis*	17	3.2	1.7	0.03	7.9	5.9
6. *Sarcocornia fruticosa*	*Arthrocnemum macrostachyum-Spergularia marina*	53	11.0	1.6	0.03	7.8	11
7. *Potamogeton pectinatus*	*Eichhornia crassipes-Ceratophyllum demersum*	87	3.3	1.1	0.11	8.1	2.9
8. *Halocnemum strobilaceum*	*Arthrocnemum macrostachyum-Spergularia marina*	21	9.4	1.7	0.03	8.0	3.5
9. *Salsola kali*	*Limbarda crithmoides-Secio galucus*	31	2.5	1.0	0.12	8.1	1.5
10. *Phragmites australis*	*Typha domingensis-Limbarda crithmoides*	18	2.3	1.7	0.03	8.0	9.0
11. *Phragmites australis-Potamogeton pectinatus*	*Eichhornia crassipes-Ceratophyllum demersum*	51	3.2	1.0	0.02	8.7	8.1
12. *Typha domingensis-Ceratophyllum demersum*	*Atriplex protulacoides- Limbarda crithmoides*	51	4.1	1.2	0.17	8.8	5.7
13. *Phragmites australis-Potamogeton pectinatus*	*Eichhornia crassipes-Ceratophyllum demersum*	15	2.3	0.9	0.10	8.9	6.6

Table 4. (Continued)

Community name	Associated species	Species diversity indices			Edaphic variables		
		Total cover of 1st dominant sp %	Sp. richness	Shannon H	Simpson C	pH	EC $mScm^{-1}$
Lake Manzala							
1. *Azolla filiculoides*	*Ludwigia stolonifera*	0.0	0.0	0.0	0.0	0.0	0.0
2. *Eichhornia crassipes*	*Azolla filiculoides-Echinochloa stagnina*	0.0	0.0	0.0	0.0	0.0	0.0
3. *Potamiogeton pectinatus*	--	0.0	0.0	0.0	0.0	0.0	0.0
4. *Ceratophyllum demersum*	*Najas marina* var *armata*	0.0	0.0	0.0	0.0	0.0	0.0
5. *Typha domingensis*	--	0.0	0.0	0.0	0.0	0.0	0.0
6. *Scirpus maritimus*	--	0.0	0.0	0.0	0.0	0.0	0.0
7. *Phragmites australis*	--	0.0	0.0	0.0	0.0	0.0	0.0
8. *Ruppia maritime*	--	0.0	0.0	0.0	0.0	0.0	0.0
Lake Bardawil							
1. *Ruppia cirrhosa-Cymodocea nodosa*	*Halodule uninervis*	17	0.3	0.4	0.40	0.0	0.0
2. *Halocnemum strobilaceum*	*Sarcocornia fruticosa-Zygophyllum album*	47	0.5	0.2	0.14	0.0	0.0
3. *Zygophyllum album*	*Arthrocnemum macrostachyum-Suaeda vera*	61	0.6	0.1	0.10	0.0	0.0
4. *Nitraria retusa*	*Juncus rigidus-Tamarix nilotica*	55	0.7	0.9	0.12	0.0	0.0
5. *Stipagrostis plumose-Retama raetum*	*Pancratium maritimum-Ononis serrata*	90	1.4	1.3	0.06	0.0	0.0
6. *Panicum turgidum-Thymelaea hirsute*	--	91	0.9	1.3	0.06	0.0	0.0
7. *Artemisia monosperma*	*Thymelaea hirsute-Moltkiopsis ciliata*	95	1.6	1.4	0.05	0.0	0.0
8. *Asparagus stipularis*	*Devera tortusa-Lycium shawii*	90	1.2	1.3	0.05	0.0	0.0
9. *Stipagrostis scoparia-Calligonum plygonoides*	*Astragalus fruticosus-Cyperus conglomeratus*	47	0.4	0.9	0.14	0.0	0.0

Table 5. The common plant communities inhabit the most habitats recorded in the five lakes (after Galal, et al., 2012)

Characteristic species	Life form	Mariut	Edku	Burullus	Manzala	Bardawil	Total
Communities of open water							
Azolla filiculoides Lam.	HH				+		1
Ceratophyllum demersum L.	HH	+	+	+	+		4
Cymodocea nodosa (Ucria) Asch.	Th					+	1
Eichhornia crassipes (C.Mast.) Solms	HH	+	+		+		3
Ludwigia stolonifera (Guill & Perr.) P. H. Raven	HH				+		1
Najas marina var. armata (H. Lindb.) Horn	HH	+			+		2
Potamogeton pectinatus L.	HH	+	+	+	+		4
Ruppia cirrhosa (Petagna) Grande	HH					+	1
Ruppia maritima L.	HH				+		1
Communities of swamps							
Arundo donax L.	GH		+				1
Cyperus articulatus L.	GH		+				1
Echinocloa stagnina (Retz.) P.Beauv.	GH	+	+		+		3
Juncus acutus L.	GH		+	+			2
Juncus rigidus Desf.	GH	+					1
Phragmites australis (Cav.) Trin.ex Steud	GH	+	+	+	+		4
Scirpus maritimus L.	GH				+		1
Typha domingensis (Pers.) Poir.ex Steud	GH	+	+	+	+		4
Communities of shoreline							
Cynanchum acutum L.	GH	+					1
Centauria calcitrapa L.	Ch		+				1
Launaea nudicaulis (L.) Hook. F.	H	+					1
Cynodon dactylon (L.) Pers.	GH		+				1
Bromus rubens L.	Th	+					1
Medicago polymorpha L.	Th		+				1
Rumex dentatus L.	Th		+				1
Volutaria tubuliflora (Murb.)Sennen.	Th	+					1
Bassia indica (Wieght) A. J. Scott	Th		+				1
Salsola kali L.	Th			+			1
Communities of saline sand formations							
Atriplex halimus L.	Ph	+					1
Nitraria retusa (Forssk.) Asch.	Ph					+	1
Tamarix amplexicaulis Ehrenb.	Ph	+					1
Arthrocnemum macrostachyum (Moric.) K. Koch	Ch	+		+			2
Halocnemum strobilaceum (Pall.) M. Bieb.	Ch	+	+	+		+	4
Limbarda crithmoides (L.) Dumort.	Ch		+	+			2
Sarcocornia fruticosa (L.) A. J. Scott.	Ch	+	+	+		+	4

Table 5. (Continued)

Characteristic species	Life form	Mariut	Edku	Burullus	Manzala	Bardawil	Total
Suaeda pruinosa Lange	Ch			+			1
Suaeda vera Forssk. Ex. J. F. Gmel.	Ch		+	+			2
Zygophyllum album L.f.	Ch					+	1
Atriplex semibaccata R.Br.	Th	+					1
Communities of non-saline sand formations							
Asparagus stipularis Forssk.	GH					+	1
Calligonum polygonoides L.	Ph					+	1
Artemisia monosperma Delile	Ch					+	1
Moltkiopsis ciliata (Forssk.) I. M. Johnst.	Ch					+	1
Thymelaea hirsuta (L.) Endl.	Ch					+	1
Stipagrostis scoparia (Trin. & Rupr.) de Winter	H					+	1
Panicum turgidum Forssk.	GH					+	1
Total		**18**	**18**	**12**	**12**	**13**	**45**

REFERENCES

Abu Al-Izz, M. S. (1971). *Landforms of Egypt*. Translated by Fayid, Y.A. American University Cairo Press, Cairo.

Al-Sodany, Y. M. (1998). Vegetation analysis of the canals, drains and lakes of the northern part of Nile Delta. Ph.D. *Thesis, Fac. Sci.*, Tanta University, Tanta., 232.

Anderson, D. M., P. M. Gilbert, J. M. & Burkholder, L. (2002). Harmful algal blooms and eutrophication: nutrient sources, composition, and consequences. *Estuaries and Coasts*, *25*, 704-746.

Anthony, A., Atwood, J., August, P., Byron, C., Cobb, S., Foster, C., Fry, C., Gold, A., Hagos, K., Heffner, L., Kellogg, D. Q., Lellis-Dibble, K., Opaluch, J. J., Oviatt, C., Pfeiffer-Herbert, A., Rohr, N., Smith, L., Smythe, T., Swift, J. & Vinhateiro, N. (2009). Coastal lagoons and climate change: ecological and social ramifications in U.S. Atlantic and Gulf coast ecosystems. *Ecology and Society*, *14*(1), 8. [online] URL: http://www.ecologyandsociety.org/vol14/iss1/art8/.

Balba, A. M. (1981). Sources and Protection of Soil and Water of the Mediterranean Coast of Egypt. Advances in Soil and Water Research. *Alexanderia*, *6*, 73.

Ben Menchem, A. (1979). Earthquake Catalogue for Middle East (92 BC-1980AD). *Bull. Geofisica Teor. Applicata*, *21*, 245-313.

Bertness, M. D. (2007). *Atlantic shorelines: natural history and ecology*. Princeton University Press, Princeton, NJ, USA.

Boulos, L. (2005). Flora of Egypt: Volume Four Monocotyledons (Alismatacea-Orchidaceae). *Al-Hadara Pupl.*, Cairo, Egypt, 617.

Canfield, R. (1941). Application of the line interception method in sampling range vegetation. *J. Forestry*, *39*, 288-294.

Demirezen, D. & Aksoy, A. (2004). Accumulation of heavy metals in Typha angustifolia (L.) and Potamegaton pectinatus (L.) living in Sultan Marsh (Kayseri, Turkey). *Chemosphere*, *56*, 685-696.

El-Hadidi, M. N. & Hosny, H. (2000). *Flora Egyptiaca*. The Palm Press, Cairo, *1*, 151.

El-Sheikh, M. A., Al-Sodany, Y. M., Eid, E. M. & Shaltout, K. H. (2012). Ten years primary succession on a newly created landfill at a lagoon of the Mediterranean Sea (Lake Burullus RAMSAR site), *Egypt. Flora*, *207* (2012) 459– 468.

Forbes, D. L. (1987). Coastal geomorphology and sediments of Newfoundland. *Geological survey of Canada*, Paper 84-1B, 11-24.

Galal, T., Shaltout, k. & Hassan, M. (2012). *The Egyptian Northern Lakes: habitat diversity, vegetation and economic importance*. Academic Publishing.

Harris, L. A., Buckley, B., Nixon, S W. & Allen, B. T. (2004). Experimental studies of predation by bluefish Pomatomus saltatrix in varying densities of sea grass and macroalgae. *Marine Ecology Progress Series*, *281*, 233-239.

Heck, K. L. & Thoman, T. A. (1984). The nursery role of seagrass meadows in the upper and lower reaches of Chesapeake Bay. *Estuaries*, *7*, 70–92.

Hill, M. O. (1979a). TWINSPAN. A Fortran program for arranging multivariate data in ordered two-way table by classification of the individuals and attributes. Cornell University, Ithaca, N.Y. 52.

Hill, M. O. (1979b). DECORANA. A Fortran program for detrended correspondence analysis and reciprocal averaging. Cornell University, Ithaca, N.Y., 90.

IUCN (1994). Guidelines for Protected Area Management Categories.

Kjerfve, B. (1986). *Comparative oceanography of coastal lagoons*. 63-81. In: Estuarine Variability. (Wolfe D.A., 4.). Academic Press, New York.

Knoppers, B, Kjerfve, B. & Carmouze, J. P. (1991). Trophic state and water turn-over time in six choked coastal lagoons in Brazil. *Biogeochemistry*, *14*, 149-166.

Magurran, A. E. (1988). Ecological diversity and its measurement. *Princeton Univ. Press. Princeton*, New Jersey, 179.

McGlathery, K. J. (2001). Macroalgal blooms contribute to the decline of seagrass in nutrient-enriched coastal waters. *Journal of Phycology*, *37*, 453-456.

Mc Lusky, D. S. (1990). *The Estuarine Ecosystem. Blackie, Chapman & Hill*, New York, USA.

Mee, L. D. (1978). *Coastal lagoons*. 441-490. In: Chemical Oceanography. Second edition. Vol. 7. (J. Riley and 0. Skirrow, eds.). Academic Press. New York.

Moore, N. H. & Slinn, D. J. (1984). The physical hydrology of a lagoon system on the Pacific coast of Mexico. stuarine, *Coastal and Shelf Science*, *19*, 413426.

National Research Council. (2000). Clean coastal waters: understanding and reducing the effects of nutrient pollution. *The National Academies Press*, Washington, D.C. USA.

Nixon, S. W. & Oviatt, C. A. (1972). Preliminary measurements of midsummer metabolism in beds of eelgrass, Zostera marina. *Ecology*, *53*(1), 150-153.

Nixon, S. W. (1982). *Nutrient dynamics, primary production and fisheries yields of lagoons*. Pages 357-371 in Oceanologica Acta, Proceedings of the International Symposium on Coastal Lagoons (Bordeaux, France, 1981). SCOR/IABO/UNESCO.

Nixon, S. W. (1995). Coastal marine eutrophication: a definition, social causes, and future concerns. *Ophelia*, *41*, 199-219.

Nixon, S. W., Granger, S., Buckley, B. A., Lamont, M. & Rowel, B. (2004). A one hundred and seventeen year coastal water temperature record from Woods Hole, *Massachusetts. Estuaries*, *27*, 397-404.

Phleger, F. B. (1969). *Some general features of coastal lagoons*. 5-26. In: Lagunas Costeras, un Simposio. (A. Ayala-Castaiieres, ed.). Universidad Nacional AuMnoma de MBxico, MBxico, DF.

Pielou, E. C. (1975). *Ecological diversity*. A Willy-Interscience Publication, 165.

Rerfve, B. & Magill, K. E. (1989). Geographic and hydrographic characteristics of shallow coastal lagoons. *Marine Geology*, *88*, 187-199.

SAS (1989–1996). SAS/STAT User's Guide. *SAS Institute Inc*., Cary, NC, USA.

Sikora, W. B. & Kjerfve, B. (1985). Factors influencing the salinity of Lake Pontchartrain, Louisiana, a shallow coastal lagoon: analysis of a long-term data set. *Estuaries*, *8* (2A), 170-180.

Shaltout, K. H., Al-Sodany, Y. M. & El-Sheikh, M. A. (2004). Phragmites australis (Cav.) Trin. ex Stud. in Lake Burullus, Egypt: is it an expanding or retreating population? *Proc. 3rd Conf. Biol. Sci., Fac. Sci., Tanta Univ. 3*, 83-96.

Shaltout, K. H., Ali, M. M., Hassan, L. M. & Galal, T. M. (2005a). Habitat and vegetation of Lake Edku, *Egypt. Taeckholmia*, *25*, 61-90.

Shaltout, K. H. & Khalil, M. T. (2005). Lake Burullus: Burullus Protected Area. Publication of National Biodiversity Unit No. 13, *Egyptian Environmental Affairs Agency (EEAA)*, Cairo.

Shaltout, K. H. & Al-Sodany, Y. M. (2008). Vegetation analysis of Burullus Wetland: a RAMSAR site. *Wetlands Ecology and Manage*, *16*, 421-439.

Täckholm, V. (1974). Student's flora of Egypt. *Cairo Univ*. Press, 888.

Ter Braak, C. F. G. & Smilauer, P. (2002). CANOCO Reference Manual and CanoDraw for Window's User's Guide: Software for Canonical Community Ordination (Version 4.5). *Microcomputer Power*, Ithaca, NY.

UNESCO (1977). *Map of the world distribution of arid regions*. MAB Technical Notes 7. w York, NY, USA.

Valiela, I., Foreman, K., Lamontagne, M., Hersh, D., Costa, J., Peckol, P., Demeo-Anderson, B., D'Avanzo, C., Babione, M., Sham, C. -H., Brawley, J. & Lajtha, K. (1992). Couplings of watersheds and coastal waters: sources and consequences of nutrient enrichment in Waquoit Bay, *Massachusetts. Estuaries*, *15*, 443-457.

Whittaker, G. H. (1972). Evolution and measurement of species diversity. *Taxon*, *21*, 213-251.

Whittaker, R. H. (1975). *Communities and ecosystems*. Second edition. MacMillan, New York, USA.

Zahran, M. A. & Willis, A. J. (1992). The vegetation of Egypt. *Chapman and Hall*, London, 424.

In: Lagoons
Editor: Mwinyikione Mwinyihija

ISBN: 978-1-62808-092-6

Chapter 8

COASTAL SAND DUNES AND FRESHWATER PONDS IN KOLKHETI: THREATS AND NEEDS FOR CONSERVATION

Izolda Matchutadze[1]*, Tamar Bakuradze[1], Mamuka Gvilava[1], Bulbuli Bolkvadze[2] and David Baratashvili[3]

[1]Society for Conservation of Wild Nature "Tchaobi"
[2]GIS and RS Consulting Centre Geographic
[3]ICZM National Focal Point for Georgia
[4,5]Shota Rustaveli State University, Tbilisi, Georgia

ABSTRACT

Black Sea coast line terraces are well developed in the coastal zone of Kolkheti, along the sand dune areas. Indeed, Sphagnum peat lands were formed within the coastal hinterland. Examples of the mentioned peat lands are the likes of Anaklia-Churia, Ureki, Grigoleti and Kobuleti areas. Initially the formation of lagoons started in these areas, later advancing into lakes which were characterized with appearance of algal cover and edifices of other aquatic organisms. Lagoon-type formations of these lake-like freshwater ponds, established behind the sand dunes, can still be found in-between the Sea and peat lands of Kolkheti area. As a good example, Lake Paliastomi and Imnati is a relict lagoon worthy of study in this area of Kolkheti. These freshwater lagoons systems are fundamental in aspects of biodiversity as they serve as critical habitats for waterfowl, plants and human water source.

The importance of the ecosystem in the Kolkheti coastal zone is based on its status as the sole and irreplaceable habitat for species of plants that thrive there such as: *Trapa colchica, Tapa maleevi, Trapa natans, Marsilea quadrifolia and Salvinia natas*. However, anthropogenic pressures are taking toil to the ecosystem with the resultant impact leading to its degradation. For example, major attributes to degradation is the extensive infrastructural development of the coastal zones where setting-up of marine terminals, construction of large coastal infrastructure and weak environmental awareness

* Contact: izo.muho@gmx.net.

initiatives are conspicuous. Indeed, the impacts derived therein are associated with the destruction and irreversible disappearance of biodiversity and at some time emergence of malarial epidemics due to ecosystem degradation. Moreover, the problems with Kholkheti coastal zones ecosystems were deeply compounded by the lack of legal framework to protect these habitats and lack of awareness in biodiversity conservation.

INTRODUCTION

There are 34 critical hotspots of biodiversity in the world (Conservation International). The Caucasus hotspot (Figure 1), which spans 532,658 km² is one of them. For instance, Georgia with the famous Kolkheti lowland surrounded by the greater and lesser Caucasus ranges belongs to this hotspot region. When the climatic and geographic dimensions are considered, the Kolkheti lowland represents (Figure 2) an important center of biodiversity but impacted by intense human activity.

The main habitats in the Kolkheti lowland include the relict forests, peat lands, wetlands, coastal sand zones, river estuaries, open fresh and salt water areas. During the last Glacial Maximum, Kolkheti lowland constituted an important refuge for the flora of Europe (Tarasov et.al. 2000). It was observed that many plant species currently preserved in Kolkheti were earlier wide-spread in Europe before their ultimate disappearance. Contributing towards the disappearance and degradation of these ecosystems were heightened anthropogenic activities in the Kolkheti forests and peatlands. These activities started to increase in the second half of the 19th century and accelerated significantly in Soviet times, without any abatement to-date. Moreover, there is great concern when the geological past and recent genesis of the dune terraces in Kolkheti are compared to the Holocene period. This is due to the later encompassing 10-12 thousand years providing glaciation as a retreat whilst currently it is the global warming threat that is eminent.

Comparatively, Black Sea terraces are well developed in the coastal zone of Kolkheti, along the sand dune areas with Sphagnum peatlands formed in adjustant coastal hinterland. In this regard peatlands of Anaklia-Churia, Ureki, Grigoleti and Kobuleti areas can be cited. It is theorized that lagoons started forming in these areas, later developing into lakes which were characterized with algal cover and appearance of group of edifices such as *Trapa colchica, Trapa natans, Phragmites australis.* Of interest was the evolvement of sphagnum species which followed, apparently marking the formation of the current peatlands in Kolkheti. Currently, the formations of freshwater lagoons are found behind the sand dunes between the Sea and the peatlands mentioned. Good examples are lagoons at Anaklia-Churia coastal areas and those at the bank near the estuary of River Chorokhi.

Lagoons like Lake Paliastomi (Figure 3) and Imnati are considered as relicts though as freshwater pond systems their importance is associated with sustainment of biodiversity. For instance, the lagoons provided critical habitats for waterfowl for a long time. Due to their importance in support of biodiversity these ponds were deemed to possibly qualify for consideration under Emerald Network and Natura 2000. Further to supporting specific birdlife, the area also provides a specialized habitat for the following species that are known to be habitant such as as: *Trapa colchica, Tapa maleevi, Trapa natans, Marsilea quadrifolia, Callitriche stagnalis* and *Salvinia natas.*

Over and above the factors mentioned there are other emerging challenges to the threat of biodiversity in Kolkheti. The most threatening are anthropogenic pressures where structures are being constructed including marine terminals and other related coastal infrastructural development. Weak environmental awareness in designing and managing of the ecosystem in these areas has culminated in the destruction and irreversible disappearance of the natural species in Kolkheti coastal zones (Figure 1). Indeed, failure of adequately addressing on habitat protection through well-defined legislation and policies to protect smaller habitats including those that are located outside the formally protected areas was significant. An example of areas under threat include Batumi area where the Ardagani Lake area with a size of 45 hectare and specific habitat of three important species of *Trapa,* the *Typhetum-Phragmitetum* and a host of many migratory birds were impacted by sanctioning of a coastal boulevard development. This lead to the destruction of this specific habitat and eventual disappearance of some of the species mentioned. The result of this impact concurrently affected the habitat as studies have validated that there is inter-relation and dependence between species and their respective habitats. However, of great interest is that with the management of the threats in some of these ecosystems, relicts and endemic plants can redevelop in their historical niche and previously established habitats. Worthy of note, is to understand that habitats are micro-ecological environment, created by the plants as such if left to regenerate they directly enrich the habitat.

In reexamining the Georgian environmental legislation, it will be observed that law identifies with the protection of fauna but rhetorically misses to include preservation of habitats. Thusly, this misnomer easily provides a leeway for habitat degradation or destruction without any legal recourse. This has been the basis of the anthropogenic interference discussed earlier leading to infrastructural development with considering the impact to the environment. Therefore, immediate reconsideration to protect the endangered species is of absolute necessity to protect the endangered habitats halt on the loss of small but valuable habitats and dependent species along the Black Sea coast of Georgia.

Figure 1. Photographic preview of Caucasus Eco-region.

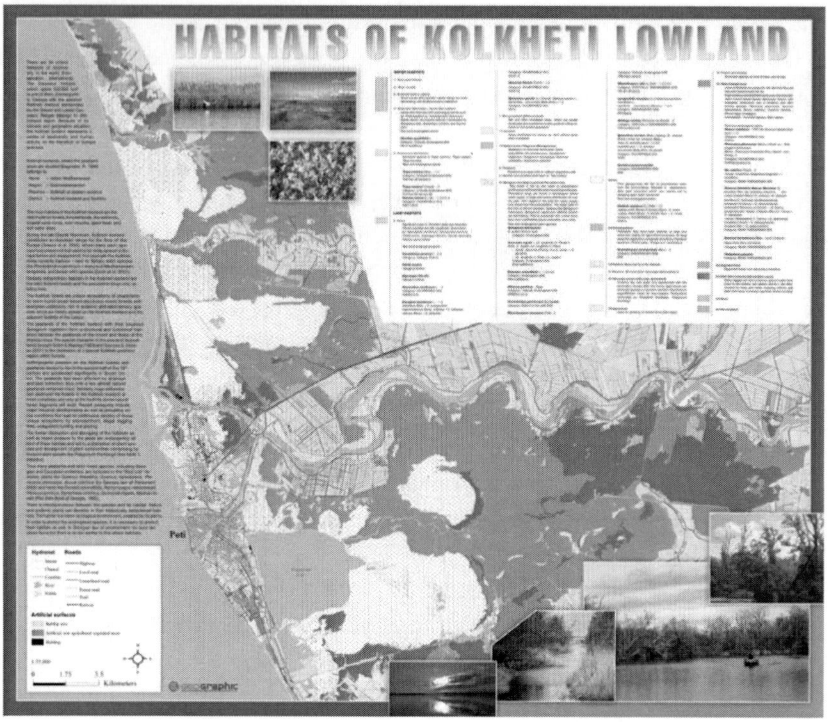

Figure 2. Illustrative information of Kolkheti Lowland habitats.

In figure 2 the main habitats in the Kolkheti lowland are illustrated depicting the relict Kolkheti forests, peatlands, the surrounding wetlands, coastal sand zones, river estuaries, open fresh and salt water areas.

Figure 3. Lake Paliastomi channel connected to the Sea.

Coastal Sand Dunes

The Coastal sand dunes are distributed along Anaklia-Churia shoreline, Grigoleti area and Kobuleti, South of Choloki River. Important coastal sand dunes include Nature 2000 and Emerald network habitat a monumental attestation of a natural living monument under the Kobuleti Protected Areas. Examples of species under this ecosystems include; *Pancratium maritimum, Cakile euxina, Glaucium flaum, Eringium maritimum, Glycyrrhiza glabra, Euphorbia pubescens, Otanthus maritimus and Convolvulus persicus.* The importance of this eco-zones includes forming critical habitats for aquatic birds that are mostly migratory. Predominantly, the plant species dominating in the coastal dunes are: *Pancratium maritimum, Eringium maritimum, Cakile euxina, Glaucium flaum and Euphorbia palustris.*

In south Kolkheti there is an observed absence of clay and presence of organic gyttja (mostly formed from decay of peat through anaerobic digestion of bacteria) between 865 and 765 cm depth in the ISP-core. This observation is indicative that Ispani 2 basin was not part of floodplain but a transformed 'lake-like' lagoon. This phenomena is also referred to through a report by Connor et al. (2007) that also demonstrated the presence of gyttja as an important feature to lagoon formation (cf. Figure 11). In this study the characterization of Ispani 2 basin was further demonstrated with the large presence of Salix (i.e Willows,Sallows, Osiers etc) species and other groups of wild grasses particularly in the lower part of zone ISP-B. Of great interest was the presence of Phragmites in the same area. Moreover, macrofossils of willow were also found at the base of the gyttja layer at a spot along the Ispani 2 cross-section about 300 m from the core (cf. Figure 3). The importance of this observation is based on the vegetation spatial restriction as indicated in other studies (Pim De Klerk et al., 2009).

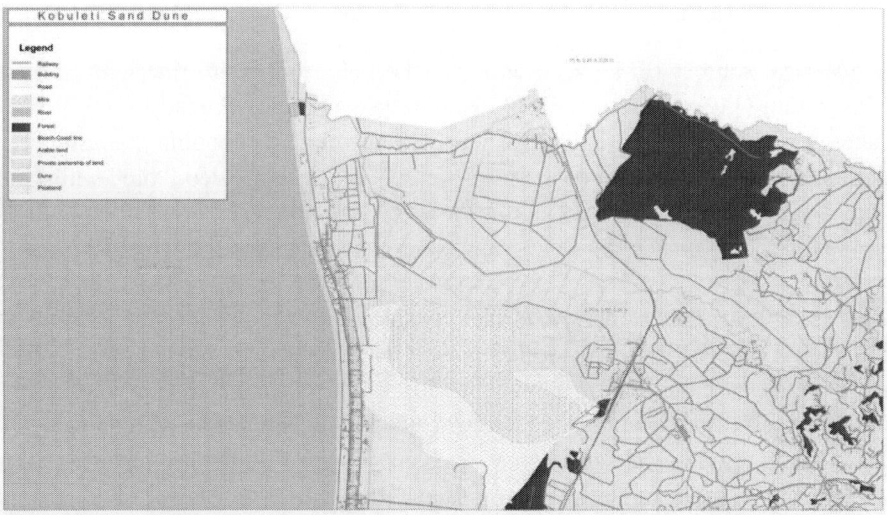

Figure 4. An outlay of the Coastal sand dunes of Kobuleti (South Kolkheti).

Freshwater Ponds

Formation of the modern Black Sea terrace of the coastal dune has resulted to the freshwater regime of these ponds. Largest one is the Paliastomi Lake, which is essentially the relict lagoon.

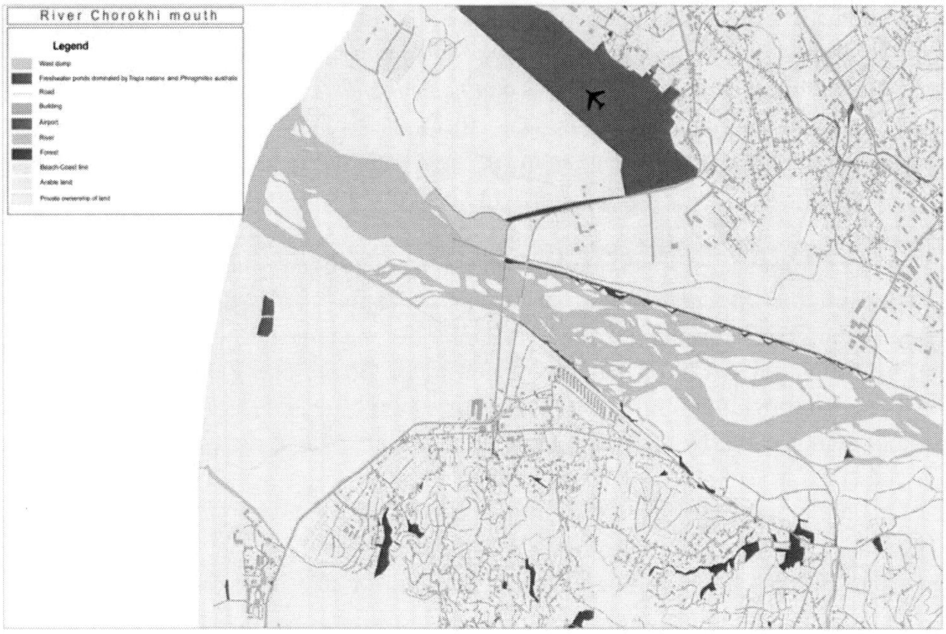

Figure 5. Freshwater ponds in the proximity of River Chorokhi.

The historical sources of the ecosytems of lake Paliastomi were first noticed by Italian authors-Pascvalini (1408), Bianca (1436), Benebicaza (1480) and Freducci (1497). The name had variations which included "Paliastoma" or "Paliastomos." Toponim "Paliastomi" which meant "old sea bay". The origins of the term were Greek where Paleo- "old" whilst stoma – meant "bay". Previously the same area drew varied description with some earlier authors (e.g. Ptolemy, Agatia, etc.) considering Lake Paliastomi areas connected to River Rioni as "small sea" or sea bay.

In describing Lake Paliastomi it is further considered as a remnant of the Black Sea basin and as a lagoon it was earlier characterized as such with the assertion based on its soil profile, geological layout and phytology of the area. Most of these characteristics concerning Lake Paliastomi's genesis were not earlier recognized till recently. Indeed geological works including drilling, lithological, geochemical, paleontological and other analyses allow the identification of the basis of its genesis and characterization (Figure 1). For example re-examination of the sediments in the bottom of Lake Paliostomi indicate the original sediments encompassed clay sapropel and peat (at monolithic horizons) with a depth of 10-12m. Conspicuously, the lake is a perfect plain, however, exhibits variations in its sediment depths

During the bay formation of the lake then later its transformation to a lake lagoon is fundamental. The fluctuating sea level, occurring in the recent geological past was critical information to the development of Lake Paliastomi. This was due to the existence of the sea bay and later its transformation to a lake basin which was influenced with prevailing sea levels. For instance, the observed average multiannual level of the lake Paliastomi exceeds the Black sea mirror by only 30-40 cm. As such changes in volume and depth over the centuries, has had an important influence on the land of the Paliastomi Lake region. If an increase of water level is experienced by a few centimetres cubic, it is enough for Lake Paliastomi to become a sea bay.

Figure 6. Ramsar site at Lake Paliastomi.

Figure 7. Kolkheti Relict forest.

The resultant impact of such conditions could have influenced the characteristics of the region around and within Lake Paliastomi. With an assumption that the sea level increases, the result would be that the channel connecting the sea to the lake will submerge, thus, widening up and allowing free flow of water between the lake and sea for the former to be considered as a bay rather than a lagoon or 'lake-like'. In contrast, if the sea level decreases (i.e. Black sea) by 2-3m, Lake Paliastomi will be relinquished due to 'drain off' and cease to exist. The principle attribute to this 'drain off' would be the build-up of the sand sediments at the bottom of the channel basin, leading to emptying of the lake.

Figure 8. A view of the Imnati mire and Lake Paliastomi.

The Imnati mire which is adjacent to Lake Paliastomi is related to its importance as a bird sanctuary (Figure 8). In the 19th Century including the early 1960's and 70's, Lake Paliastomi together with the adjoining Imnati Sphagnum mire, the forested peatlands and the adjacent sections of the Black Sea coastal land-water areas served as one of the most important rest areas for migratory birds. This sanctuary was particularly crucial during winter when the birds from Africa migrated towards the moderate and polar belts of Eurasia and vice versa. Moreover, the bogs and the marsh forests surrounding Lake Paliastomi served as a shelter for even greater diversity of animals such as wild-boars, roe-dears, martens, otters etc.

Figure 9. View of the Imnati mire and Lake Imnati.

The Imnati mire is located near Lake Imnati whose ecological importance has been previously discussed. In this particular ecosystem potrays an important association of the mire and the lake with certain species found to be predominant such as, *nymphaea colchica, Nymphaea alba, Potampogeton natans, Potamogeton pectibnatus.*

Lake Imnati provides fundamental geological evidence where the indication is that the peat formation in Imnati began approximately 6000 BP (Nejstadt et al. 1965). Prior to the peat formation the area was a lagoon separated from the Black Sea by a spit build-up of sediments emanating from River Rioni. This lagoon was characterized by a peat body with a depth of 3-12m overlying clay and silt sediment at the bottom (Figure 10). Furthermore, the peat stratigraphy reveals two main phases of mire development. First, a fen phase which initiated with the territorialisation of the lagoon with reed plants (*Phragmites australis*) growing under enriched nutrient conditions. The fen phase is estimated to have lasted approximately 4000 years and was associated with the continuous tectonic sinking of the area. Finally, at about 2000 BP, peat accumulation rates exceeded the rate of sea level. This increased rise was particularly influenced by rain water and the bog phase become dominated with emergent species such as Sphagnum mosses, *Molinia litoralis* and *Cladium mariscus* ultimately signified the development of Inmati mire .

In support to the lagoon systems in the Kolkheti area there still exist smaller sized freshwater ponds which provide habitat to specific species such as; *Trapa colchica, Trapa natans, Trapa hyrcana, Salvinia natans, Potamogeton natans* etc. However, with the advent of these specific species, presence of the invasive *Egeria* denca in the mire adjacent Adjara-Guria are observed. In particular, within the the Kolkheti area, Salvinia covers are only found forming a habitat to species located in the Emerald Network Small niches such as *Typha minima.*

Figure 10. Representation of Imnati mire in front of Adjara-Guria Mountains.

This type of freshwater lagoons some found near the coastline of the Tskaltsminda village are still pristine surrounded by pine tree patches and sand dunes. The importance of these lagoons is the existence therein of critical habitat that host's a specific specie known as *Marsilea quadrifolia*. Indeed, as per IUCN categorization this specific specie is considered as Critically Endangered (EN) and represents species of Emerald Network.

It is worthy of note to observe the presence of other freshwater lagoons which are spreading along the left bank of River Chorokhi. These ecosystems provide habitat to species such as *Trapa colchica, Phragmites australis, Typha angustifolia* whilst on the riparian areas species such as *Polygonum thunbergii, Pancratium maritimum, Juncus, Eringium maritimum* flourish mainly on the dunes.

Figure 11. Lake Paliastomi viewed with the mire

Anthropogenic activities at present are in the rise within the Anaklia-Churia areas which is an area situated in the 'outskirts' of the virgin forest in Kolkheti. In retrospect, the destruction of the forest by felling of trees and the construction of drainage canals for Kulevi oil terminal has impacted negatively to the lagoon ecosystem. Consequently ground waters have been most affected by not being recharged with fresh water resulting to the neighbouring community lacking portable source of drinking water. In Anklia-Churia there is a well-defined biodiversity and only the anthropogenic activities will impact on them. For example, the species that can be found there include the habitat for endemic flora such as *Hibiscus ponticus, Ruscus colchica, Ptericarya pterocarpa* with birds such as *Colchicus phasanus*, found to be well established in this area. However, with the infringement of human activity as mentioned earlier the biodiversity is in danger including humans themselves. Alongside,

Ramsar conservation has been affected also with coastal area pristine environment and Bronze Age settlements and archaeological excavations damaged.

Figure 12. Lake Paliastomi viewed with a background of anthropogenic activities.

Anthropogenic Impact

Anthropogenic pressure on the Kolkheti forests and peatlands started to rise in the second half of the 19th century and accelerated significantly in Soviet times, without any abatement to date.

The pretentious socialist agricultural development in Kolkheti was accompanied by massive deforestation activities. Some of these activities involved depletion of the virgin forests where they were replaced with plantations of tea, citrus, and tung trees. Furthermore, the construction of canals as mentioned earlier drastically affected the lagoons ecosystems. In retrospect peats from Imnati and Ispani I were extracted to be used as soil improvers in the plantations. This destabilized the ecosystems in these areas and invader species emerged to replace the damaged original species.

Certain interventions have previously been undertaken to manage environmental calamities in the area. Some of these interventions had negative consequences to both the aquatic and terrestrial ecosystems of Kolkheti region. For example, in 1924 for purposes controlling floods, channels were constructed such as the Maltakva channel which connected the freshwater of lake Paliastomi with the Black Sea. However, salinization of the lake started to occur transforming the lagoon from fresh to slightly saline water. The water salt content othe lagoon before the channel was constructed did not exceed 1% promile. However, after the operationalization of the Maltakva channel salt content increase to between 5-14 %. In retrospect higher salinity levels are observed during sea storms when injection of more salty water into the lagoons occur. Moreover, infiltration of the Kolkheti forests through anthropogenic activities as discussed earlier has affected certain species of tress such as oak, beech, hornbeam, and wing-nut which have become rare and only survive in places that are difficult to access. Other noted repercussions are as discussed earlier.

Figure 13. A glimpse of Grigoleti sand dune.

Figure 14. Destruction of the coastal sand dunes of Anaklia.

Another anthropogenic related event that emerged at the end of the 19th and the beginning of the 20th century was the malaria epidemics which broke out in Kolkheti. This happened with the resident population attributing the occurrence of the epidemic to the mire rather than the destruction of the Kolkhet forest. However, the degraded ecosystems through the deforestation favoured the breeding of *Anopheles* mosquito that bred in the emergent water puddles. The regrowth of the mire plants and its re-colonisation of the puddles and subsequent restoration of the ecosystem resulted to the elimination of mosquito habitat. This eventually lead to eradication of malarial epidemics in the area.

In conclusion, anthropogenic pressures, such as extensive infrastructural development of the coastal zones, building of marine terminals, other large coastal infrastructure development, and weak legislative, policy and lack of environmental awareness contributed to the degradation of the ecosystems and emergent of epidemics in Kolkheti lagoons region. Further evidence indicate that in Batumi area, an area of 45 acres surrounding Lake Ardagani were surrendered for housing projects leading to destruction of habitat and disappearance of important species with ramifications discussed earlier.

Anaklia-Churia

Anaklia-Churia areas were considered since 2002 as part of the Strict Protection Zone of Kolkheti National Park. However, since 2009 infrastructural development in this area has continued but disregarded to conduct environmental impact assessment to ascertain the underlying threats to the biodiversity. As a result, freshwater lagoons located at Anaklia-Churia were destroyed with all the inhabitant species, while these areas were potential candidates for Ramsar site status in terms of the area being recognized as a sanctuary for migratory birds. Earlier, since 1999, Kulevi area's habitat was destroyed with the construction of the oil terminal and the railway line heavily impacting the shoreline all along to the south including the Poti region.

Estuary of River Chorokhi

Selected examples of important species in the estuary region of river Chorokh threatened with extinction due to anthropogenic pressures are depicted. Some of these species from the freshwater and lagoons in this area have their photographic illustration and respective categories are provided:

1. *Marsilea quadrifolia* L.

    Category: Critically Endangered (EN)
    EN A1acdB1(i, ii)

2. *Trapa colchica*

    Category: Critically Endangered (EN)
    EN A1acdB1(i, ii)

Figure 15. Trapacolchica.

3. *Trapa natans*

    Category: Critically Endangered (EN)
    EN A1acdB1(i, ii)

4. *Salvinia natans*

    Category: Critically Endangered (EN)
    EN A1acdB1(i, ii)

Figure 16. *Salvinia natans* and *Trapa natans* dominating the freshwater pond along the Anaklia shoreline (photo by Francis Muller).

5. *Ceratophyllum demersum*

   Category: Critically Endangered (EN)
   EN A1acdB1(i, ii)

6. *Typha minima*

   Category: Critically Endangered (EN)
   EN A1acdB1(i, ii)

7. *Callitriche stagnalis*

   Category: Critically Endangered (EN)
   EN A1acdB1(i, ii)

8. *Sagittaria sagitiifolia*

   Category: Critically Endangered (EN)
   EN A1acdB1(i, ii)

Figure 17. *Sagittaria sagitiifolia.*

1. Coastal Sand Dune Habitat

9. *Otanthus maritimus*

Category: Critically Endangered (EN)
EN A1acdB1(i, ii)

Figure 18. *Otanthus maritimus.*

10. *Cakile maritime*

Category: Critically Endangered (EN)
EN A1acdB1(i, ii)

Figure 19. *Cakile maritime.*

11. *Convolvulus persicus*

Category: Critically Endangered (EN)
EN A1acdB1(i, ii)

12. *Glaucium flaum*

Category: Critically Endangered (EN)
EN A1acdB1(i, ii)

13. *Asparagus litoralis*

Category: Critically Endangered (EN)
EN A1acdB1(i, ii)

14. *Pancratium maritimumL.* – 2

Category: VULNERABLE (VU)
VUB2c (i, ii, iii)

15. *Spiranthes spiralis* (L.) Chevall. (*Ophrys spiralis* L., *Spiranthes autumnalis* (Balb.)Rich.) –1,2

Category: VULNERABLE (VU)
Criteria. A1a

Figure 20. *Pancratium maritimum.*

16. *Eringium amritimum*

Category: VULNERABLE (VU)
Criteria. A1a

2. *Peatland Habitat*

17. *Hibiscus ponticus*

Category: Critically Endangered (EN)
Criteria: A1acdB1(i, ii)

Figure 21. *Eringium amritimum.*

Figure 22. *Hibiscus ponticus.*

18. *Kosteletzkya pentacarpa* (L.) Ledeb.

    Category: Critically Endangered (EN)
    Criteria: A1acdB1(i, ii)

19. *Ilex colchica* –P.EF.Yeo (*Ruscus hypophyllum* auct.) –1,2
    Category: VULNERABLE (VU)
    Criteria: A1a

Figure 23. *Kosteletzkya pentacarpa (*Photo by Hans Joosten).

Figure 24. *Ilex colchica*.

20. *Pterocarya pterocarpa*

Category: Endangered (EN)
Criteria: A1a

21. *Quercus hartwissiana*

Category: Endangered (EN)
Criteria: A1a

22. *Buxus colchica*

Category: Endangered (EN)
Criteria: A1a

References

[1] Haberl, A., Kahrmann, M., Krebs, M., Matchutadze, I. and Joosten, H. (2006): *The Imnati mire in the Kolkheti lowland* (Georgia), Peatlands International 2006/1.

[2] Integrated Plan for Sustainable Development of Tskaltsminda Coastal Community, 2009., Tamar Bakuradze, Amiran Gigineishvili, Mamuka Gvilava, Kakhaber Kheladze, Inesa Kuzanova, Irene Lucius, Izolda Matchutadze. EU funded ECBSea Project. 44 p. www.iczm.ge.

[3] Matchutadze I. 2009, *Anthropogenic impact on Kolkheti mires and succession,* IMCG scientific congress.

[4] "Paliastomi Lake Environmental Study - Final Report", Georgia ICZM Program, Financed by the World Bank / GEF Tbilisi, 1998.

[5] Pim de Klerk, Andreas Haberl, Andreas Kaffke, Matthias Krebs, Izolda Matchutadze, Merten Minke, Jenny Schulz, Hans Joosten, 2009, Vegetation history and environmental development since ca. 6000 calyr BP in and around Ispani 2 Kolkheti lowlands, Georgia.

[6] The mires of the Kolkheti - a paradigm example for global ecosystem [a- and g-] biodiversity 2009., Matthias Krebs, Hans Joosten, Andreas Haberl, Andreas Kaffke, Izolda Matchutadze., Rustaveli State University, *Proceedings of the international conference "Biodiversity of Southern Kolkheti"*, p. 39-45.

In: Lagoons
Editor: Mwinyikione Mwinyihija
ISBN: 978-1-62808-092-6

Chapter 9

REVIEW OF MAN-MADE LAGOONS FOR TANNERY WASTE MANAGEMENT

Mwinyikione Mwinyihija*
COMESA/Leather and Leather Products Institute,
Addis Ababa, Ethiopia

ABSTRACT

Lagoons worldwide are phenomena worth discussing as they appear naturally or anthropogenically to serve various purposes. In the leather sector, lagoons are constructed to facilitate in managing tannery effluent emanating from associated industrial processes. For instance tannery effluent is characterized by both organic and inorganic pollutants. Whilst the biogeochemistry of lagoons could be looked at as aerobic or anaerobic, the ultimate benefits is what matters Indeed for the tanning industry the anaerobic lagoons has traditionally been used as an important phase of effluent treatments. Biotechnology as an important aspect in the present world was used in this study to dissect the pollutant load and toxicity levels. Five interconnected anaerobic lagoons in a tannery site were determined using both chemical and bioassay techniques. The evaluation of the lagoon samples indicated the presence of organic and inorganic pollutants. The most conspicuous aspect in this study was increased anaerobism in the lagoons accelerated with tanning effluent. Moreover through biosensor analysis, it was demonstrated that depletion of oxygen levels resulted to increased anaerobism which induced the conversion of sulphates to the highly toxic sulphide levels. Secondly, the study attempted to ascertain the efficacy of anaerobic lagoons where the chemistry of the effluent and pre-treatment of the waste was fundamental. Thus the use of man-made lagoons for the management of tannery waste should be supported by mechanized aeration mechanism to avert increase in toxicity in the final effluent.

Keywords: Anaerobic lagoons, Efficacy, Toxicity, Tannery effluents, Biosensors

* Corresponding author: COMESA/LLPI, P O Box 2358, Addis Ababa, Ethiopia. Email: m_ mwinyi@ hotmail.co.uk.

INTRODUCTION

Tannery waste emanating from leather processing (attained from the hides and skins industry) activities worldwide is considered the main source of contamination and pollution (Khwaja, 1998; Mwinyihija, 2012a). Indeed the sector is associated with a negative long term impact on the economic growth potential of a country irrespective of the immediate profit accruals intended (Mwinyihija and Killham, 2006). Tanning is therefore a major contaminant and environmental pollutant (Ros and Ganter, 1998; Mwinyihija et al, 2006; Mwinyihija, 2012) can cause an irreparable environmental impact on aquatic systems with its high oxygen demand, discolouration and toxic chemical constituents (Song et al., 2000; Mwinyihija et al., 2005a; Mwinyihija, 2011).

Thusly, the major public phobia is related to the tanning industry is directed towards the production of odours and water pollution from untreated or poorly treated effluent. Critical sources associated with leather processing include chlorides, tannins, chromium, sulphate and sulphides. Moreover addition to trace organic chemicals and increasing use of synthetic chemicals such as pesticides, dyes and finishing agents, as well as from the use of newer processing chemical solvents complicates the chemical composition of the discharged effluent even further (Anon., 2003; Mwinyihija, 2011a). These substances are frequently toxic and persistent, and affect both human health and the environment (UNEP, 1994; Mwinyihija et al., 2005a).

This contamination impacts negatively with related agronomical activities and structures with a higher seepage potential to ground water systems. Resulting groundwater contamination is envisaged to occur when the tannery waste water laded with polluting chemical seep through the soil from unlined waste lagoons, water ponds, pipes and drains, or from dumps and spills (Mwinyihija, 2011b). Furthermore deposition of solids, raw unsettled tannery wastewaters can cause encrustation (of calcium carbonate) to waste pathways leading to and from the lagoons. Moreover serious corrosion of metals as well as concrete sewers due to H_2S biological oxidation to H_2SO_4 (Balusubramanian and Pugalenthi, 2000) could occur. The adoption of cleaner technology in the short term and drastic measures to reduce on environmental impacts will require expenditure of funds, which literally could have promoted positive and sustainable development to the leather sector (Mwinyihija and Killham, 2006).

The biogeochemistry of the effluent emanating from the tanning sector and in certain instances are deemed recalcitrant and has actually resulted to the notion that leather processing is the largest source of hazardous wastes production (Figure 1) (Mwinyihija, 2012b). However, to date there are no reliable estimates of the quantity and types of hazardous waste generated in most developing countries. Chabaan (2001) indicates that an estimate of 10-15% of the wastes from the leather processing industry are potentially hazardous with a per annum increase of 2-5%.

Due to the complexities of tannery waste identified earlier (including chromium and chlorinated phenols), it is essential to dissect the waste water composition to determine the toxic nature of such effluent. This will further implore on the issue to create an understanding of the impacts and provide an opportunity to identify both environmental managemental aspects and potential remediation strategies (Mwinyihija, 2012). The approach towards managing tannery wastewater is generally treated by various physico-chemical or biological methods in the anaerobic lagoons (Figure 2)) or by a combination of both techniques

(Reemste and Jekel, 1997; Mwinyihija, 2011a). In addition physico-chemical processes are applied as interventional strategies towards treatment of contaminated sites (this includes use of ameliorants), but often do not destroy or degrade the contaminants (Bouwer et al., 1994: Gatima et al., 2006).

The aim of the chapter is therefore to ascertain the efficacy of manmade lagoons in the treatment of resultant effluent from leather processing tanneries. In carrying out this task, application of chemical techniques and bioassay to assess and determine the biogeochemistry of the anaerobic lagoons efficacy in reducing the toxic nature of effluents from the tanning industry will be attempted. For purposes of generating data a site was identified in Kenya. During the initial evaluation it was ascertained that the movement of the effluent from one lagoon to the other was estimated to take six days in each lagoon before being discharged to a nearby river (Figure 1).

2. MATERIALS AND METHOD

All areas of the study site were marked using a twelve channel GPS (*e*Trex™) personal navigator, part number 190-00205-00 Rev. B (Sept. 2000) GARMIN International Inc. USA). This was carried out to obtain waypoints and altitude at identified sampling points. Sediments and effluent samples from the anaerobic lagoons were collected in 20 ml super polyethylene vials (in triplicates) in a tannery site in Kenya. Bottles were filled to the brim and carefully capped without air bubbles. During storage and transportation from the field, they were stored in the dark at 4°C and tested within four weeks of sampling.

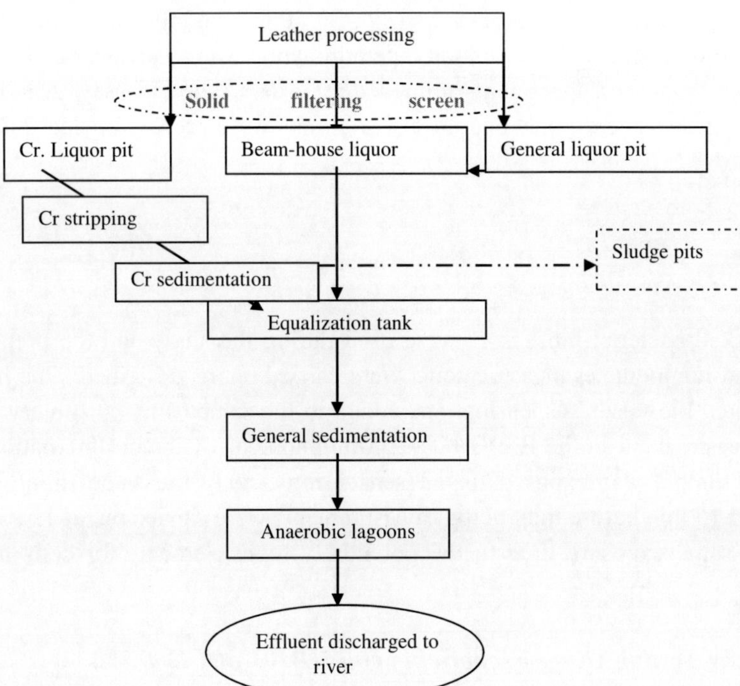

Figure 1. Flow diagram illustrating effluent treatment in a typical leather processing site in Kenya (Mwinyihija, 2011a).

Figure 2. Layout of the tannery waste lagoons leading to the discharge points.

2.1. Preparation of Sediment and Effluent Samples

Sediment extraction was performed using a method adapted from that described by Mathews and Hastings, 1987. Samples (10 g) of sediment were weighed into centrifuge tubes and 20 ml aliquots of deionized water (Millipore-Q, Quantum™ EX ultra pure organex cartridge) added (in triplicate). The pH of the samples was measured using a portable, standard, pH electrode (Hanna HI 8424, Norlab Instrument Ltd., Aberdeen, U.K). The samples were placed in an end-over-end shaker for 30 min and then centrifuged at 1100 g for 15 min. Parallel extractions for toxicity determination were carried out in triplicate and samples taken from the supernatants immediately prior to the bioassay. All chemicals used were of analytical-reagent grade and all test solutions were prepared using double deionized water.

2.2. Bioassay

Lux-marked bacterial biosensors were used during the study and the preparation of the biosensor and luminometer measurements were carried out as described. One hundred μL of the resuscitated biosensor suspension was added to the samples at 15 s intervals, accurately timed for measurement in the Bio Orbit 1253 luminometer (Labtech International, Uckfield, U.K). Each sample was exposed to the sensor for exactly the same time. Samples were incubated for 15 min before light output measurements were carried out at 15 s intervals. This ensured the same exposure time to the potentially toxic elements for cells in each of the cuvettes.

2.2.1. Toxicity Testing Using Escherichia coli HB101 pUCD607

Determination of toxicity was based on the bioluminescence response of the *lux*-modified biosensor, *E. coli* HB101 pUCD607, which had previously been, marked with the *lux*

CDABE genes, (isolated from *Vibrio fischeri)* using the multi-copy plasmid pucd607 (Amin-Hanjani *et al.*, 1993). The biosensor was stored at -20°C and resuscitated from freeze dried prior to bioassay.

2.2.2. Resuscitation of Freeze Dried Cultures

Freeze dried cultures of *Escherichia coli* HB101 pUCD607 were resuscitated in 10 ml of sterile 0.1 M KCl (contained in a Universal). 1ml of KCl was added and the culture resuspended by mixing (drawing up and down 5 times into a PI000 Gilson pipette). The resuspended culture was transferred back to the universal and the culture placed in a shaking (200 rpm) incubator (25° C) for 1 h.

2.2.3 Sample Addition and Luminometry Measurements

One hundred µL of the resuscitated biosensor suspension was added to the samples at 15 s intervals, accurately timed for measurement in the Bio Orbit 1253 luminometer (Labtech International, Uckfield, U.K). Each sample was exposed to the sensor for exactly the same time. Samples were incubated for 15 min before light output measurements were carried out at 15 s intervals. This ensured the same exposure time to the potentially toxic elements for cells in each of the cuvettes.

2.2.4. Data Analysis

The output from the luminometer resulting from each assay carried out was recorded in relative light units (RLU's) (equating to mV/10 s/ml). The light output was then converted to percentage maximum bioluminescence. This was calculated against a blank of double deionised water adjusted to pH 5.5, the optimum pH for bioluminescence.

$$\%\ \text{maximum bioluminescence} = I_S/I_{C*100} \quad (1)$$

where

I_S = RLU's emitted by the cells exposed to the sample
I_C = RLU's emitted by the cells exposed to the control

The percentage (%) maximum bioluminescence was determined for the three sample replicates. A mean of this determination was then calculated. The assay performance was monitored by reference to the response to the control, the reproducibility of the response to the three replicates and the response to a standard of trichlorophenol (TCP) (Figure 3). Effect of exposure time on toxicity to a range of standard solutions of Zn and Cu were prepared by dilution with double deionised water at pH 5.5 (Figure 4 (i&ii).

NB: R^2 =0.99, Sigmoid, 5 parameter, equation - f=y0+a/(1+exp (-(x-x0)/b))^c (All error bars (Figure 3 & 4 (i & ii) < 17.6% of the largest bar show as points in the graph). All standards were prepared by dilution with double deionised water at pH 5.5. Curves were fitted using Sigma plot 9.0 and an equation and r-squared determined.

The experimental data were analysed by a one-way analysis of variance (ANOVA) and regression analysis using Minitab.

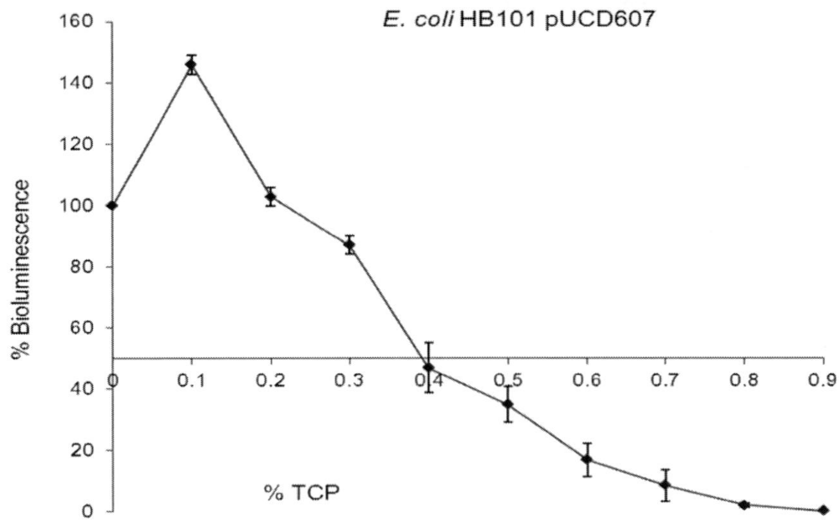

Figure 3. The assay performance monitored by the response to the control, the reproducibility of the response to the three replicates and the response to a standard of trichlorophenol (TCP).

2.2.5. Toxicity Dissection

2.2.5.1. Sparging

Aliquots (900 μl) of effluent and sediment extract were pipetted into luminometer cuvettes and sparged with air (N_2) for 10 min (at 1650 ml/min). The high flow of air allowed for rapid removal of volatiles present.

2.2.5.2 Activated Charcoal Treatment

Charcoal was first conditioned by placing 100 g of charcoal in a Duran bottle filled with double deionised water and allowing it to stand for 48 h, prior to multiple (x 10) deionised water rinsing and recovery by filtration.

Ten ml of the effluent or sediment extract was placed in a centrifuge tube. Charcoal (0.1g) was added and shaken for 30 min. Samples were centrifuged (3000 g) for 20 min (MSE Coolspin 2) at 48C and toxicity of the supernatant tested by transferring 900 μl of the sample into a luminometer cuvette (Clinicon, Petworth, West Sussex, U.K.) and adding 100 μl of resuscitated biosensor suspension.

2.2.5.3. Filtration

Aliquots of water samples and sediment extracts (50 ml) were filtered through a cellulose acetate membrane filter (0.22μm pore diameter) in order to determine toxicity related to suspended solids. Triplicates of each sample were submitted to the bioassay immediately after filtration.

2.2.5.4. pH Adjustment

Aliquots of the water samples (pH 5.5) and effluent/sediment extracts were adjusted to pH 4.0, 6.0 and 8.0 with 0.1M sodium hydroxide and 0.1M hydrochloric acid. Triplicates of the samples were submitted to the bioassay immediately after pH adjustment.

i.) Zn toxicity test

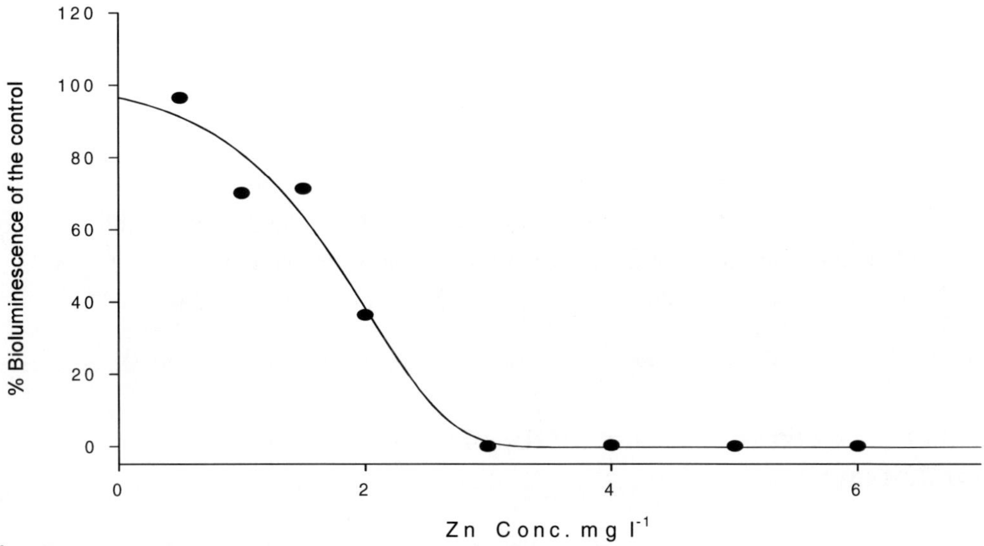

R^2 =0.98, Sigmoid, 5 parameter, equation - f=y0+a/(1+exp (-(x-x0)/b))^c.

ii.) Cu toxicity test

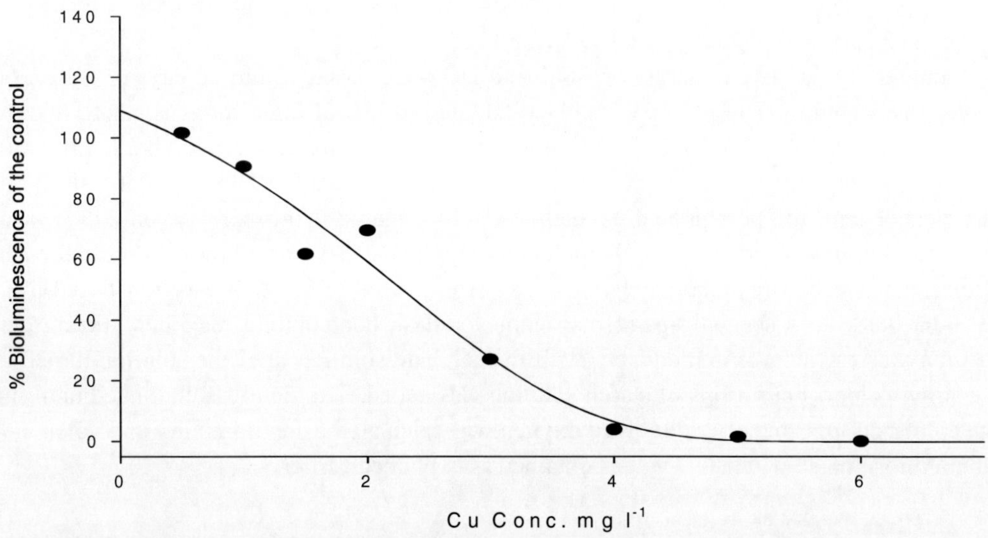

Figure 4 (i & ii). Quality control of *E. coli* HB101 pUCD607 IMS (15 min bioassay) on (i) Zn standard solutions (ii) Cu standard solutions.

2.3. Chemical Analysis

A total concentration of Cr, Ni, Cu, Zn, Cd, and Fe in each tannery effluent sample was determined (acidified with 1% HNO_3) by Atomic Absorption spectrometry (Perkins Elmer Analyst 100).

2.4. Sample Digestion (Sediments Samples)

Aliquots (2.5 mL) of concentrated HNO_3 (69% Analar grade) were added to 200 mg dried sediments samples obtained from the anaerobic lagoons, which had been weighed into 75 mL digestion tubes. The mixture was then allowed to stand overnight at 15°C. The following day, the digestion tubes were placed on a heating block and the temperature was gradually raised to 100°C for 8 h. The samples were allowed to digest for 3 h, after which the volume was reduced to 3-4 mL. The digest was cooled at room temperature, and diluted to 10 mL with double deionised water in graduated tubes. Total concentrations of Pb, Cu, Zn, Fe, Ni, Cd, and Cr were determined using atomic absorption spectrometry (Perkin Elmer Analyst 100).

2.5. Biological Oxygen Demand (BOD)/Dissolved Oxygen (DO) Determination

Levels of BOD and DO were determined using standard protocols (APHA, 1965) and carried out in an accredited laboratory (Inveresk laboratory, Scotland). Samples of water were incubated at 20°C for 5 days in a dark water bath. Every day for five days, DO was determined. The difference between initial value and the value at each time ($_t$) period (i.e. Oxygen demand) was plotted as the BOD_t (mg l^{-1}).

Samples for the determination of BOD and DO were tested within 10 days of collection to avoid degradation. The protocol included adding 1.0 mL of manganous sulphate reagent followed immediately by 1.0 ml of alkaline-iodide-azide solution to the BOD bottle (300 ml). The bottle was restoppered immediately and the contents mixed by shaking vigorously for at least 20 s or until the precipitated manganous and manganic hydroxide is evenly dispersed. After 2 to 3 min of shaking again, the precipitate in the sample was allowed to settle for 1 h. By means of a two-way pipette and vacuum system, 100 ml of solution was transferred from the BOD bottle to a specially-painted Erlenmeyer flask containing a magnetic stirring bar. Titration was carried out immediately with thiosulphate solution until the solution turned to pale straw colour. Four drops of starch solution was added. Titration was continued until the blue colour disappeared. The dissolved oxygen was calculated using normality and volume of sodium thiosulphate with BOD values obtained as explained earlier.

2.6. Sulphate and Chloride Determination

Appropriate standards for the determination of sulphate and chloride using ion exchange chromatography (Dionex, series 4500i – Autosampler AS40) were prepared in various concentration ranges (0, 2, 5 & 10 mg L^{-1}) and a calibration curve obtained. For sulphate determination, a stock solution of 1000 mg L^{-1} SO_4 was prepared by dissolving 1.818 g of potassium sulphate (K_2SO_4) in 1 L of deionized water. A chloride stock standard solution of 1000 mg L^{-1} Cl was prepared by dissolving 1.648 g of sodium chloride (NaCl) in 1 L of deionized water. The eluent was prepared by weighing out 0.95 g of Na_2CO_3 and 0.71 g of $NaHCO_3$. The preparation was then dissolved and made up to 5 L with deionized water. The regenerant was prepared by adding 3.5 ml of concentrated H_2SO_4 to approximately 300 ml of deionized water and making up to 5 L with deionized water. The following instrumentation conditions were maintained:

i.)	Eluent	
	Eluent flow rate	: 1.2 ml min^{-1}
	Suppressor	: $H_2\ SO_4$
	Background conductivity	: 17.3 μS
ii.)	Analyte	
	Analyte flow rate	: 1.2 ml min^{-1}
	Temperature compensation	: Selectable compensation between 0.0% and 3.0% per 1.7 °C
	Pressure	: 2900 psi
	Limit	: 5000 psi

2.7. Total Phenols

A spectrophotometric determination of phenols based on a multicommuted flow system with a 100 cm optical path flow cell was used to determine total phenols. All solutions were prepared with distilled and deionized water; and used analytical grade chemicals. Phenol reference solutions within 10.0 and 100 g L^{-1} were prepared by appropriate dilutions of a 1.00 g L^{-1} stock solution. Reagent R1 was prepared by dissolving 50.0 mg of 4-aminoantipyrine (4-AAP) in 50 ml of a buffer containing 5.2 g L^{-1} $NaHCO_3$, 5.8 g L^{-1} H_3BO_3 and 6.2 g L^{-1} KOH (pH 10.0). Reagent R2 was a 0.20% *m*/*v* $K_3[Fe(CN)_6]$ solution prepared in water. Water was used as the carrier. The analysis was carried out by an accredited laboratory (Robertson laboratories, U.K).

Where; V*i*, three-way solenoid valves; B, reaction coil (80 cm); D, long pathlength flow cell (100 cm optical path); C, water carrier (5.4 ml min^{-1}); S, sample (6.0 ml min^{-1}); R1, 0.10% (*m*/*v*) 4-AAP buffered at pH 10.0 (0.8 ml min^{-1}); R2, 0.2% (*m*/*v*) $K_3[Fe(CN)_6]$ (0.6 ml min^{-1}); W, waste.

Solutions containing 7.0 mol l^{-1} phenol, *m*-cresol, *p*-cresol, *p*-chlorophenol, catechol, hydroquinone, *p*-aminophenol and *p*-nitrophenol were employed for evaluation of the relative response for different phenols.

These data were then collated using Microsoft Excel. Statistical analysis involved two way analysis of variance (ANOVA) and treatment effects determined by least significant difference (LSD).

3. Results and Discussion

Generally, before any manipulations were carried out (untreated samples), high toxicity ($p \leq 0.05$) was observed for the samples obtained from the effluent treatment pits (Table 1) leading to the anaerobic lagoons. This was indicative of the tannery samples from the beam house, chrome stripping and general sedimentation pit which showed extreme toxicity.

Results from the different effluent treatment pits were all significantly different ($p \leq 0.05$). This observation was similar for the analyzed samples obtained from the anaerobic lagoons (except for lagoon 3) (Table 2). When the organic matter contents was compared to the bioluminescence values, a reduction in the value of the biosensor (indicating increased toxicity) similarly demonstrated a general reduction in the organic matter content as indicated in Table 2. Studies elsewhere suggested that strong binding of the pollutants (e.g. Chromium, Mercury etc.) capability towards humic acid (a major component of the organic matter content) is most probably due to sulfhydryl and thiophenolic groups (Miretzky et al., 2005). Therefore, the presence of natural humic substances in sediments reduces the availability (thus the toxicity) of the pollutants leached from the sediments to the aquatic systems. The environmental dynamic mechanism could involve both complexation and reduction and most probably occurring simultaneously. The overall effect of this is the reduction of the sediment strength in the affected lagoons as shown in Table 2. Moreover Schumacher, 2002 reported that the quality of organic matter in sediments is critical to the partitioning and bioavailability of sediments-associated contaminants. Specifically Talley et al., 2002 found out that PAH's clay/silt sediment fraction were more mobile and available and thus posed a great concern. The observation was critical as it provided the basis of evaluating the efficacy of man-made lagoons.

Table 1. Percentage maximum bioluminescence of effluent treatment pits. Figures in parentheses are standard errors of the mean (sem) (n=9)

Samples	Means (%)	SEM
Beam house	0.06	0.04)
General Sedimentation	2.40	(1.02)
Strip Chrome Tank	0.004	(0.004)
Chrome sedimentation	26.26	(12.70)
Equalization. Tank	6.83	(2.60)
Reference (Ddw)	104.85	(5.36)
LSD (5%)		18.27

Ddw - Double deionized water.
Figures in parentheses are standard errors of the mean (sem) (n=9).

Table 2. Percentage maximum bioluminescence, Total Organic Compounds (TOC) and Organic Matter (OM) of anaerobic effluent treatment lagoons

Samples	Bioluminescence Means (%)	SEM	TOC(ppm)	OM%
Lagoon1	27.86	(7.0)	274	520.6
Lagoon2	34.07	(21.2)	81.7	155.23
Lagoon3	77.08	(23.1)	30.4	57.76
Lagoon4	8.39	(0.90)	34.35	181
Lagoon5	1.83	(0.50)	52.55	99.845
Reference (Ddw)	104.85	(5.36)	n.a	n.a
LSD (5%)				45.37

Ddw –Double deionised water.
n.a – not available.
Figures in parentheses are standard errors of the mean (sem) (n=9).

3.1. Bioassay

In determining the efficiency of the lagoons as a treatment phase for the tannery waste, it was crucial to evaluate the bioavailability potential of the pollutants and contaminants passing through the identified man-made lagoons. As such use of bioassay becomes fundamental and especially through the application of biosensors. Biosensors have been used to monitor environmental contamination (Steinberg et al., 1995; van de Lelie et al., 1994; Mwinyihija et al., 2005b; 2006) by heavy metals (Paton et al., 1995) and organic contaminants, as well as toxicity in soils and water contaminated by industrial effluents (Brown *et al.*, 1996), BTEX (benzene, toluene, ethlybenzene, xylene) compounds (Sousa et al., 1998) and chlorinated aromatics (Steiert and Crawford, 1985). Toxicity is assessed by examining a decrease in light output from the bacterium when it is exposed to environmental samples or chemicals (Steinberg et al., 1995). The reduction (due to a toxic substance) in light output is proportional to the bioavailable concentration present (Sousa et al., 1998). Therefore, maximum percentage bioluminescence for all the analysed samples calculated against a blank of double deionised water at pH 5.5. The choice of the *lux*-marked biosensor in this work offered great environmental relevance in determining the bioavailability of various groups of the pollutants and the toxic nature of the effluent from the Kenyan tanning industry. Samples from the effluent treatment pits (Table 1) and anaerobic lagoons (lagoons 1, 2 and 3) (Table 2) were associated with this type of toxicity. The increase of luminescence in this study suggested that considerable toxicity was caused by volatile organics. However lagoons 3 and 4 indicated extreme toxicity with bioluminescence values of 8.39% and 1.83% respectively. Available metals are generally in the form of soluble cations and their tendency to be present in ionic form increases with increasing acidity (Sposito, 1989). In addition the high toxicity values could possibly be linked with the increased sulphide levels (Table 4). The toxicity relationship with the organic matter content and potential impact has been discussed earlier (para 3.0). Sarin (2000) reported the toxicity response of *lux*-marked *E. coli* HB101 to a range of metals thus its suitability in the present study.

Toxicity in samples such as treatment effluents, anaerobic lagoons and downstream riverine sampling points (Table 1and 2) was attributed to high concentrations of chromium and phenols (Table 3) (Mwinyihija, 2012b). For example, although in tannery wastewater Cr^{3+} is the most expected Cr form, the Redox reactions occurring in the sludge can increase the concentration of the hexavalent form (Kotas and Stascika, 2000). Most metals show increased solubility with decreased pH (Artiola, 1996), indicating increased bioavailability (chemical assimilation and possible toxicity) of organic/inorganic compounds (Steinberg et al, 1995; Shaw et al., 2000; Alexander, 2000). Under slightly acidic or neutral pH conditions in this type of wastewater, the poorly soluble ($Cr(OH)_3$ aq should be the preferred form, but a high content of organic matter originating from the hide material processing is effective in forming soluble organic Cr^{3+} complexes (Stein and Schwedt, 1994; Walsh and O'Halloran, 1996a&b). Other related studies investigating the fractionation of chromium toxicity in water using *E. coli* HB101 pUCD607 showed that speciation of chromium at different pH levels and a synergistic effect with other metals (e.g. copper and zinc) contributed to its toxicity (Wararatananurak, 2000). This observation suggested that chromium is frequently a constraint to bioremediation in contaminated environments (Killham Pers. Comm. 200b).

The choice of the organism (*E. coli* HB101 pUCD607) was suitable for the study as it is found in a variety of habitats including sewage sludge and contaminated river water, and it functions over a wide pH range. However, the future use of indigenous organisms mostly associated with tannery waste management as biosensors should complement this study and enables exploration of adaptability and acquired pollutant resistance over time. Indeed when measuring the bioluminescence the resultant data obtained was similar for the analyzed samples obtained from the anaerobic lagoons (except for lagoon 3) (Table 3). Therefore this depicted that a toxic substance generates a reduction in light output proportional to the concentration present. In this study, the percentage maximum bioluminescence results for all the treatments (sparging, activated charcoal, filtration and pH adjustment) were calculated against a blank of double deionised water at pH 5.5. The response to sparging suggests that toxicity could be due to VOC's (volatile organic compounds), activated charcoal to certain inorganic and organic compounds (particularly chlorinated hydrocarbons), and filtration to particulate and colloidal fractions

3.1.2. Chemical Analysis

Total phenols were present at elevated concentrations within most of the effluent treatment pits, lagoons and at the discharge point. However, a gradual decrease in phenol concentration was observed downstream (Table 4). The samples with the highest total phenols were the general sedimentation, chrome sedimentation, equalisation tank, lagoons 1 and 3, and the discharge point at the river (> 30 mg L^{-1}). The presence of high total phenols within the sampled areas was demonstrated by increased (except lagoon 3) bioluminescence (Table 2). Increase in bioluminescence is related to the uncoupling of the proton gradient (release of the fatty acid substrate for the luciferase enzyme) by chlorinated phenols, causing an increase in electron transfer rate and therefore an increase in the respiration rate, resulting in an inclined light output.

To ascertain the impact of the pollutants all the treatment phases before the anaerobic lagoons were determined. Indeed metals such as chromium and iron showed high concentrations as shown in Table 4. Chromium stripping (22.58 mg L^{-1}) and chrome sedimentation (191.47 mg L^{-1}) samples were, not surprisingly, associated with high chromium concentrations, hence linked to decreased luminescence of the sensor (Table 3) calculated against a blank of double deionized water at pH 5.5. When table 4 was further scrutinized and the effluent treatment pits compared to the anaerobic lagoons, chromium unlike the other parameters was mostly affected. The chrome stripping (22.58 mg L^{-1}) and chrome sedimentation (191.47 mg L^{-1}) pits demonstrated high concentration values whereas within the anaerobic lagoons very low concentration levels ($\leq$ 0.13 mg L^{-1}) were observed. On the other hand total phenol did not necessarily show any difference between effluent treatment pits and the anaerobic lagoons (the pH values were at most > 7.8). This could be attributed to the behavior of chlorinated phenols where in conditions of ambient pH constant pKa, chlorophenols can donate protons to form phenolate ions (Schellenberg et al., 1984). Therefore the effect is based on the tendency of donating the protons to increase along with increasing chlorination of the benzene ring. Thus the phenolate form is more water soluble than its non-ionized counterpart (Lyytikäinen, 2004). Furthermore the tendency of phenolate ions to sorb on soil or sediment is 15-30 times lower compared to their conjugated acids (Lagas, 1988). This overall result (as shown in Table 4) was critical in understanding the role of indigenous and habitat microorganism in breaking down of metals and organic contaminants within the anaerobic lagoons sediment matrices.

Table 3. Percentage maximum bioluminescence of anaerobic effluent treatment lagoons untreated and after treating with sparging, filtration and pH adjustment. Figures in parentheses are standard errors of the mean (sem) (n=9)

Samples	No Treatment Means	N_2 Sparged Means	Activated Charcoal Means	Filtration Means	pH adjustment		
					4.00	6.00	8.00
Lagoon1	27.86 (7.0)	90.89 (9.93)	70.92 (3.60)	149.32 (1.70)	2.10 (0.78)	4.86 (0.11)	115.11 (21.24)
Lagoon2	34.07 (21.2)	87.07 (8.00)	122.43 (22.55)	139.58 (6.64)	1.40 (0.36)	56.00 (1.45)	116.49 (23.14)
Lagoon3	77.08 (23.1)	89.39 (13.5)	154.07 (3.44)	145.62 (1.67)	1.02 (0.35)	4.89 (0.49)	91.76 (0.88)
Lagoon4	8.39 (0.90)	100.54 (7.70)	104.92 (20.30)	94.56 (1.50)	15.17 (0.91)	182.65 (0.70)	138.23 (0.50)
Lagoon5	1.83 (0.50)	95.59 (11.70)	69.88 (10)	104.77 (1.22)	0.62 (0.62)	3.16 (0.70)	110.74 (0.39)
Reference (Ddw)	104.85 (5.36)	94.75 (0.49)	101 (1.61)	99.97 (1.13)	102.87 (3.82)	107.29 (4.06)	114.24 (2.70)
LSD (5%)	45.37	32.72	45.52	10.29	1.85	2.43	44.29

Table 4. Total Phenols, pH, and metal concentrations of tannery effluent treatment pits ($mg\ L^{-1}$), anaerobic lagoons ($mg\ L^{-1}$) and riverine sediments ($mg\ kg^{-1}$) (Mwinyihija et al., 2005a)

Samples	pH	Cr $mg\ L^{-1}$	Pb $mg\ L^{-1}$	Fe $mg\ L^{-1}$	Cu $mg\ L^{-1}$	Cd $mg\ L^{-1}$	Zn $mg\ L^{-1}$	Ni $mg\ L^{-1}$	Total Phenol ($mg\ L^{-1}$)
Effluent treatment pits ($mg\ L^{-1}$)									
Beam-house	12	0.07	0	1.75	0.02	0.01	0.07	0.05	ND
General sedimentation	8.34	0.31	0	0.15	0.01	0.01	0..07	0	72
Chrome stripping	9.6	22.58	0.06	1.39	0.03	0.01	0	0.15	ND
Chrome Sedimentation	8.25	191.47	0	7.42	0.06	0.01	0.71	0.53	52.9
Equalization tank	8.05	0.3	0	0.5	0	0.01	0	0.02	36.8
Anaerobic lagoons ($mg\ L^{-1}$)									
Lagoon 1	7.8	0.1	0	0.17	0	0.01	0.02	0.03	30
Lagoon 2	7.92	0.06	0	0.09	0.01	0.01	0	0.04	NA
Lagoon 3	8.3	0.07	0.01	0.03	0	0.01	0	0.04	48.1
Lagoon 4	7.82	0.13	0.04	0.06	0.01	0.01	0.05	0.01	24.5
Lagoon 5	8.40	0.03	0	0	0	0.01	0.01	0.06	16.6
Riverine sediment sample ($mg\ kg^{-1}$)									
0 m discharge point	8.01	1.41	0.39	4237	0.57	0.02	1.63	0.72	30

The limit of detection (LOD) was determined on the basis of five blank samples at average blank signal plus three and ten times the standard deviation.

NA - Not Available.

ND – None detected.

Table 5. Results depicting analysis of various ecotoxicological parameters related to the three main stages of tannery waste water treatment flow; Raw effluent, sedimentation tank and final effluent at the discharge point (mg L^{-1}) (n=3)

	p H	COD	BOD	Chloride	Sulphide	Susp. Slds	Tot. Cr	oil/ grease
Raw effluent	7.90 (0.36)	3080.90 (1140.4)	1625.00 (240.7)	1816.67 (1092.7)	77.33 (5.0)	621.67 (177.4)	36.50 (42.1)	458.33 (192.7)
Sedimentation tank	7.10 (0.2)	8735.93 (11118.3)	4500.00 (6011.7)	1708.33 (405.4)	55.33 (36.5)	508.00 (256.4)	1.55 (0.8)	193.17 (150.5)
Final effluent	7.37 (0.2)	1179.98 (550.3)	514.17 (485.0)	1800.00 (1383.1)	1102.77 (1448.4)	330.67 (79.0)	0.93 (0.3)	94.17 (61.6)

Susp. Slds – Suspended solids.
Figures in parentheses are standard errors of the mean (sem) (n=9).

In table 5 a three point (Raw effluent, sedimentation tank and final effluent) analysis was determined to comprehend the trends of COD, BOD, Chlorides and the suspended solids in the treatment pits mentioned. The results showed that all the investigated parameters were quite erratic as depicted through the standard errors of the mean (sem) most probably due to the different types of cured raw material and diverse processing regimes. In particular, the Sulphide content increased (from 77.33 ± 5.0 to 1102.77 ± 1448.4 mg L^{-1}) sequentially through the effluent treatment pits as BOD levels reduced (except for the sedimentation pits), likewise demonstrating a depletion of oxygen levels within the treatment pits. In addition the depletion of dissolved oxygen encourages anaerobic activity, which leads to release of noxious gases (Pepper et al., 1996, Mwinyihija, 2011b).

Sulphates are reversed to sulphides through reduction by anaerobic bacteria thus increased anaerobic conditions within the lagoons resulted to an increase in Sulphide contents (Table 5). This observation showing increases in sulphide concentration in correlation to increased anaerobic conditions has been observed in earlier studies by Balusubramanian and Pugalenthi, 2000. However sulphide is not only toxic for higher organisms, it is also known as an inhibiting substance in anaerobic microbial processes (Weimann et al., 1998).

CONCLUSION

It is important to note that anaerobic lagoons cause sedimentation of suspended solids, partial conversion of COD to methane gas, conversion of sulphides into sulphates and production of sludge. In retrospect, it was demonstrated in this study that depleted levels of oxygen occurred due to biological decomposition of organic matter and other tannery waste, imparting anaerobic conditions observed. Thus the study demonstrated that anaerobic lagoons by themselves have very little impact and could be the potential source of Hydrogen sulphide re-generation and methylation. The lagoons efficacy can only be assured if the waste is pretreated effectively and the passage of effluent doesn't exceed more than six days in moving from one pit to another to avert anaerobic conditions setting in. Furthermore the study successfully demonstrated the potential of *lux*-marked biosensors to determine toxicity of tannery effluent. Indeed appropriate chemical analysis was used to further confirm the biosensor-based identification of 'cause and effect' of toxicity and potential constraints to remediation. The success of identifying the primary contaminants within the anaerobic lagoons prompts the need to further investigate the impact of effluent emanating from the tanning industry to either terrestrial or aquatic ecosystem when discharged or the dynamics of methylation within the anaerobic lagoons.

REFERENCES

Artiola, J. F. (1996). *Waste disposal*, chapter 10, 142. In: Pepper, L.I., Gerba, P.C., Brusseau, M.L Pollution Science, Academic Press (Elsevier Science, USA), San Diego, California, USA.

Alexander, M. (2000). Ageing, bioavailability, and overestimation of risk from environmental pollutants: Critical review. *Environmental Toxicology & Chemistry*, *34*, 4259-4265.

American Public health Association (APHA), (1965). Standard methods for the examination of water and wastewater, twelfth Ed., New York.

Balusubramanian, S. & Pugalenthi, V. (2000). A Comparative study of the determination of sulphide in tannery wastewater by ion selective electrode (ISE) and Iodimetry. *Water Research*, *34*, 4201-4206.

Bouwer, E., Durant, N., Wilson, L., Zhang, W. & Cunningham, A., (1994). Degradation of xenobiotic compounds *in situ*: Capabilities and limits. *FEMS Microbiology Reviews*, *15*, 307-317.

Brown, J. S., Rattray, E. A. S., Paton, G. I., Reid, G., Caffoor, I. & Killham, K. (1996). Comparative assessment of the toxicity of a paper mill effluent by respiratory and luminescence based bacteria assay. *Chemosphere*, *32*, 1553 – 1561.

Chaaban, M. A., (2001). Hazardous waste source reduction in materials and processing technologies. *Journal of Materials Processing Technology*, *119*, 336-343.

Gatima, E., Mwinyihija, M. & Killham, K. (2006). Determination of adsorption efficiency based on cation exchange capacity related to red earth, bone meal and pulverized fly ash as ameliorants to lead contaminated soils. *International Journal of Environmental Science and Technology*, *3*(3), 269-380.

Khwaja, A. R. (1998). *Studies on pollution abatement of wastes from leather Industries*, PhD thesis, University of Roorkee, India.

Killham, K. (2006). Personal communications.

Kotaś, J. & Stasicka, Z. (2000). Chromium occurrence in the environment and methods of its speciation. *Environmental Pollution*, *107*, 263 - 283.

Lagas, P. (1988). Sorption of chlorophenols in the soil. *Chemosphere* 17(2): 205-216.

Lyytikäinen, M. (2004). Transport, bioavailability and effects of Ky-5 and CCA wood preservative components in aquatic environment. University of Joensuu, Ph.D. Dissertations in Biology, No 26. ISSN 1457-2486, ISBN 952-458-524-3, 53.

Matthews, J. E. & Hastings, L. (1987). Evaluation of toxicity test procedure for screening treatability potential of waste in soil. *Toxicology Assessment*, *2*, 265 – 281.

Miretzky, P., Jardim, W. F., Bisnoti, C. M. & Rocha, C. J. (2005). Factors affecting Hg(II) Adsorption in soils from the Rio Negro basin (Amazon). *Quim. Nova*, Vol. *28*(3), 438 – 443.

Anonymous (2003). *Unione Nazionale Industria Conciaria.* Environmental report 2003, www.unic.it.

Mwinyihija, M. (2012a). Pollution control and remediation of the tanning effluent. *The Open Environmental Pollution and Toxicology Journal*, *3*, 55-64.

Mwinyihija, M. (2012b). Application and relevance of biosensors in the Tanning Industry. *Journal of Environmental Research*, *6*(4), 969-984.

Mwinyihija, M. (2011a). Ecological risk assessment (ERA) as a tool to pollution control of the tanning industry. *Resource and Environment*, *1*(1), 1-12.

Mwinyihija, M. (2011b). Ecotoxicology in the tanning industry. *Journal of Environmental Chemistry and Ecotoxicology*. Vol. *3*(13), 323-331.

Mwinyihija, M. & Killham, K. (2006). Is the Kenyan Tanning Industry integral to prioritized environmental sustainability targets set in the quest to Industrialization by 2020? *Environmental Sciences*, *3*(2), 113-134.

Mwinyihija, M., Meharg, A., Dawson, J., Strachan, NJC. & Killham, K. (2006). An ecotoxicological approach to assessing the Impact of Tanning Industry effluent on river health. *Arch. Environ. Contam. Toxicol. 50*, 316-324.

Mwinyihija, M., Meharg, A., Standing, D., Strachan, NJC. & Killham, K. (2005a). Ecological Screening of Kenyan Tannery dust using a luminescent-based bacterial biosensors. *International Journal of Environmental Health Research. 16*(1),47 – 58.

Mwinyihija, M., Meharg, Strachan, NJC. & Killham, K. (2005b). Ecological Risk Assessment of The Kenyan Tanning Industry. *Journal of American Leather Chemist Association*. Vol.*100*, 380-395.

Paton, G. I. (1995b). The development and application of a bioassay using *lux*-marked microorganisms to assess terrestrial ecotoxicity. PhD-thesis, *Department of Plant and Soil Science*, University of Aberdeen

Pepper, I. L., Gerba, C. P. & Brussean, M. L. (1996). *Pollution Science*, Academic press Inc., pp 194.

Ros, M. & Ganter, A. (1998). Possibilities of reduction of recipient loading of tannery waste Slovenia. *Water Science & Technology*, *37*, 145-152.

Sarin, C. (2000). A *lux*-based bioassay of heavy metal contamination of wastes. Ph.D Thesis, University of Aberdeen, U.K.

Schellenberg, K., Leuenberger. C. & Schwarzenbach, R. P. (1984). Sorption of chlorinated phenols by natural sediments and aquifer materials. *Environmental Science & Technology*, *18* (9), 652-657.

Schumacher, B. A.(2002). *Methods for the determination of Total Organic Carbon (TOC) in soils and sediments.* NCEA-C-1282 EMASC-001. United States Environmental Protection Agency Report.

Shaw, L. J. Y., Glover, L. A., Killham, K., Osborn, D. & Meharg, A. A. (2000). Bioavailability of 2,4-dichlorophenol associated with soil water-soluble humic material. *Environmental Science and Technology*, *34*, 4721-4726.

Song, Z., Williams, C. J. & Edyvean, R. J. (2000). Sedimentation of tannery wastewater. *Water Research*, *34*, 2171-2176.

Sousa, S., Duffy, C., Weitz, H., Glover, A. L., Bar, E., Henkler, R. & Killham, K. (1998). Use of a *lux*-modified bacterial biosensor to identify constraints to bioremediation of BTEX-contaminated sites. *Environmental Toxicology & Chemistry*, *17*, 1039-1045.

Sposito, G. (1989). *The Chemistry of Soils*. Oxford University Press, London

Steiert, J. G. & Crawford, R. L. (1985). Microbial degradation of chlorinated phenols. *Trends in Biotechnology*, *3*, 300 – 305.

Stein, K. & Schwedt, G. (1994). Chromium speciation in the wastewater from a tannery. *Fresenius Journal of Analytical Chemistry*, *350*, 38-41.

Steinberg, S. M., Poziomek, E. J., Engelman, W. H. & Rogers, K. R. (1995). A review of environmental applications of bioluminescence measurements. *Chemosphere*, *30*, 2155-2195.

Talley, J. W., Ghosh, U., Tucker, S. G., Furey, J. S. & Luthy, R. G. (2002). Particle-scale understanding of the bioavailability of PAHs in sediment. *Environ. Sci. Technol. 36*, 477-483.

van de Lelie, D., Corbisier, P., Baeyens, W., Wuertz, S., Diels, L. & Mergeay, M., (1994). The use of biosensors for environmental monitoring. *Research in Microbiology*, *145*, 67 – 74.

Walsh, A. R. & O'Halloran, J. (1996a). Chromium speciation in tannery effluent -1: An assessment of techniques and role of organic Cr(III) complexes. *Water Research*, *30*, 2393 – 2400.

Walsh, A. R. & O'Halloran, J. (1996b). Chromium speciation in Tannery effluents – II: Speciation in effluent and in a receiving estuary. *Water Research*, *30*, 2401 – 2412.

Wararatananurak, P. (2000). *Fractionation of chromium toxicity in water*. PhD Thesis, University of Aberdeen, U.K.

Wiemann, M., Schenk, H. & Hegemann, W. (1998). Anaerobic treatment of tannery wastewater with simultaneous sulphide elimination. *Water Research*, *32*(3), 774-780.

UNEP IE/PAC (1994). Tanneries and the Environment – A Technical Guide, Technical Report (2nd Print) Series No 4, ISBN 92 807 1276 4.

Reemste, T. & Jekel, M. (1997). Dissolved organics in tannery wastewaters and their alteration by a combined anaerobic and aerobic treatment. *Water Research*, *31*, 1035-1046.

INDEX

C

D

E

F

M

N

O

P

Q

R

T